New Technologies for Power System Operation and Analysis

New Technologies for Power System Operation and Analysis

Edited by

HUAIGUANG JIANG

National Renewable Energy Laboratory, Golden, CO, United States

YINGCHEN ZHANG

Chief Engineer and Group Manager of National Renewable Energy Laboratory, Golden, CO, United States

EDUARD MULJADI

Department of Electrical and Computer Engineering, Auburn University, Auburn, AL 36849, United States

ELSEVIER

ACADEMIC PRESS

An imprint of Elsevier

Academic Press is an imprint of Elsevier
125 London Wall, London EC2Y 5AS, United Kingdom
525 B Street, Suite 1650, San Diego, CA 92101, United States
50 Hampshire Street, 5th Floor, Cambridge, MA 02139, United States
The Boulevard, Langford Lane, Kidlington, Oxford OX5 1GB, United Kingdom

British Library Cataloguing-in-Publication Data

A catalogue record for this book is available from the British Library

Library of Congress Cataloging-in-Publication Data

A catalog record for this book is available from the Library of Congress

ISBN: 978-0-12-820168-8

For Information on all Academic Press publications
visit our website at https://www.elsevier.com/books-and-journals

Publisher: Joe Hayton
Acquisition Editor: Lisa Reading
Editorial Project Manager: Ali Afzal Khan
Production Project Manager: R. Vijay Bharath
Cover Designer: Greg Harris

Typeset by MPS Limited, Chennai, India

Contents

List of contributors *ix*

1. Introduction **1**

Yan Li, Changhong Zhao, Huaiguang Jiang, Yingchen Zhang and Eduard Muljadi

1.1 Overview of power systems 1
1.2 The development history of power systems in the United States 4
1.3 Distributed energy resource units 5
1.4 Steady-state conditions 8
1.5 AI and machine learning 10
1.6 Network dynamic operation 11
1.7 Multisector coupling 15
1.8 Structure of this book 17
References 17

2. Decoupled linear AC power flow models with accurate estimation of voltage magnitude in transmission and distribution systems **23**

Ning Zhang, Jingwei Yang, Yi Wang, Hai Li and Chongqing Kang

Nomenclature 23
2.1 Introduction 24
2.2 Linear power flow models for the meshed transmission systems 26
2.3 Linear three-phase power flow models of the unbalanced distribution systems 37
2.4 Case study 40
2.5 Conclusion 47
References 47

3. Renewable energy integration and system operation challenge: control and optimization of millions of devices **49**

Ying Xu, Wei Sun and Zhihua Qu

3.1 Introduction 50
3.2 Distribution system model with high penetration of renewables 52
3.3 Autonomous distributed voltage control 57
3.4 Hierarchical multiagent control of large-scale distribution system 68

3.5 Islanded microgrid with high penetration of distributed generations 74
3.6 Grid-edge situational awareness: enhanced observability by voltage inference 79
3.7 Control-enabled dynamic hosting allowance: *P* and *Q* control capacity and impact analysis 86
3.8 Cosimulation of integrated transmission and distribution systems 91
3.9 Conclusion 95
References 97

4. Advances of wholesale and retail electricity market development in the context of distributed energy resources 99

Qin Wang

4.1 Introduction 99
4.2 Modern wholesale electricity market 100
4.3 Modern retail electricity market 130
4.4 Conclusion 140
References 140

5. Wide-area monitoring and anomaly analysis based on synchrophasor measurement 143

Shutang You, Yu Su, Yong Liu and Yilu Liu

5.1 Synchrophasor measurement technology introduction 144
5.2 Wide-area measurement system example—FNET/GridEye 147
5.3 FNET/GridEye wide-area measurement system applications overview 151
References 160
Further reading 161

6. Advanced grid operational tools based on state estimation 163

Yu Liu, Yuzhang Lin and Junbo Zhao

6.1 Introduction 163
6.2 Model validation 165
6.3 System monitoring 178
6.4 Protective relaying 192
6.5 Conclusion remarks 204
References 205
Further reading 208

7. Advanced machine learning applications to modern power systems 209

Cong Feng, Mucun Sun, Morteza Dabbaghjamanesh, Yuanzhi Liu and Jie Zhang

7.1 Introduction 209
7.2 Modern forecasting technology 212
7.3 Machine learning—based control and optimization 228
7.4 Advanced artificial intelligence and machine learning applications to building occupancy detection 235
7.5 Conclusion 251
References 252

8. Power system operation with power electronic inverter—dominated microgrids 259

Yuhua Du, Xiaonan Lu and Xiongfei Wang

Nomenclature 259
8.1 Power system evolution toward modernization 260
8.2 Networked microgrids with parallel inverters 261
8.3 Parallel inverter operation in microgrids 264
8.4 Conclusion 273
References 274

9. Automated optimal control in energy systems: the reinforcement learning approach 275

Xiangyu Zhang and Huaiguang Jiang

9.1 Introduction 276
9.2 Deep reinforcement learning 286
9.3 Reinforcement learning in energy systems 297
References 314

10. Power, buildings, and other critical networks: Integrated multisystem operation 319

Kyri Baker

10.1 Introduction 320
10.2 Grid-interactive buildings 327
10.3 Interdependent critical networks 338
10.4 Electrification of the transportation sector 347
10.5 Considerations for future power systems 352
Acknowledgments 354
References 354

Index *359*

List of contributors

Kyri Baker
University of Colorado Boulder, Boulder, CO, United States

Morteza Dabbaghjamanesh
Department of Mechanical Engineering, The University of Texas at Dallas, Richardson, TX, United States

Yuhua Du
Temple University College of Engineering, Philadelphia, PA, United States

Cong Feng
Department of Mechanical Engineering, The University of Texas at Dallas, Richardson, TX, United States

Huaiguang Jiang
National Renewable Energy Laboratory, Golden, CO, United States

Chongqing Kang
Department of Electrical Engineering, Tsinghua University, Beijing, P.R. China

Hai Li
Department of Electrical Engineering, Tsinghua University, Beijing, P.R. China

Yan Li
Pennsylvania State University, University Park, State College, PA, United States

Yuzhang Lin
Department of Electrical and Computer Engineering, University of Massachusetts, Lowell, MA, United States

Yilu Liu
University of Tennessee, Knoxville, TN, United States; Oak Ridge National Laboratory, Oak Ridge, TN, United States

Yong Liu
Pacific Gas and Electric Company, San Francisco, CA, United States

Yu Liu
School of Information Science and Technology, ShanghaiTech University, Shanghai, P.R. China

Yuanzhi Liu
Department of Mechanical Engineering, The University of Texas at Dallas, Richardson, TX, United States

Xiaonan Lu
Temple University College of Engineering, Philadelphia, PA, United States

Eduard Muljadi
Auburn University, Auburn, AL, United, States

Zhihua Qu
Department of Electrical and Computer Engineering, University of Central Florida, FL, United States

Yu Su
University of Tennessee, Knoxville, TN, United States

Mucun Sun
Department of Mechanical Engineering, The University of Texas at Dallas, Richardson, TX, United States

Wei Sun
Department of Electrical and Computer Engineering, University of Central Florida, FL, United States

Qin Wang
Department of Power Delivery & Utilization, Electric Power Research Institute, Palo Alto, CA, United States

Xiongfei Wang
Department of Energy Technology, Aalborg University, Aalborg, Denmark

Yi Wang
Department of Electrical Engineering, Tsinghua University, Beijing, P.R. China

Ying Xu
Department of Electrical and Computer Engineering, University of Central Florida, FL, United States

Jingwei Yang
Department of Electrical Engineering, Tsinghua University, Beijing, P.R. China

Shutang You
University of Tennessee, Knoxville, TN, United States

Jie Zhang
Department of Mechanical Engineering, The University of Texas at Dallas, Richardson, TX, United States

Ning Zhang
Department of Electrical Engineering, Tsinghua University, Beijing, P.R. China

Xiangyu Zhang
National Renewable Energy Laboratory, Golden, CO, United States

Yingchen Zhang
National Renewable Energy Laboratory, Golden, CO, United States

Changhong Zhao
Chinese University of Hong Kong, Hong Kong, P.R. China

Junbo Zhao
Department of Electrical and Computer Engineering, Mississippi State University, Starkville, MS, United States

CHAPTER ONE

Introduction

Yan Li[1], Changhong Zhao[2], Huaiguang Jiang[3], Yingchen Zhang[3] and Eduard Muljadi[4]
[1]Pennsylvania State University, University Park, State College, PA, United States
[2]Chinese University of Hong Kong, Hong Kong, P.R. China
[3]National Renewable Energy Laboratory, Golden, CO, United States
[4]Auburn University, Auburn, AL, United, States

Contents

1.1 Overview of power systems 1
1.2 The development history of power systems in the United States 4
1.3 Distributed energy resource units 5
1.4 Steady-state conditions 8
1.5 AI and machine learning 10
1.6 Network dynamic operation 11
1.7 Multisector coupling 15
1.8 Structure of this book 17
References 17

1.1 Overview of power systems

The modern power grid is one of the most complex human-made engineering systems. In power grids, there are three subsystems, that is, generation system, transmission system, and distribution system, as shown in Fig. 1.1. In the generation system, electricity is produced in power plants through burning fossil fuels (e.g., coal, gas, and oil), nuclear material, or using hydroelectric power plants. The function of generation systems is to convert chemical energy to electric energy. Then the electrical energy is converted to a higher voltage level through transformers and transmitted through the power lines in the transmission system. Once the energy has traveled through the transmission system, the voltage is brought back down. Then, the energy reaches the distribution system, where we deliver energy to residential customers, commercial customers, industry, and other power loads.

New Technologies for Power System Operation and Analysis.
DOI: https://doi.org/10.1016/B978-0-12-820168-8.00001-8

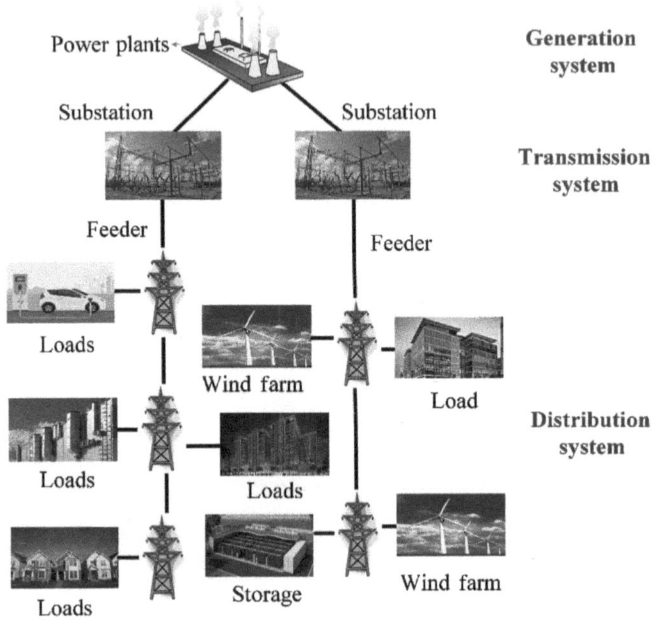

Figure 1.1 Overview of power systems.

Power generation in the traditional power grid is highly centralized, with power and energy flowing unidirectionally from generators in the generation system through a transmission/distribution network to end users. However, technological issues of traditional electric utilities as well as environmental problems caused by the combustion of fossil fuels have prompted the research and development of new pattern of power systems. With the emergence of distributed energy resource (DER) units —for example, wind, photovoltaic (PV), battery, biomass, micro-turbine, and fuel cell — microgrids have attracted increasing attention as an effective means of integrating such DER units into power systems.

The major constituents in microgrid are [1] DER units for power generation and energy storage, control systems for power/energy control and dispatch, and controllable loads for demand response. Microgrid research and development is currently propelled by technology advances in renewable energy, power electronics (PEs), control theory, data management, interconnected systems, wireless networks, as well as advances in interoperability and high-performance computing. The abovementioned technologies have been increasing the complexity of operation and control of microgrids. On one hand, microgrids will create a complicated interconnected power grid with various partners as shown in Fig. 1.2; on the other hand, the presence of microgrids will transform the traditional centralized system in to a distributed active one.

A typical microgrid prototype is given in Fig. 1.3, where PV, battery, PE device, and controller are installed in a house. This microgrid has two operational modes: a grid-connected mode and islanded mode. This means "my house" can either purchase electricity from "power utility" or directly get power energy from PV or battery. When the sunlight is strong, the electricity power generated by PV can be stored in the battery. When the sunlight is weak or during night, the power energy stored in the battery can be released to support customers' electricity consumption. Thus, it significantly improves the reliability of electricity. Moreover, the controller is playing a critical role in coordinating and operating this microgrid prototype. Although this microgrid system is very promising, it also induces several challenging problems. For instance, how can it dispatch power between roof top solar panel and battery? How can we sell electricity to power utility?

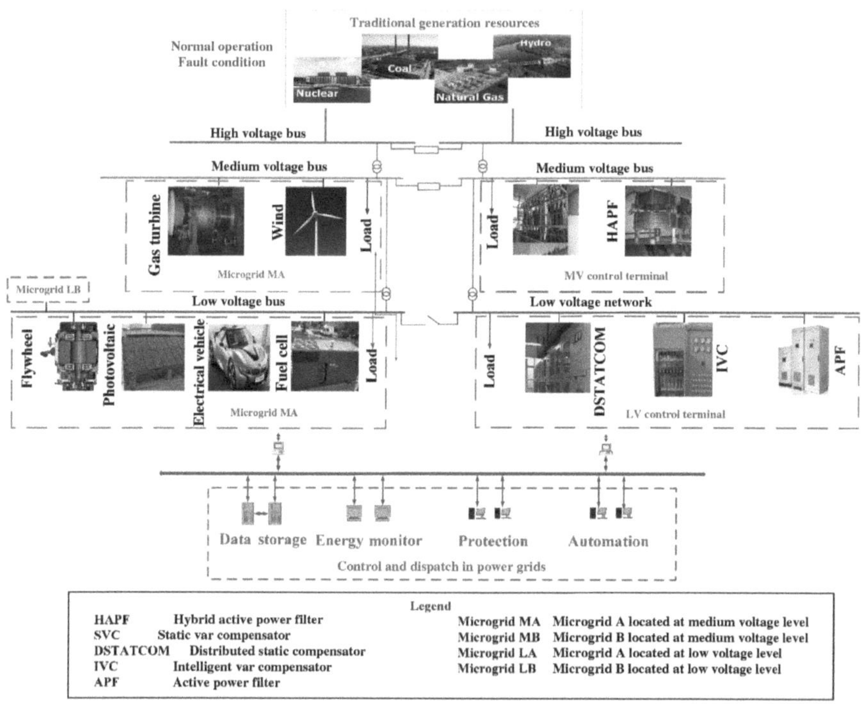

Figure 1.2 The future power system.

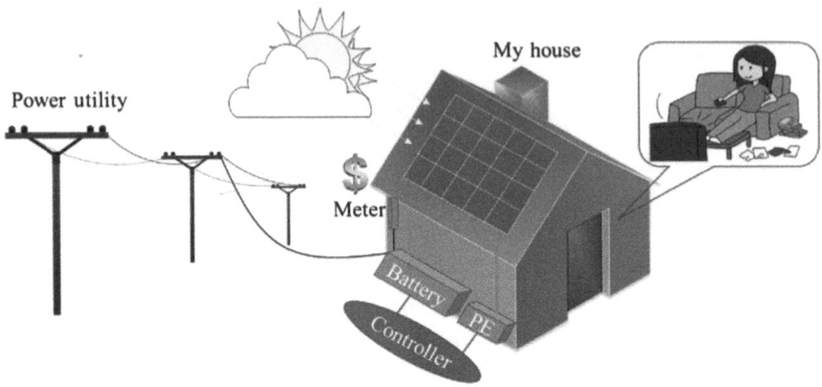

Figure 1.3 An example of community microgrid.

1.2 The development history of power systems in the United States

The development of power systems started in the 18th century.

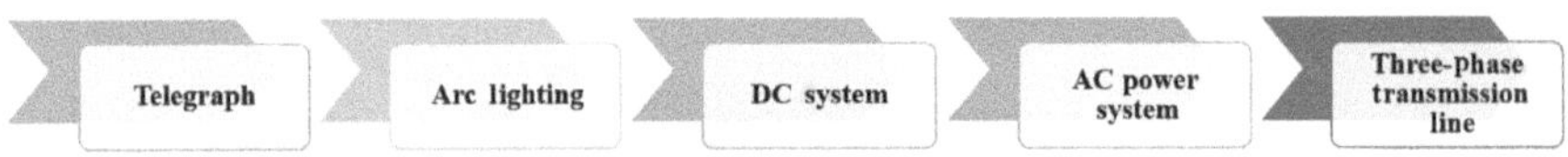

- The first real practical uses of electricity began with the telegraph in the 1860s.
- The use of electricity continued with arc lighting in the 1870s.
- In the early 1880s, Edison introduced Pearl Street DC system in Manhattan, which supplied 59 customers. In 1884, Sprague produced practical DC motor. In 1885, the transformer was invented to change voltage levels. In the mid-1880s, AC power system was introduced by Westinghouse/Tesla. In late 1880s, the AC induction motor was invented by Tesla.
- In 1893, the first three-phase transmission line was produced, which was at 2.3 kV. In 1896, AC lines started to deliver electricity. For instance, electricity was delivered from hydro generation at Niagara Falls to Buffalo, which was 20 miles away.

In the 19th century, the power system is further developed.

- In the early 1900s, private utilities supplied all customers in area (city), which was recognized as a natural monopoly. At that time, states stepped in to begin regulating the system.
- By the 1920s, large interstate holding companies control most electricity systems.
- In 1935 US Congress the passed Public Utility Holding Company Act to establish national regulation, which broke up large interstate utilities. This gave rise to electric utilities that only operated in one state. In 1935/36, the Rural Electrification Act brought electricity to rural areas. In the 1930s, electric utilities were established as vertical monopolies. In the 1930s, the frequency of power systems was standardized as 60 Hz in the United States.
- The 1970s brought inflation, increased fossil-fuel prices, and called for conservation and growing environmental concerns. In 1978, the Public Utilities Regulator Policies Act (PURPA) was passed by US Congress, which mandated that utilities must purchase power from independent generators located in their service territory. Therefore, PURPA started to introduce competition.
- Then in 1992, National Energy Policy Act was passed to boost the major opening of industry to competition. This act mandated that utilities provide "nondiscriminatory" access to the high voltage transmission, which means the goal was to set up true competition in generation. Over the last few years the electric utility industry has been dramatically restructured.

1.3 Distributed energy resource units

DER units have attracted increasing attention due to the worldwide problem of energy and the rapid development of PEs. DER units can be either rotating or stationary (PE-based DER units) and either dispatchable or nondispatchable. Moreover, some DER units may operate in plug-and-play fashion. In microgrids, the dispatchable DER units can be divided into three categories: (1) the DER unit itself can be dispatchable, for example, fuel cell, micro-turbine, and battery; (2) these units can be operated with revised MPPT making them dispatchable, although the outputs of DER units fluctuate, which are usually controlled by maximum power point tracking (MPPT), for example, wind and PV; and (3) for the abovementioned DER units with fluctuating outputs, they can be combined with energy storage units. The combination of these various types of DER units makes microgrid control a challenging task. Furthermore, the increasing penetration of DER units changes the operation and management of power systems. The main challenges include but are not limited to the bidirectional power flow, the variations in voltage and frequency profiles, the variations in short circuit current, etc. All of these play an important role in the system operations; and thus, it is important to investigate and understand the operations of DER units. Taking PV as an example, Fig. 1.4 shows the typical structure of connecting PV to the grid.

PV is a direct means to convert sunlight to electrical energy. The efficiency of energy conversion is usually in the range of 10%–20%. One important feature of PV is that it outputs DC power instead of AC power. The DC output of PV is converted by a PE interface (e.g., the inverter in Fig. 1.4) to AC output and then integrated into the physical system through an interface circuit. Phase lock loop (PLL), outer controller, inner controller, and pulse width modulation (PWM) are the typical control systems of PV. Specifically, the PLL is adopted to identify the phase of input signals, usually the three-phase voltages. This phase is then used to generate a control signal for the double-loop controller, specifically the outer controller and inner controller. Finally, the signals generated by the double-loop controller are transformed from the *dq* frame to the *abc* frame, and the PWM technique is then used to generate signals to

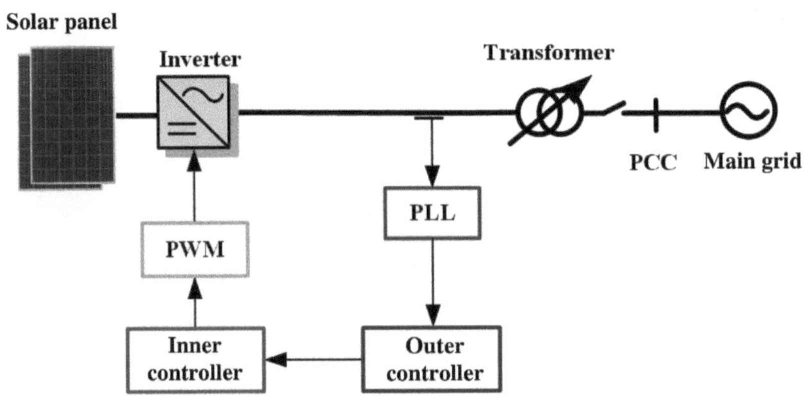

Figure 1.4 Connecting photovoltaic with grids.

control the switches, such as insulated-gate bipolar transistors (IGBTs), in the inverter.

The output of PV highly depends on the environment condition, such as irradiance and temperature. Fig. 1.5 gives PV curves that show typical output of PV under different conditions. Fig. 1.5 shows that there is a maximum output for each specific temperature or irradiance value. Therefore, the MPPT strategy is usually adopted to allow PV to generate the maximum power output, in order to increase the efficiency of energy conversion. Fig. 1.5 also shows that when the temperature increases, the maximum values of PV curves decrease; when the irradiance increases, the maximum values of PV curves increase as well.

Integrating DER into the power grids has been a challenge for utilities. Recently, microgrids have been studied from various aspects, including control strategies for DER units in a microgrid; stability analysis for a single microgrid, and control and operation of microgrids.

For the control strategies, optimal control methods based on different algorithms (e.g., particle swarm optimization [2–5] and sequential quadratic programing [6]) from different operation perspectives (e.g., stability and economics) were proposed for the optimal dispatch and control of DER units in a microgrid.

For the stability research of a single microgrid, optimal design of novel controllers (e.g., feedback controllers [7], mode-adaptive controller [8,9], hierarchical control system [10–12], adaptive decentralized controller [13], advanced droop control loop [14,15], nonlinear stabilizer [16], and distribution static compensator [17]) were proposed for the effective integration of DER units and stable operation of microgrids [18]. Besides, a synchronverter was developed to mimic synchronous generators, based on which DER units within microgrids can be

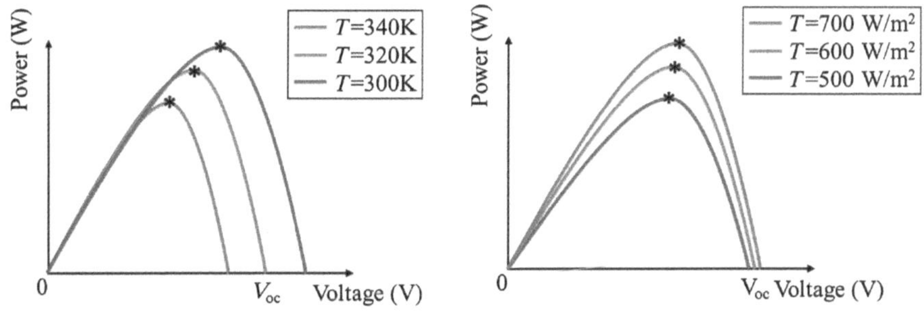

Figure 1.5 PV curves. *PV*, Photovoltaic.

automatically synchronized with main grid without the traditional PLL and then controlled to participate frequency regulation and voltage regulation [19,20].

For the multiple microgrid system, research has been conducted for the control and operation. New methodologies exploiting optimization tools (e.g., based on a metaheuristic approach [21]) were developed for the voltage stability in the multiple microgrid system. In addition, frequency control issue was investigated in Ref. [21] based on hierarchical control structure of multiple microgrids during mode transitions and load following in islanded operation. Furthermore, frequency control reserves in multiple microgrids were researched in Refs. [22,23] under different control strategies and different dynamic loads. Also state estimation and assessment of multiple microgrids were investigated in Refs. [21,24] based on the weighted least squares algorithm, real-time measurements, and fuzzy state estimation approach for different operation modes. In addition, other research has been focused on impact of high DER penetration and system parameters and structures [25–27], optimal power flow (OPF) [28], demand response [29,30], stability analysis [31–38], power quality issue [39], modeling and evaluation of system reliability [39], optimized operation [40], market mechanism [41], service restoration for black start [42], etc.

1.4 Steady-state conditions

Due to system interconnections and integration of DERs, the modern electrical power systems have become more and more complex. The steady-state operation is playing a fundamental role in running power [36]. Steady-state operation refers to the ability of the power system to maintain synchronism after small and slow disturbances, such as gradual power changes and fluctuations of PV. From the frequency perspective, in the United States, all synchronous generators are required to rotate with the same speed and produce 60 Hz voltage. The permitted frequency deviation is less than ± 0.5 Hz. From the voltage perspective, any increase or decrease in power load will change the angle between the induced voltage and terminal voltage. So, to maintain the stable operation, the voltage on each bus must be within the range of $\pm 5\% - \pm 8\%$ of the rated value.

Power flow calculation is usually used to investigate the steady-state operation of power grids. In power flow calculation, there are three typical buses as shown in Fig. 1.6: a slack bus, a generator bus, and a load bus. The slack bus provides the reference for the whole system, which means its voltage magnitude and angle are predefined. For the generator buses, their active power output and voltage magnitude are usually known, which means the voltage angle and reactive power need to be determined via power flow calculation. For the load buses, their active and reactive power outputs are usually known, which means the voltage magnitude and angle need to be determined via power flow calculation.

The goal of power calculation is to get the voltage and power output in the power system by solving the following algebraic equations:

$$\boldsymbol{I} = \boldsymbol{Y}\boldsymbol{U}, \tag{1.1}$$

where $\boldsymbol{Y}$ is the system admittance matrix, $\boldsymbol{U}$ is the voltage vector, and $\boldsymbol{I}$ is the injection current vector which represents the power output of generators and power consumption of loads.

There are several algorithms to solve the previous power flow issue; for instance, Newton–Raphson solution method and its derivatives, Gauss–Seidel method, Fast-decoupled-load-flow method, and DC power flow. The appropriate algorithm should be selected on the basis of study system.

AC OPF is a fundamental problem in power system operations. At the distribution level, OPF underlies (and possibly unifies) many applications, such as Volt/VAr/Watt control, dispatch of renewable energy sources, and demand response. With the rapid growth of DERs—including renewables, energy storage devices, and flexible loads—it is crucial for distribution systems to solve OPF in a fast and scalable way over a large number of active nodes. Toward this end, nonconvexity of AC OPF is a

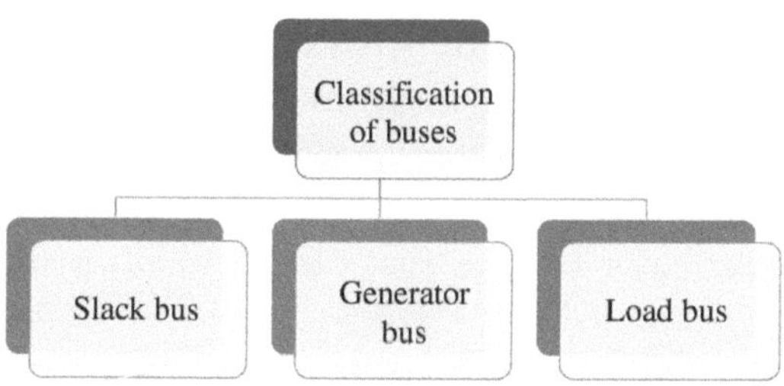

Figure 1.6 Classification of buses in power grids.

major hurdle, and recent efforts have looked at centralized and distributed OPF solution methods based on convex approximations or relaxations; see, for example, Refs. [43–46].

A popular convex approximation is obtained through the linearization of the power flow equations [47,48]. Semidefinite relaxation is another commonly taken approach to convexify OPF problems. To the best of our knowledge, it was first proposed in Ref. [49] to solve OPF as a semidefinite program (SDP) in single-phase networks with general topologies, and it was first studied in Ref. [50] whether and when this SDP relaxation is exact, meaning will the optimal solution of the SDP be feasible and globally optimal for the original nonconvex OPF as well. Sparsity of power networks was exploited to simplify the SDP relaxation [51,52], and relaxation to a more efficiently solvable second-order cone program (SOCP) is available in radial (tree) networks [53,54]. Techniques such as quadratic convex relaxation [55], moment/the sum-of-squares hierarchy [56], and strong SOCP relaxation [57] have also been explored to strengthen the SDP relaxation. However, power distribution networks in practice are multiphase and radial and composed of both wye- and delta-connected power sources and loads. In multiphase radial networks, Ref. [58] was the first that we know of that applied SDP relaxation; later, Ref. [59] illustrated SDP relaxation on a numerically more stable branch flow model. Both Refs. [58,59] considered wye connections only. In Ref. [60], a first semidefinite relaxation was proposed to incorporate delta-connected sources and loads.

1.5 AI and machine learning

With the increasing number of the smart sensors such as synchronized sensors and unsynchronized sensors in power systems, massive heterogeneous data are collected: structured data such as voltage, frequency, wind speed, solar irradiance, and real-time electrical price in addition to unstructured data such as customer service (voice and text), sky image, and satellite image.

There are three major features that can be found in the collected big data from power systems [61,62]:

1. Heterogeneous: The collected data is from different areas and contains different characteristics such as different sample speeds, different volumes, different scales, and different physical meanings.

2. Large volume and high velocity: The data volume can reach PB level with various kinds of high-speed sensors, and the natural language (customer service) and images also bring in a large volume of data.
3. Low value density: With this huge volume of data, the useful information is very sparse in the collected data and requires fast and accurate methods to process.

As discussed previously, it is imperative to build an artificial intelligence (AI) and machine learning based on the data processing and information extraction methods developed for power system operation and control. Generally, there are four major areas:

1. Perception: In this area, the goal is to monitor the power system, estimate the real system state, and analyze different scenarios (including the various faults) that appeared in the power system.
2. Forecasting: In this area, the goal is to track and forecast several variables in power systems such as load and system states. Furthermore, some basic inferences, such as basic decisions can also be based on the forecasting or regression results.
3. Operation and control: In this area, the AI and machine learning approaches are used to assist power system optimization, control, schedule, operation, and complex decision-making based on the results of perception and forecasting.

In Refs. [63–66], based on the machine learning methods, the smart sensor data are used to detect, locate, and identify different scenarios or faults in the power systems. In Refs. [67–70], various machine learning methods are implemented to forecast the power system's future information, such as load and system states. In Refs. [71–73], with the perception and forecast results, various machine learning–based methods are investigated for power system scheduling, control, operation, and complex decision-making. Furthermore, with advanced machine learning approaches, such as reinforcement learning and other approaches, the performance of power system operation, schedule, and control can be further improved and can be integrated with other systems [74–78].

1.6 Network dynamic operation

Although steady state is fundamentally important for power system operation, only studying the steady-state operation is not enough to run a practical power system. In order to operate an interconnected power grid, all the generators (and other synchronous machines) must remain in synchronism, which requires the rotors of synchronous machines turning at the same speed. Loss of synchronism will result in several severe conditions where power cannot be successfully delivered to customers. If one or more generators lose synchronism under a disturbance, this scenario is transiently unstable. In order to study the transient response of a power system, we need to develop system models during the transient time frame of several seconds following a system disturbance, where both electrical and mechanical models will be developed. Fig. 1.7 summarizes the timescale of different power system studies.

For different studies, we may need to consider different models. For instance, when we analyze transient stability, three systems usually need to be investigated, that is, prefault, faulted, and postfault. In the prefault condition, the system is assumed to be at an equilibrium point before the fault occurs. During fault, the system equations are changed, moving the system away from the above equilibrium point. Postfault shows system performance after the fault is cleared, where the system might or might not move to a new stable operating point.

Power system stability is another important concept in analyzing the dynamics of power grids. Fig. 1.8 summarizes the typical power system stability studies.

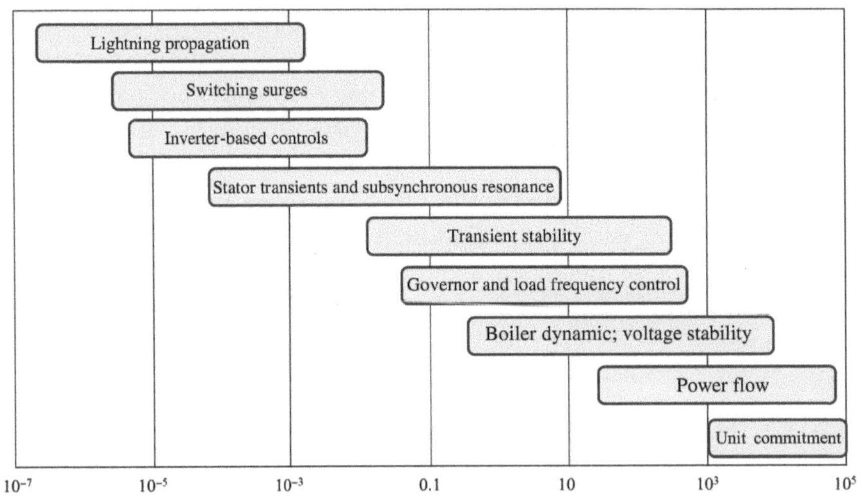

Figure 1.7 Timescale of power system studies [79].

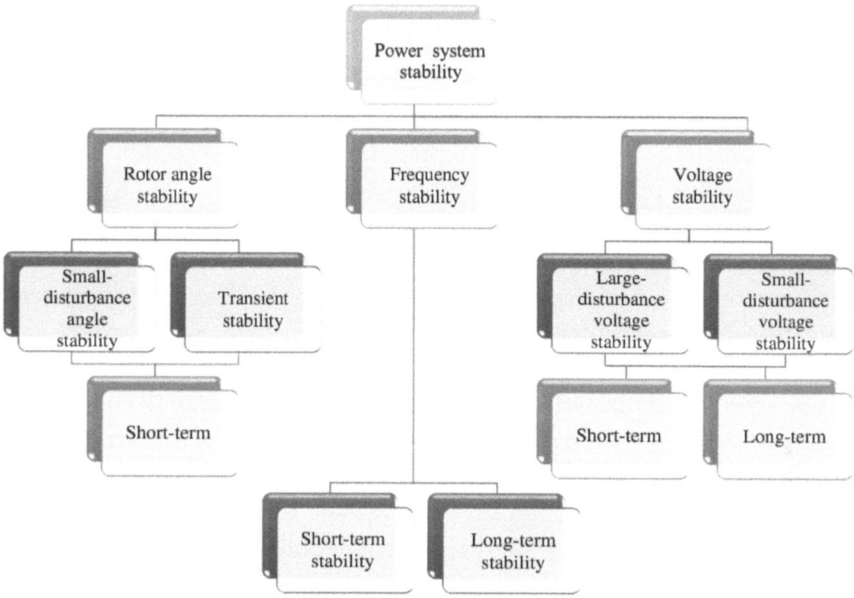

Figure 1.8 Categories of power system stability.

There are several methods for each stability analysis. Taking transient stability as an example, time-domain numerical integration and direct method are two typical approaches.

Time-domain numerical integration is used to determine following a contingency whether the power system can return to a steady-state operating point. The goal of numerical integration is to solve a set of differential and algebraic equations (DAE) which are used to model the system. The DAE is given as follows:

$$\begin{cases} \frac{d\boldsymbol{x}}{dt} = \boldsymbol{f}(\boldsymbol{x}, \boldsymbol{y}) \\ \boldsymbol{0} = \boldsymbol{g}(\boldsymbol{x}, \boldsymbol{y}), \end{cases} \tag{1.2}$$

where $\boldsymbol{x}$ is the set of dynamic state variables and $\boldsymbol{y}$ is the set of algebraic variables.

Numerical integration is by far the most common technique to analyze system dynamics, particularly for large power systems. During the fault and after the fault, the power system DAE models are updated and solved using numerical methods.

In the numerical integration, when the values of some variables change very slowly, such as the chemical action in the fuel cell, they can be treated as constant variables during the integration, whereas when the values of other variables change very quickly, such as PE devices, they can be treated as algebraic variables.

The second approach of system transient stability analysis is direct method, which is mostly used to provide an intuitive insight into the transient stability problem. The regions of attraction can be provided by direct method computation, which is unattainable with time-domain numerical integration. Moreover, direct method can be used to quickly check whether a specific control action is able to stabilize a power system. For a two-bus power system, this method is known as the equal area criteria.

The continued growth of variable renewable energy brings major challenges to the grid: the increased volatility of power injections from renewables and the reduced number of synchronous generators providing mechanical inertial support can severely degrade the dynamic performance and jeopardize the stability of the grid, especially in terms of *frequency*. Frequency control is one of the most critical grid services to maintain the safety of the electrical infrastructure and the quality of power delivery. Traditionally, large synchronous generators have been responsible for keeping the frequency tightly around its nominal value (e.g., 60 Hz in the United States) through coordinated actions across multiple timescales: inertial response of rotating masses to limit the rate of change of frequency in the first few seconds after a disturbance; droop control or primary frequency control of speed governors to stabilize frequency to a steady state within the low tens of seconds; and automatic generation control or secondary frequency control, which is a centralized paradigm run by the transmission system operator, to restore the nominal frequency within a few minutes. This type of hierarchical control architecture featuring a clear timescale separation and relying on slow and infrequent actions of a small number of generators might be sufficient for today's grid; but it lacks the responsiveness and flexibility to meet the future requirements for grid resilience, reliability, and efficiency at higher penetration levels of variable renewable generation.

A tremendous opportunity to address this rising concern lies in (1) the proliferation of DERs, for example, controllable loads, electric vehicles, energy storage devices, and PV systems, which are equipped with advanced PEs such as DC/AC inverters that can push or pull power at a much faster timescale and with a much finer resolution than synchronous generators and (2) the recent and continued advances in sensing, control, and communication technologies and the large-scale deployment of these technologies in the grid. Based on these trends, we envision a future grid with millions of interconnected active endpoints—DERs—as well as our increased capability to coordinate and optimize the operations of these endpoints. With this in mind the following natural question arises: how shall we exploit the control capabilities of DERs to improve grid dynamics, stability, and resilience in response to severer generation variations and contingencies under higher penetration levels of renewables? This question can be answered in different contexts (transient stability, voltage regulation, frequency control, economic dispatch, etc.). One series of work [80–82] tried to answer this question from a perspective of jointly developing dynamic feedback control and steady-state optimization of power networks, with application in distributed, load-side, optimal frequency control. Specifically, scalable distributed feedback control algorithms are developed and deployed to steer the power network frequency, power flow, generation, and load to a steady-state operating point that not only fulfills all the stability and security requirements but also achieves economic efficiency measured by generation cost, user disutility, and so on. From another perspective, feedback-based algorithms have been developed to save computation effort in pursing the optimal steady state and facilitate online tracking of OPF solutions that may vary on a timescale of seconds [83–85].

1.7 Multisector coupling

Subsystems constituting our interconnected critical infrastructure, such as electricity, natural gas, municipal water, district heating, and transportation systems, are traditionally operated in isolation. Further, due to nonlinearity and nonconvexity of their underlying physics, off-the-shelf "black-box" tools are used to solve challenging optimization and control problems associated with each subsystem separately. Such separation of architecture and limitation in optimization tools brings, a huge gap between the benefit that can be achieved theoretically and that has been gained today. Overcoming these limitations can bring significant benefits from operational-efficiency, socioeconomic, and environmental perspectives, as recognized by the literature. Therefore, in recent years, researchers have been pursuing the goal of developing distributed and computationally efficient optimization algorithms for interconnected power, gas, water, heating, and transportation networks.

For instance, Ref. [86] formulated an optimal power—water flow problem to jointly control DERs, tanks, and pumps in an interconnected power distribution and municipal water network and used a successive convex inner approximation method to identify a locally optimal solution. In addition, it developed a distributed optimization solver based on alternating the direction method of multipliers, which enables power and water operators to pursue individual objectives, while respecting intrinsic couplings between them. Another facet of research on the coupling of multiple sectors is to integrate transportation network with power network, on which Ref. [87] delivers a first version of an analytic model and distributed algorithm for optimal routing and charging scheduling of electric vehicles to simultaneously shave the peak load in the power distribution system and the peak traffic in the urban transportation system.

Future research challenges along this direction will include (1) deriving optimization models for large-scale comprehensive urban infrastructure that strikes a good balance between accuracy and computational affordability; (2) developing convex relaxation techniques that can produce globally optimal solutions or characterize optimality gaps for optimization problems containing nonconvex constraints associated with the intrinsic

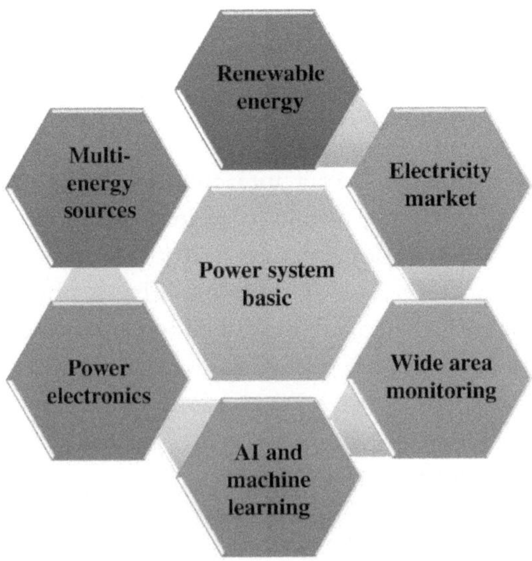

Figure 1.9 Structure of this book.

physics of different sectors; (3) developing fast, scalable, and privacy-preserving distributed optimization and control algorithms for the synergic and economically efficient operations of different system operators and utilities in different sectors; and (4) exploiting learning approaches applied on the big volume of urban data to aid complex and coupled system modeling and controller design.

1.8 Structure of this book

The structure of this book is given in Fig. 1.9, where all the chapters briefly introduced previously will be discussed.

References

[1] Y. Li, Y. Qin, P. Zhang, A. Herzberg, SDN-enabled cyber-physical security in networked microgrids, in: IEEE Transactions on Sustainable Energy, accepted, 2018.
[2] I.-Y. Chung, W.X. Liu, D.A. Cartes, E.G. Collins, S.-I. Moon, Control methods of inverter-interfaced distributed generators in a microgrid system, IEEE Trans. Ind. Appl. 46 (3) (2010) 1078–1088.
[3] I.-Y. Chung, W.X. Liu, D.A. Cartes, S.-I. Moon, Control parameter optimization for multiple distributed generators in a microgrid using particle swarm optimization, Eur. Trans. Electr. Power 21 (2) (2011) 1200–1216.
[4] A. Nasri, M.E. Hamedani Golshan, S. Mortaza Saghaina Nejad, Optimal planning of dispatchable and non-dispatchable distributed generation units for minimizing distribution system's energy loss using particle swarm optimization, Int. Trans. Electr. Energy Syst. 24 (2014) 504–519. Available from: https://doi.org/10.1002/etep.1708.
[5] M. Akbari, M.A. Golkar, S.M.M. Tafreshi, A PSO solution for improved voltage stability of a hybrid ac-dc microgrid, in: IEEE PES Innovative Smart Grid Technology (ISGT India), Dec. 2011, pp. 352–357.
[6] C.S. Wang, Y. Li, K. Peng, B.W. Hong, Z. Wu, C.B. Sun, Coordinated optimal design of inverter controllers in a micro-grid with multiple distributed generation units, IEEE Trans. Power Syst. 28 (3) (2013) 2679–2687.
[7] R.D. Robinett III, D.G. Wilson, S.Y. Goldsmith, Collective control of networked microgrids with high penetration of variable resources Part I: Theory, in: Proceedings of the IEEE International Conference on Cyber Technology in Automation, Control, and Intelligent Systems, May. 2012, pp. 1–4.
[8] J. Kim, J.M. Guerrero, P. Rodriguez, P. Rodriguez, R. Teodorescu, K. Nam, Mode adaptive droop control with virtual output impedances for an inverter-based flexible AC microgrid, IEEE Trans. Power Electron. 26 (3) (2011) 289–701.
[9] C.N. Rowe, T.J. Summers, R.E. Betz, T.G. Moore, An adaptive sliding mode controller for enhanced Q′-V droop in a microgrid, in: IEEE Energy Conversion Congress and Exposition (ECCE), Sept. 2012, pp. 2727–2734.
[10] Y.A.-R.I. Mohamed, A.A. Radwan, Hierarchical control system for robust microgrid operation and seamless mode transfer in active distribution systems, IEEE Trans. Smart Grid 2 (2) (2011) 352–362.

[11] S. Su, Y.T. Yan, M. Jin, Q.G. Wang, N.C. Zhou, Hierarchical coordination control of multi-micro grids system in series and parallel structure, in: China International Conference on Electricity Distribution (CICED), Sept. 2012, pp. 1–5.

[12] N.J. Gil, J.A. Peas Lopes, Hierarchical frequency control scheme for islanded multi-microgrids operation, in: IEEE Lausanne Power Tech, Jul. 2007, pp. 473–478.

[13] Y.A.-R.I. Mohamed, E.F. El-Saadany, Adaptive decentralized droop controller to preserve power sharing stability of paralleled inverters in distributed generation microgrids, IEEE Trans. Power Electron. 23 (6) (2008) 2806–2816.

[14] R. Majumder, B. Chaudhuri, A. Ghosh, R. Majumder, G. Ledwich, F. Zare, Improvement of stability and load sharing in an autonomous microgrid using supplementary droop control loop, IEEE Trans. Power Syst. 25 (2) (2010) 796–808.

[15] C.N. Rowe, T.J. Summers, R.E. Betz, D.J. Cornforth, T.G. Moore, Arctan power-frequency droop for improved microgrid stability, IEEE Trans. Power Electron. 28 (8) (2013) 3747–3759.

[16] S.M. Ashabani, Y.I. Mohamed, A flexible control strategy for grid-connected and islanded microgrids with enhanced stability using nonlinear microgrid stabilizer, IEEE Trans. Smart Grid 3 (3) (2012) 1291–1301.

[17] R. Majumder, A. Ghosh, G. Ledwich, F. Zare, Power sharing and stability enhancement of an autonomous microgrid with inertial and non-inertial DGs with DSTATCOM, in: International Conference on Power Systems (ICPS), Dec. 2009, pp. 1–6.

[18] Y. Li, W. Gao, J. Jiang, C. Wang, E. Muljadi, Unified calculation of eigen-solutions in power systems based on matrix perturbation theory, Sci. China Technol. Sci. 57 (8) (2014) 1594–1601.

[19] Q.C. Zhong, P.L. Nguyen, Z.Y. Ma, W.X. Sheng, Self-synchronised synchronverters: inverters without a dedicated synchronisation unit, IEEE Trans. Power Electron. 29 (2) (2014) 617–630.

[20] Q.C. Zhong, G. Weiss, Synchronverters: inverters that mimic synchronous generators, IEEE Trans. Ind. Electron. 58 (4) (2011) 1259–1267.

[21] A.G. Madureira, J.C. Pereira, N.J. Gil, J.A. Pecas Lopes, G.N. Korres, N.D. Hatziargyriou, Advanced control and management functionalities for multi-microgrids, Eur. Trans. Electr. Power 21 (2) (2011) 1159–1177.

[22] C. Yuen, A. Oudalov, A. Timbus, The provision of frequency control reserves from multiple microgrids, IEEE Trans. Ind. Electron. 58 (1) (2011) 173–183.

[23] A. Kahrobaeian, Y.A.-R. Mohamed, Analysis and mitigation of low-frequency instabilities in autonomous medium-voltage converter-based microgrids with dynamic loads, IEEE Trans. Ind. Electron. 61 (4) (2014) 1643–1658.

[24] G.N. Korres, N.D. Hatziargyriou, P.J. Katsikas, State estimation in multi-microgrids, Eur. Trans. Electr. Power 21 (2) (2010) 1178–1199.

[25] N.D. Hatziargyriou, A.G. Anastasiadis, J.Vasiljevska, A.G. Tsikalakis, Quantification of economic, environmental and operational benefits of microgrids, in: Proceedings of the IEEE Power Tech, Jun./Jul. 2009, pp. 1–8.

[26] M.A. Sofla, R. King, Control method for multi-microgrid systems in smart grid environment-stability, optimization and smart demand participation, in: IEEE PES Innovative Smart Grid Technology (ISGT), Jan. 2012, pp. 1–5.

[27] S. Tabatabaee, H.R. Karshenas, A. Bakhshai, P. Jain, Investigation of droop characteristics and X/R ratio on small-signal stability of autonomous microgrid, in: Second Annual Power Electronics, Drive Systems & Technologies Conference (PEDSTC), Feb. 2011, pp. 223–228.

[28] A. Kargarian, B. Falahati, F. Yong, M. Baradar, Multiobjective optimal power flow algorithm to enhance multi-microgrids performance incorporating IPFC, in: IEEE Power and Energy Society General Meeting, Jul. 2012, pp.1–6.

[29] A.G. Tsikalakis, N.D. Hatziargyriou, Operation of microgrids with demand side bidding and continuity of supply for critical loads, Eur. Trans. Electr. Power 21 (2) (2011) 1238–1254.
[30] H.S.V.S.K. Nunna, S. Doolla, Demand response in smart distribution system with multiple microgrids, IEEE Trans. Smart Grid 3 (4) (2012) 1641–1649.
[31] Y. Li, P. Zhang, M. Althoff, M. Yue, Distributed formal analysis for power networks with deep integration of distributed energy resources, in: IEEE Transactions on Power Systems, Accepted, 2018.
[32] Y. Li, P. Zhang, M. Yue, Networked microgrid stability through distributed formal analysis, Appl. Energy 228 (2018) 279–288.
[33] Y. Li, P. Zhang, P.B. Luh, Formal analysis of networked microgrids dynamics, IEEE Trans. Power Syst. 33 (3) (2018) 3418–3427.
[34] Y. Li, P. Zhang, W. Li, J.N. Debs, D.A. Ferrante, D.J. Kane, et al., Non-detection zone analytics for unintentional islanding in distribution grid integrated with distributed energy resources, in: IEEE Transactions on Sustainable Energy, Accepted, 2018.
[35] Y. Li, P. Zhang, J.N. Debs, D.A. Ferrante, D.J. Kane, S.N. Woolard, et al., A generic method for the determination of non-detection zones in DER-dominated distribution grids, in: IEEE Power and Energy Society General Meeting (PESGM), Portland, OR, Aug. 5–9, 2018, pp. 1–5.
[36] Y. Li, P. Zhang, Reachable set calculation and analysis of microgrids with power-electronic interfaced renewables and loads, in: IEEE Power and Energy Society General Meeting (PESGM), Portland, OR, Aug. 5–9, 2018, pp. 1–5.
[37] Y. Li, P. Zhang, L. Ren, T. Orekan, A Geršgorin theory for robust microgrid stability analysis, in: IEEE Power and Energy Society General Meeting (PESGM), Boston, MA, Jul. 17–21, 2016, pp. 1–5.
[38] Y. Li, W. Gao, Stability analysis on microgrids with variable DER units and loads based on matrix perturbation theory, in: IEEE Power and Energy Society General Meeting (PESGM), National Harbor, MD, Jul. 27–31, 2014, pp. 1–5.
[39] E.N. Dialynas, L. Daoutis, Modelling and evaluation of microgrids reliability and operational performance and its impact on service quality, Eur. Trans. Electr. Power 21 (2) (2011) 1255–1270.
[40] C. Schwaegerl, L. Tao, P. Mancarella, G. Strbac, A multi-objective optimization approach for assessment of technical, commercial and environmental performance of microgrids, Eur. Trans. Electr. Power 21 (2) (2011) 1269–1288.
[41] G.S. Kasbekar, S. Sarkar, Pricing games among interconnected microgrids, in: IEEE Power and Energy Society General Meeting, Jul. 2012, pp. 1–8.
[42] J.W. Li, J.H. Su, X.Z. Yang, T. Zhao, Study on microgrid operation control and black start, in: Fourth International Conference on Electric Utility Deregulation and Restructuring and Power Technologies (DRPT), Jul. 2011, pp. 1652–1655.
[43] M. Kraning, E. Chu, J. Lavaei, S. Boyd, Dynamic network energy management via proximal message passing, Found. Trends Optim. 1 (2) (2013) 70–122.
[44] B. Zhang, A.Y. Lam, A.D. Dominguez-Garcia, D. Tse, An optimal and distributed method for voltage regulation in power distribution systems, IEEE Trans. Power Syst. 30 (4) (2015) 1714–1726.
[45] Q. Peng, S.H. Low, Distributed algorithm for optimal power flow on an unbalanced radial network, in: Proceedings of IEEE Conference on Decision and Control, Osaka, Japan, 2015, pp. 6915–6920.
[46] Q. Peng, S.H. Low, Distributed optimal power flow algorithm for radial networks, I: Balanced single phase case, IEEE Trans. Smart Grid 9 (1) (2018) 111–121.
[47] M.E. Baran, F.F. Wu, Network reconfiguration in distribution systems for loss reduction and load balancing, IEEE Trans. Power Deliv. 4 (2) (1989) 1401–1407.

[48] C. Coffrin, P. Van Hentenryck, A linear-programming approximation of AC power flows, INFORMS J. Comput. 26 (4) (2014) 718–734.
[49] X. Bai, H. Wei, K. Fujisawa, Y. Wang, Semidefinite programming for optimal power flow problems, Int. J. Electr. Power Energy Syst. 30 (6) (2008) 383–392.
[50] J. Lavaei, S.H. Low, Zero duality gap in optimal power flow problem, IEEE Trans. Power Syst. 27 (1) (2012) 92–107.
[51] D.K. Molzahn, J.T. Holzer, B.C. Lesieutre, C.L. DeMarco, Implementation of a large-scale optimal power flow solver based on semidefinite programming, IEEE Trans. Power Syst. 28 (4) (2013) 3987–3998.
[52] S. Bose, S.H. Low, T. Teeraratkul, B. Hassibi, Equivalent relaxations of optimal power flow, IEEE Trans. Autom. Control 60 (3) (2015) 729–742.
[53] R.A. Jabr, Radial distribution load flow using conic programming, IEEE Trans. Power Syst. 21 (3) (2006) 1458–1459.
[54] M. Farivar, S.H. Low, Branch flow model: relaxations and convexification (parts I, II), IEEE Trans. Power Syst. 28 (3) (2013) 2554–2572.
[55] C. Coffrin, H.L. Hijazi, P. Van Hentenryck, The QC relaxation: a theoretical and computational study on optimal power flow, IEEE Trans. Power Syst. 31 (4) (2016) 3008–3018.
[56] C. Josz, D.K. Molzahn, Moment/sum-of-squares hierarchy for complex polynomial optimization, arXiv (2015) preprint arXiv:1508.02068.
[57] B. Kocuk, S.S. Dey, X.A. Sun, Strong SOCP relaxations for the optimal power flow problem, Oper. Res. 64 (6) (2016) 1177–1196.
[58] E. Dall'Anese, H. Zhu, G.B. Giannakis, Distributed optimal power flow for smart microgrids, IEEE Trans. Smart Grid 4 (3) (2013) 1464–1475.
[59] L. Gan, S.H. Low, Convex relaxations and linear approximation for optimal power flow in multiphase radial networks, in: Power Systems Computation Conference, 2014.
[60] C. Zhao, E. Dall'Anese, S.H. Low, Convex relaxation of OPF in multiphase radial networks with delta connection, in: IREP Symposium on Bulk Power System Dynamics and Control, 2017.
[61] Y. Gu, H. Jiang, Y. Zhang, J. Zhang, T. Gao, E. Muljadi, Knowledge discovery for smart grid operation, control, and situation awareness a big data visualization platform, in: 2016 North American Power Symposium, 2016.
[62] Y. Zhang, R. Yang, K. Zhang, H. Jiang, J.J. Zhang, Consumption behavior analytics-aided energy forecasting and dispatch, IEEE Intell. Syst. 32 (4) (2017) 59–63.
[63] H. Jiang, X. Dai, Y. Zhang, J.J. Zhang, E. Muljadi, Spatial-temporal synchrophasor data characterization and analytics in smart grid fault detection, identification, and impact causal analysis, IEEE Trans. Smart Grid 7 (2016) 2525–2536.
[64] H. Jiang, Y. Li, Y. Zhang, J.J. Zhang, D.W. Gao, E. Muljadi, et al., A big data based automatic security approach to detect, locate, and enhance stability for an unplanned islanding microgrid, J. Energy Eng. 143 (5) (2017) 04017045.
[65] H. Jiang, Y. Zhang, J. Zhang, E. Muljadi, PMU-aided voltage security assessment for a wind power plant, in: IEEE Power and energy Society General Meeting, 2015, pp. 1–5.
[66] H. Jiang, J.J. Zhang, W. Gao, Z. Wu, Fault detection, identification, and location in smart grid based on data-driven computational methods, IEEE Trans. Smart Grid 5 (6) (2014) 2947–2956.
[67] H. Jiang, Y. Zhang, E. Muljadi, J. Zhang, W. Gao, A short-term and high-resolution distribution system load forecasting approach using support vector regression with hybrid parameters optimization, in: IEEE Transactions on Smart Grid, 2016.
[68] H. Jiang, Y. Zhang, Short-term distribution system state forecast based on optimal synchrophasor sensor placement and extreme learning machine, in: 2016 IEEE Power & Energy Society General Meeting, 2016, pp. 1–5.

[69] H. Jiang, F. Ding, Y. Zhang, Short-term load forecasting based automatic distribution network reconfiguration, in: IEEE Power and energy Society General Meeting, 2017.
[70] F. Ding, H. Jiang, J. Tan, Automatic distribution network reconfiguration: an event-driven approach, in: IEEE Power and energy Society General Meeting, 2016.
[71] Y. Gu, H. Jiang, J.J. Zhang, Y. Zhang, E. Muljadi, Multi-timescale three-phase unbalanced distribution system operation with variable renewable generations, in: IEEE Transactions on Smart Grid, Jul. 2018.
[72] Y. Gu, H. Jiang, J.J. Zhang, Y. Zhang, E. Muljadi, F.J. Solis, Chance-constrained day-ahead hourly scheduling in distribution system operation, in: 2017 IEEE. 51st Asilomar Conference on Signals, Systems, and Computers, 2017, pp. 1363–1367.
[73] Y. Gu, H. Jiang, J.J. Zhang, Y. Zhang, E. Muljadi, F.J. Solis, Load forecasting based distribution system network reconfiguration—a distributed data-driven approach, in: 2017 IEEE. 51st Asilomar Conference on Signals, Systems, and Computers, 2017, pp. 1358–1362.
[74] Q. Lu, Safe and efficient intersection control of connected and autonomous intersection traffic, arXiv (2018) preprint arXiv:1801.09361.
[75] Q. Lu, K.-D. Kim, A genetic algorithm approach for expedited crossing of emergency vehicles in connected and autonomous intersection traffic, J. Adv. Transp. 2017 (2017).
[76] Q. Lu, K.-D. Kim, Autonomous and connected intersection crossing traffic management using discrete-time occupancies trajectory, Appl. Intell. 49 (5) (2019) 1621–1635.
[77] Q. Lu, K.-D. Kim, Intelligent intersection management of autonomous traffic using discrete-time occupancies trajectory, J. Traffic Logist. Eng. 4 (1) (2016) 1–6.
[78] H. Xu, X. Li, X. Zhang, J. Zhang, Arbitrage of energy storage in electricity markets with deep reinforcement learning, arXiv (2019) 1904.12232v1.
[79] V. Vijay, J.D. McCalley, P.A. Anderson, A.A. Fouad, Power system control and stability. John Wiley & Sons, 2019.
[80] C. Zhao, U. Topcu, N. Li, S.H. Low, Design and stability of load-side primary frequency control in power systems, IEEE Trans. Autom. Control 59 (5) (2014) 1177–1189.
[81] E. Mallada, C. Zhao, S.H. Low, Optimal load-side control for frequency regulation in smart grids, IEEE Trans. Autom. Control 62 (12) (2017) 6294–6309.
[82] C. Zhao, E. Mallada, S.H. Low, J. Bialek, Distributed plug-and-play optimal generator and load control for power system frequency regulation, Int. J. Electr. Power Energy Syst. 101 (2018) 1–12.
[83] L. Gan, S.H. Low, An online gradient algorithm for optimal power flow on radial networks, IEEE J. Sel. Areas Commun. 34 (3) (2016) 625–638.
[84] Y. Tang, K. Dvijotham, S.H. Low, Real-time optimal power flow, IEEE Trans. Smart Grid 8 (6) (2017) 2963–2973.
[85] E. Dall'Anese, A. Simonetto, Optimal power flow pursuit, IEEE Trans. Smart Grid 9 (2) (2018) 942–952.
[86] A.S. Zamzam, E. Dall'Anese, C. Zhao, J.A. Taylor, N.D. Sidiropoulos, Optimal water-power flow problem: formulation and distributed optimal solution, IEEE Trans. Control Netw. Syst. 6 (1) (2019) 37–47.
[87] H. Jiang, Y. Zhang, Y. Chen, C. Zhao, J. Tan, Power-traffic coordinated operation for bi-peak shaving and bi-ramp smoothing—a hierarchical data-driven approach, Appl. Energy 229 (2018) 756–766.

CHAPTER TWO

Decoupled linear AC power flow models with accurate estimation of voltage magnitude in transmission and distribution systems

Ning Zhang, Jingwei Yang, Yi Wang, Hai Li and Chongqing Kang
Department of Electrical Engineering, Tsinghua University, Beijing, P.R. China

Contents

Nomenclature 23
2.1 Introduction 24
2.2 Linear power flow models for the meshed transmission systems 26
2.2.1 Decoupling of voltage magnitude and phase angle 26
2.2.2 Matrix formulation of the decoupled linearized power flow model 30
2.2.3 Transformers and phase shifters 31
2.2.4 Derivation and justification of the fast decoupled linearized power flow 33
2.3 Linear three-phase power flow models of the unbalanced distribution systems 37
2.4 Case study 40
2.4.1 Meshed transmission systems 40
2.4.2 Balanced distribution systems 41
2.4.3 Unbalanced distribution systems 44
2.5 Conclusion 47
References 47

Nomenclature

The main notation used in this chapter is provided later; other symbols are defined as required.

$\boldsymbol{n}$ number of power system buses
$\boldsymbol{m}$ number of power system lines
$\mathscr{R}$ set of $V\theta$ buses (slack buses)
$\mathscr{S}$ set of PV buses

New Technologies for Power System Operation and Analysis.
DOI: https://doi.org/10.1016/B978-0-12-820168-8.00002-X

$\mathcal{L}$ set of PQ buses
$\boldsymbol{Y}=\boldsymbol{G}+j\boldsymbol{B}$ admittance matrix of the power system
$\boldsymbol{Y}'=\boldsymbol{G}'+j\boldsymbol{B}'$ admittance matrix without shunt elements
$\boldsymbol{Y}_{ij}=\boldsymbol{G}_{ij}+j\boldsymbol{B}_{ij}$ element in the ith row and jth column of $\boldsymbol{Y}$
$\boldsymbol{Y}'_{ij}=\boldsymbol{G}'_{ij}+jB'_{ij}$ element in the ith row and jth column of $\boldsymbol{Y}$
$\boldsymbol{V}, \boldsymbol{\theta}$ magnitudes and phase angles of bus voltages
$\boldsymbol{V}_i, \boldsymbol{\theta}_i$ voltage magnitude and phase angle of bus i
$\boldsymbol{P}, \boldsymbol{Q}$ bus injected active and reactive powers
$\boldsymbol{P}_i, \boldsymbol{Q}_i$ injected active and reactive power at bus i
$\boldsymbol{P}_{ij}$ lossless MW flow from bus i to bus j
$z_{ij}=r_{ij}+jx_{ij}$ impedance of line (i, j)
$y_{ij}=g_{ij}+jb_{ij}$ admittance of line (i, j)
$y_{ii}=g_{ii}+jb_{ii}$ shunt admittance at bus i

2.1 Introduction

Power flow analysis is the backbone of power system studies and is widely applied in daily operations as well as short-term and long-term planning. Direct calculations of power flow are of great significance for contingency analysis [1], reliability assessment [2], and probabilistic load flow analysis [3], if they can be performed accurately and efficiently. Although conventional AC power flow calculations yield accurate results, their nonlinearity, difficulty in convergence, and low computational efficiency limit their application in the power system industry, especially for large-scale systems [4]. Therefore a fast and linear power flow model would be of great interest, provided it could offer reasonable accuracy and robustness for all types of grids. Such a linear model could also be beneficial for solving optimization problems, such as locational marginal pricing [5] and security-constrained unit commitment (SCUC) [6], by allowing these problems to be transformed into linear programming problems.

The DC power flow (DCPF) model is one of the most widely used linear power flow models for power systems. Because it is a linear, noniterative model with reasonable accuracy in terms of MW flow, it has considerable analytical and computational appeal compared with the AC power flow model [7]. The classical DCPF model is derived based on the assumptions of a lossless MW flow and a flat bus voltage profile. More general versions of the DCPF model have been proposed, including hot- and cold-start models, for broader applications.

Hot-start models correct the bus power injections according to base points obtained from initial AC power flow solutions in various ways, such as net loss dispersal and base-point matching [7].

However, in situations such as DC-model-based SCUC [6], financial transmission rights allocations [8] and power system planning, no reliable AC base points are available. Therefore it is necessary to use cold-start models. As alternatives to the classical DCPF model, some cold-start models modify the bus injections by estimating loss as a percentage of net load, whereas others apply an empirical fixed voltage magnitude. Several tests have shown, however, that there is no large difference in MW-flow error between the classical DCPF model and other cold-start models [7].

Although cold-start DCPF models are widely used in many fields, it is worth noting that they can neither consider reactive power nor shed light on bus voltage magnitude. In fact, the lack of a voltage magnitude consideration leads to many problems in system planning and operation when cold-start DCPF models are applied. For instance, it is a common practice to simply consider the MW flows in reliability evaluations. The bus voltage quality is inevitably ignored and, moreover, ZIP loads are modeled at nominal and typical voltages [2]. In terms of electricity markets, only active generation is priced in markets such as PJM. The reactive power offered by generators is paid for merely as an ancillary service [9]. Probabilistic load flow calculations, which are an increasingly important tool in power system analysis, typically face the dilemma that DCPF models are incapable of accounting for voltage magnitude, whereas AC power flow models that are linearized around the operating point lead to errors in scenarios with large signal perturbations [10].

However, it is not easy to consider bus voltage magnitudes in a decoupled linear flow model, because they are sensitive to both active and reactive power injections. In Ref. [11] a linear model was proposed for radial distribution systems. However, this model is dependent on the radial topology of the distribution system and cannot be extended to meshed transmission systems. The authors of Ref. [12] have achieved a great advancement in formulating a linear load equation that considers reactive power. This power flow model reflects both active and reactive power balance with respect to the square of the voltage magnitude, V^2, and the modified phase angle, $V^2\theta$.

Nevertheless, the voltage magnitude and phase angle are not completely decoupled, giving rise to a quadratic programming problem if this model is applied for optimization.

This chapter makes a further step toward a $V-\theta$ decoupled linear power flow model. We propose a decoupled linearized power flow (DLPF) model with respect to the voltage magnitude V and the phase angle θ. The model is distinguished by its accuracy in voltage magnitude and robustness for application to different types of systems. The error in MW flow is comparable to that of cold-start DCPF models, which is much smaller in practice than is predicted, as mentioned in Ref. [7]. We also design a fast DLPF (FDLPF) model by developing a fast approximation of the matrices used in the DLPF model, which is as fast as classical DCPF calculations while still as accurate as DLPF calculations. The approximation is justified by theoretical derivation and numerical tests. Furthermore, we generalize this model to the three-phase power flow in unbalanced distribution systems.

The remainder of the chapter is organized as follows: Section 2.2 formulates the DLPF model in transmission grid based on the initial AC power flow equation and derives the expressions for voltage magnitude, phase angle, and branch MW flow. Then the FDLPF model is formulated by making a subtle change to the DLPF model, which is justified by theoretical derivation and numerical examples. The specific steps for utilizing the FDLPF model are presented. In Section 2.3 a linear three-phase power flow of unbalanced distribution systems is proposed, which is a generalization of the DLPF model. The linearization of ZIP characteristics of the loads is also discussed in this section. In Section 2.4, several cases, including radial distribution systems, meshed large-scale transmission systems, and ill-conditioned systems, are analyzed to demonstrate the accuracy and robustness of the DLPF/FDLPF approach and the computational efficiency of the FDLPF model. Finally, conclusions are drawn and further discussions are presented.

2.2 Linear power flow models for the meshed transmission systems

2.2.1 Decoupling of voltage magnitude and phase angle

For a power system with n buses the polar power–voltage AC power flow model is well known:

$$P_i = \sum_{j=1}^{n} G_{ij} V_i V_j \cos\theta_{ij} + \sum_{j=1}^{n} B_{ij} V_i V_j \sin\theta_{ij} \tag{2.1}$$

$$Q_i = \sum_{j=1}^{n} B_{ij} V_i V_j \cos\theta_{ij} + \sum_{j=1}^{n} G_{ij} V_i V_j \sin\theta_{ij} \quad i = 1, 2, \ldots, n \tag{2.2}$$

For most scenarios in real power systems the bus voltage magnitudes are approximately 1 p.u. The absolute value of phase angle differences across lines rarely exceed 30 degrees [12], and most of them are within 10 degrees [13]. Moreover, the admittance matrix of the power system has a special structure, in which the diagonal elements are the sum of the non-diagonal elements in the corresponding rows, with the contribution of shunt elements as well:

$$Y_{ij} = \begin{cases} -y_{ij} & \text{if } j \neq i \\ y_{ii} + \sum\limits_{k=1,\, k\neq i}^{n} y_{ik} & \text{if } j = i \end{cases} \tag{2.3}$$

Note that in this chapter, the "shunt elements" y_{ii} denote not only the contribution of shunt capacitors but also the line-charging susceptance and the equivalent admittance of transformers and phase shifters (Section 2.2.3 of this section provides the details of the consideration of transformers and phase shifters).

Considering these facts, the following linear approximations are applied to (2.1) to decouple the voltage magnitudes and the phase angles:

$$\begin{aligned} P_i &= \sum_{j=1}^{n} G_{ij} V_i V_j \cos\theta_{ij} + \sum_{j=1}^{n} B_{ij} V_i V_j \sin\theta_{ij} \\ &= g_{ii} V_i^2 + \sum_{j=1,\, j\neq 1}^{n} \left(g_{ij} V_i \left(V_i - V_j \cos\theta_{ij} \right) - b_{ij} V_i V_j \sin\theta_{ij} \right) \\ &\approx g_{ii} V_i + \sum_{j=1,\, j\neq 1}^{n} g_{ij} (V_i - V_j) - \sum_{j=1,\, j\neq 1}^{n} b_{ij} (\theta_i - \theta_j) \\ &= \left(V_i \sum_{j=1}^{n} g_{ij} + \sum_{j=1,\, j\neq 1}^{n} (-g_{ij}) V_j \right) - \left(\theta_i \sum_{j=1,\, j\neq 1}^{n} b_{ij} + \sum_{j=1,\, j\neq 1}^{n} (-b_{ij}) \theta_j \right) \\ &= \sum_{j=1}^{n} G_{ij} V_j - \sum_{j=1}^{n} B'_{ij} \theta_j \end{aligned} \tag{2.4}$$

Similarly, Eq. (2.2) for reactive power can be approximated as follows:

$$Q_i = -\sum_{j=1}^{n} B_{ij}V_j - \sum_{j=1}^{n} G_{ij}\theta_j \tag{2.5}$$

Note that $G'_{ij} \approx G_{ij}$ is assumed in (2.5) because the shunt conductance is negligible compared with the shunt susceptance.

The approximations made in (2.4) and (2.5) are not the same as those made in other linear power flow models. Consider the following three alternative approximations in the expression of branch MW flow: corresponding to the DCPF, the approximation made by [12], and the proposed approximation, respectively,

- Approximation 1: $g_{ij}\ V_i(V_i - V_j \cos\theta_{ij}) \approx 0$
- Approximation 2: $g_{ij}\ V_i(V_i - V_j \cos\theta_{ij}) \approx g_{ij}(V_i^2 - V_j^2)$
- Approximation 3: $g_{ij}\ V_i(V_i - V_j \cos\theta_{ij}) \approx g_{ij}\ (V_i - V_j)$

The common assumption is that $\cos\theta_{ij} \approx 1$. However, the methods dealing with voltage magnitude are not identical. In the DC model the voltage differences across branches are neglected. However, these are considered in Approximation 3, which is the key approximation step in the derivation of the proposed DLPF model. Although Ref. [12] takes voltage differences into account by using Approximation 2, the result is not satisfactory. Fig. 2.1 illustrates the effects of the three approximations on branch MW flow in two typical scenarios (a high and a low r/x ratio). It is evident that the approximation proposed in this chapter (Approximation 3) performs the best in both cases.

In fact, a theoretical mechanism exists underlying the proposed approximation:

$$\begin{aligned} g_{ij}V_i\left(V_i - V_j\cos\theta_{ij}\right) &\approx g_{ij}V_i\left(V_i - V_j\right) \\ &= g_{ij}\left(1+\Delta V_j\right)\left(\Delta V_i - \Delta V_j\right) \\ &\approx g_{ij}\left(\Delta V_i - \Delta V_j\right) \\ &= g_{ij}\left(1+\Delta V_i - 1 - \Delta V_j\right) \\ &= g_{ij}\left(V_i - V_j\right) \end{aligned} \tag{2.6}$$

Note that V_i is decomposed into $(1 + \Delta V_i)$, where ΔV_i is typically one order of magnitude smaller than V_i. In the second step of (2.6), ΔV_i^2 and $\Delta V_i \Delta V_j$ can be neglected without excessive error because they are two orders of magnitude smaller than V_i and V_j.

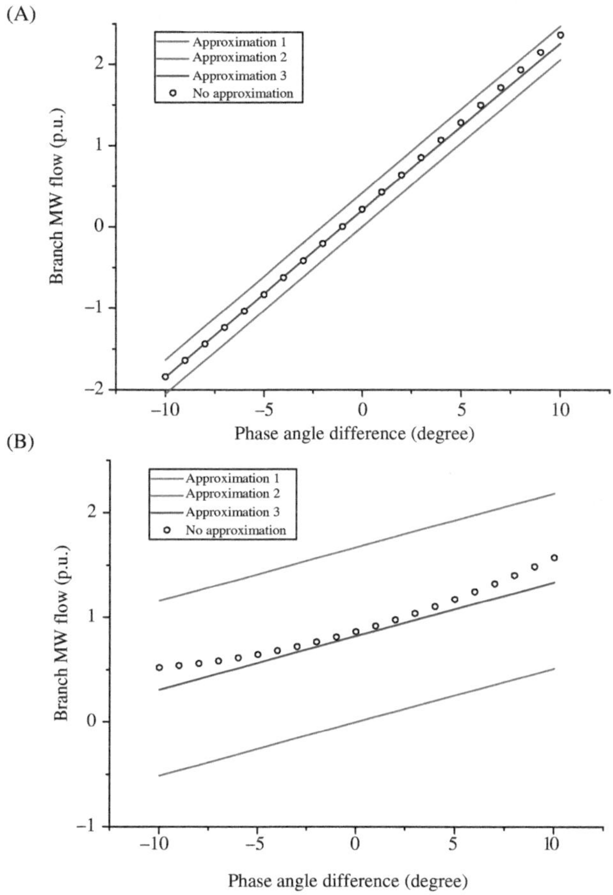

Figure 2.1 An illustration of the accuracy of different approximations with $V_i = 1.05$ p.u. and $V_j = 0.98$ p.u. (A) Case I: $r/x = 1/4$ with $r = 0.02$ p.u. and $x = 0.08$ p.u. and (B) case II: $r/x = 4/1$ with $r = 0.08$ p.u. and $x = 0.02$ p.u.

A linear expression for the average branch MW flow P_{ij} can also be easily obtained from the first step of (2.4):

$$P_{ij} = g_{ij}\left(V_i - V_j\right) - b_{ij}\left(\theta_i - \theta_j\right) \tag{2.7}$$

Because the line loss is ignored for linearity, we have $P_{ij} = -P_{ji}$, where P_{ij} denotes the power flow from bus i to bus j. It should be noted that Eq. (2.7) has to be modified for lines with transformers and phase shifters. The details are also provided in Section 2.2.3.

2.2.2 Matrix formulation of the decoupled linearized power flow model

Equations (2.4) and (2.5) represent the basic formulation of the DLPF model. The matrix form is as follows:

$$\begin{bmatrix} \boldsymbol{P} \\ \boldsymbol{Q} \end{bmatrix} = -\begin{bmatrix} B' & \boldsymbol{G} \\ \boldsymbol{G} & B \end{bmatrix} \begin{bmatrix} \boldsymbol{\theta} \\ \boldsymbol{V} \end{bmatrix} \tag{2.8}$$

It is evident that both $\boldsymbol{\theta}$ and $\boldsymbol{V}$ consist of three subvectors, corresponding to the $V\theta$, PV, and PQ buses. Without loss of generality, we arrange these buses in the following sequence: $V\theta$, PV, and PQ.

$$\begin{aligned} \boldsymbol{\theta} &= [\boldsymbol{\theta}_{\mathcal{R}}^T, \boldsymbol{\theta}_{\mathcal{S}}^T, \boldsymbol{\theta}_{\mathcal{L}}^T]^T \\ \boldsymbol{V} &= [\boldsymbol{V}_{\mathcal{R}}^T, \boldsymbol{V}_{\mathcal{S}}^T, \boldsymbol{V}_{\mathcal{L}}^T]^T \end{aligned} \tag{2.9}$$

The admittance matrix $\boldsymbol{Y}$ can also be partitioned in the same manner:

$$Y = \begin{bmatrix} \boldsymbol{Y}_{\mathcal{R}\mathcal{R}} & \boldsymbol{Y}_{\mathcal{R}\mathcal{S}} & \boldsymbol{Y}_{\mathcal{R}\mathcal{L}} \\ \boldsymbol{Y}_{\mathcal{S}\mathcal{R}} & \boldsymbol{Y}_{\mathcal{S}\mathcal{S}} & \boldsymbol{Y}_{\mathcal{S}\mathcal{L}} \\ \boldsymbol{Y}_{\mathcal{L}\mathcal{R}} & \boldsymbol{Y}_{\mathcal{L}\mathcal{S}} & \boldsymbol{Y}_{\mathcal{L}\mathcal{L}} \end{bmatrix} \tag{2.10}$$

With the aid of the known subvectors $\boldsymbol{\theta}_{\mathcal{R}}$, $\boldsymbol{V}_{\mathcal{R}}$, and $\boldsymbol{V}_{\mathcal{S}}$, we transform (2.8) into the following equation:

$$\begin{bmatrix} \tilde{\boldsymbol{P}} \\ \tilde{\boldsymbol{Q}} \end{bmatrix} = \begin{bmatrix} \boldsymbol{H} & \boldsymbol{N} \\ \boldsymbol{M} & \boldsymbol{L} \end{bmatrix} \begin{bmatrix} \tilde{\boldsymbol{\theta}} \\ \tilde{\boldsymbol{V}} \end{bmatrix} \tag{2.11}$$

where

$$\begin{bmatrix} \tilde{\boldsymbol{P}} \\ \tilde{\boldsymbol{Q}} \end{bmatrix} = \begin{bmatrix} \boldsymbol{P}_{\mathcal{S}} \\ \boldsymbol{P}_{\mathcal{L}} \\ \boldsymbol{Q}_{\mathcal{L}} \end{bmatrix} \begin{bmatrix} \boldsymbol{B'}_{\mathcal{S}\mathcal{R}} - \boldsymbol{G}_{\mathcal{S}\mathcal{R}} - \boldsymbol{G}_{\mathcal{S}\mathcal{S}} \\ \boldsymbol{B'}_{\mathcal{L}\mathcal{R}} - \boldsymbol{G}_{\mathcal{L}\mathcal{R}} - \boldsymbol{G}_{\mathcal{L}\mathcal{S}} \\ \boldsymbol{B'}_{\mathcal{L}\mathcal{R}} - \boldsymbol{G}_{\mathcal{L}\mathcal{R}} - \boldsymbol{G}_{\mathcal{L}\mathcal{S}} \end{bmatrix} \begin{bmatrix} \boldsymbol{\theta}_{\mathcal{R}} \\ \boldsymbol{V}_{\mathcal{R}} \\ \boldsymbol{V}_{\mathcal{S}} \end{bmatrix} \tag{2.12}$$

$$\begin{bmatrix} \boldsymbol{H} & \boldsymbol{N} \\ \boldsymbol{M} & \boldsymbol{L} \end{bmatrix} = -\begin{bmatrix} \boldsymbol{B'}_{\mathcal{S}\mathcal{S}} - \boldsymbol{B'}_{\mathcal{S}\mathcal{L}} - \boldsymbol{G}_{\mathcal{S}\mathcal{L}} \\ \boldsymbol{B'}_{\mathcal{L}\mathcal{S}} - \boldsymbol{B'}_{\mathcal{L}\mathcal{L}} - \boldsymbol{G}_{\mathcal{L}\mathcal{L}} \\ \boldsymbol{G}_{\mathcal{L}\mathcal{S}} - \boldsymbol{G}_{\mathcal{L}\mathcal{L}} - \boldsymbol{B}_{\mathcal{L}\mathcal{L}} \end{bmatrix} \tag{2.13}$$

$$\begin{aligned} \tilde{\boldsymbol{\theta}} &= \left[\boldsymbol{\theta}_{\mathcal{S}}^T, \boldsymbol{\theta}_{\mathcal{L}}^T\right]^T \\ \tilde{\boldsymbol{V}} &= \boldsymbol{V}_{\mathcal{L}} \end{aligned} \tag{2.14}$$

Performing elementary row operations on both sides of (2.11) leads to the following equations:

$$\begin{bmatrix} \tilde{\boldsymbol{P}} - \boldsymbol{N}\boldsymbol{L}^{-1}\tilde{\boldsymbol{Q}} \\ \tilde{\boldsymbol{Q}} \end{bmatrix} = \begin{bmatrix} \boldsymbol{H} - \boldsymbol{N}\boldsymbol{L}^{-1}\boldsymbol{M} & 0 \\ \boldsymbol{M} & \boldsymbol{L} \end{bmatrix} \begin{bmatrix} \tilde{\boldsymbol{\theta}} \\ \tilde{\boldsymbol{V}} \end{bmatrix} \tag{2.15}$$

and

$$\begin{bmatrix} \tilde{\boldsymbol{P}} \\ \tilde{\boldsymbol{Q}} - \boldsymbol{M}\boldsymbol{H}^{-1}\tilde{\boldsymbol{P}} \end{bmatrix} = \begin{bmatrix} \boldsymbol{H} & \boldsymbol{N} \\ \boldsymbol{0} & \boldsymbol{L} - \boldsymbol{M}\boldsymbol{H}^{-1}\boldsymbol{N} \end{bmatrix} \begin{bmatrix} \tilde{\boldsymbol{\theta}} \\ \tilde{\boldsymbol{V}} \end{bmatrix} \tag{2.16}$$

Combining the first part of (2.15) and the second part of (2.16) into one equation leads to the decoupling of voltage magnitudes and phase angles:

$$\begin{bmatrix} \tilde{\boldsymbol{P}} - \boldsymbol{N}\boldsymbol{L}^{-1}\tilde{\boldsymbol{Q}} \\ \tilde{\boldsymbol{Q}} - \boldsymbol{M}\boldsymbol{H}^{-1}\tilde{\boldsymbol{P}} \end{bmatrix} = \begin{bmatrix} \tilde{\boldsymbol{H}} & 0 \\ 0 & \tilde{\boldsymbol{L}} \end{bmatrix} \begin{bmatrix} \tilde{\boldsymbol{\theta}} \\ \tilde{\boldsymbol{V}} \end{bmatrix} \tag{2.17}$$

where

$$\tilde{\boldsymbol{H}} = \boldsymbol{H} - \boldsymbol{N}\boldsymbol{L}^{-1}\boldsymbol{M} \tag{2.18}$$

$$\tilde{\boldsymbol{L}} = \boldsymbol{L} - \boldsymbol{M}\boldsymbol{H}^{-1}\boldsymbol{N} \tag{2.19}$$

According to (2.17), the expressions for calculating the unknown $\tilde{\boldsymbol{V}}$ and $\tilde{\boldsymbol{\theta}}$ with respect to the known bus injections $\tilde{\boldsymbol{P}}$ and $\tilde{\boldsymbol{Q}}$ are as follows:

$$\tilde{\boldsymbol{\theta}} = \tilde{\boldsymbol{H}}^{-1}\tilde{\boldsymbol{P}} - \tilde{\boldsymbol{H}}^{-1}\boldsymbol{N}\boldsymbol{L}^{-1}\tilde{\boldsymbol{Q}} \tag{2.20}$$

$$\tilde{\boldsymbol{V}} = \tilde{\boldsymbol{L}}^{-1}\tilde{\boldsymbol{Q}} - \tilde{\boldsymbol{L}}^{-1}\boldsymbol{M}\boldsymbol{H}^{-1}\tilde{\boldsymbol{P}} \tag{2.21}$$

Note that the decoupling of the voltage magnitudes and phase angles is not achieved by simply ignoring the relatively small conductance matrix $\boldsymbol{G}$. On the contrary, the coupling between the active and reactive powers is considered in Eqs. (2.20) and (2.21).

2.2.3 Transformers and phase shifters

This section describes the changes to the admittance matrix and MW flow approximations needed for branches with transformers or phase shifters.

2.2.3.1 Contributions to the admittance matrix

According to the formation of nodal admittance matrix $\boldsymbol{Y}$ [14], the contribution of the branch with a phase-shifting transformer (Fig. 2.2, t for tap ratio and θ_s for phase-shifting angle) to the nodal admittance matrix is

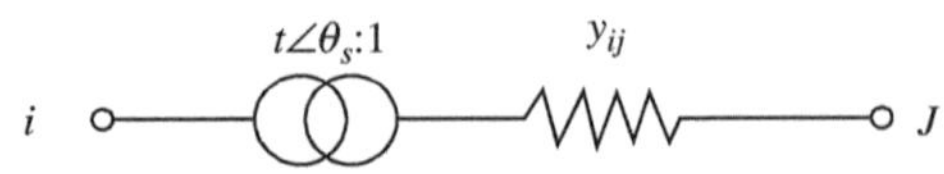

Figure 2.2 A general diagram for a branch with phase-shifting transformers.

$$\Delta \boldsymbol{Y} = \boldsymbol{C} y_{ij} \overline{\boldsymbol{C}}^T \tag{2.22}$$

where $\boldsymbol{C}$ is an $n \times 1$ vector of all zeros except $\boldsymbol{C}_i = 1/te^{-j\theta_s}$ and $\boldsymbol{C}_j = -1$. $\overline{\boldsymbol{C}}$ is the conjugate of $\boldsymbol{C}$.

$\Delta \boldsymbol{Y}$ can be split into two parts:

$$\Delta \boldsymbol{Y} = \Delta \boldsymbol{Y}_1 + \Delta \boldsymbol{Y}_2 \tag{2.23}$$

where $\Delta \boldsymbol{Y}_1$ has four nonzero elements:

$$\begin{aligned} \Delta \boldsymbol{Y}_{1,ii} &= -\Delta \boldsymbol{Y}_{1,ij} = \frac{y_{ij}}{te^{-j\theta_s}} \\ \Delta \boldsymbol{Y}_{1,jj} &= -\Delta \boldsymbol{Y}_{1,ji} = \frac{y_{ij}}{te^{j\theta_s}} \end{aligned} \tag{2.24}$$

while $\Delta \boldsymbol{Y}_2$ is a diagonal matrix with only two nonzero elements:

$$\begin{aligned} \Delta \boldsymbol{Y}_{2,ii} &= \frac{y_{ij}}{t^2} - \frac{y_{ij}}{te^{-j\theta_s}} \\ \Delta \boldsymbol{Y}_{2,jj} &= y_{ij} = \frac{y_{ij}}{te^{j\theta_s}} \end{aligned} \tag{2.25}$$

In fact, $\Delta \boldsymbol{Y}_1$ represents the contribution of equivalent line admittance to $\boldsymbol{Y}$. $\Delta \boldsymbol{Y}_2$ can be regarded as the equivalent shunt elements of phase-shifting transformers and is considered in y_{ii} as stated in Section 2.2.1.

2.2.3.2 Influence on the branch MW flow

For the branch shown in Fig. 2.2, the branch complex power has the following relationship with bus voltage and phase angle:

$$\begin{bmatrix} \overline{S}_{ij} \\ \overline{S}_{ji} \end{bmatrix} = \begin{bmatrix} V_i e^{-j\theta_i} & \\ & V_j e^{-j\theta_j} \end{bmatrix} \begin{bmatrix} \dfrac{y_{ij}}{t^2} & -\dfrac{y_{ij}}{te^{-j\theta_s}} \\ -\dfrac{y_{ij}}{te^{j\theta_s}} & y_{ij} \end{bmatrix} \begin{bmatrix} V_i e^{j\theta_i} \\ V_j e^{j\theta_j} \end{bmatrix} \tag{2.26}$$

The branch MW flow P_{ij} is the real part of S_{ij}:

$$
\begin{aligned}
P_{ij} &= \mathrm{Re}\left(V_i e^{-j\theta_i}\left(\frac{y_{ij}}{t^2}V_i e^{j\theta_i} - \frac{y_{ij}}{te^{-j\theta_s}}V_j e^{j\theta_j}\right)\right) \\
&= \mathrm{Re}\left(\frac{y_{ij}}{t^2}V_i^2 - \frac{y_{ij}}{t}V_i V_j e^{-j(\theta_i-\theta_j-\theta_s)}\right) \\
&= \frac{g_{ij}}{t}V_i\left(\frac{V_i}{t} - V_i V_j\cos(\theta_i-\theta_j-\theta_s)\right) \\
&\quad - \frac{b_{ij}}{t}V_i V_j\sin(\theta_i-\theta_j-\theta_s) \\
&\approx \frac{g_{ij}}{t}\left(\frac{V_i}{t} - V_j\right) - \frac{b_{ij}}{t}(\theta_i-\theta_j-\theta_s)
\end{aligned} \tag{2.27}
$$

The approximation made in the last step of (2.27) is similar to that in (2.4), and this provides a linear expression for branch MW flow with transformers or phase shifters. Because the line loss is ignored for linearity, we have $P_{ij} = -P_{ji}$. It should be noted that (2.27) and (2.7) have the same pattern. Eq. (2.7) can be seen as a special case of (2.27) where $t=1$ and $\theta_s = 0$.

2.2.4 Derivation and justification of the fast decoupled linearized power flow

A fast computation speed is one of the key advantages of DCPF models. The classical DCPF model achieves this speed because the admittance matrix is formulated using $1/x_{ij}$–directly from the system information. For the DLPF model proposed in Section 2.2, its solution requires the factorization of four matrices: $\boldsymbol{H}$, $\boldsymbol{L}$, $\tilde{\boldsymbol{H}}$, and $\tilde{\boldsymbol{L}}$. Although $\boldsymbol{H}$ and $\boldsymbol{L}$ are sparse matrices, $\tilde{\boldsymbol{H}}$ and $\tilde{\boldsymbol{L}}$ maybe nonsparse matrices, for which factorization is computationally intractable. This section, however, shows that $\tilde{\boldsymbol{H}}$ and $\tilde{\boldsymbol{L}}$ can be approximated by the sparse matrices $\tilde{\boldsymbol{H}}'$ and $\tilde{\boldsymbol{L}}'$, which have the same structure as $\boldsymbol{H}$ and $\boldsymbol{L}$:

$$
\tilde{H}'_{ij}\begin{cases} -\dfrac{1}{x_{ij}} & \text{if } j \neq i \\ \displaystyle\sum_{k=1,\, k\neq i}^{n} \frac{1}{x_{ik}} & \text{if } j = i \end{cases} \tag{2.28}
$$

$$\tilde{L}'_{ij}\begin{cases} -\dfrac{1}{x_{ij}} & \text{if } j \neq i \\ \displaystyle\sum_{k=1}^{n} \dfrac{1}{x_{ik}} & \text{if } j = i \end{cases} \tag{2.29}$$

Expressions (2.20) and (2.21) for bus voltage magnitude and phase angle are thus transformed into

$$\tilde{\boldsymbol{\theta}} = \tilde{\boldsymbol{H}}'^{-1}\tilde{\boldsymbol{P}} - \tilde{\boldsymbol{H}}'^{-1}\boldsymbol{N}\boldsymbol{L}^{-1}\tilde{\boldsymbol{Q}} \tag{2.30}$$

$$\tilde{\boldsymbol{V}} = \tilde{\boldsymbol{L}}'^{-1}\tilde{\boldsymbol{Q}} - \tilde{\boldsymbol{L}}'^{-1}\boldsymbol{M}\boldsymbol{H}^{-1}\tilde{\boldsymbol{P}} \tag{2.31}$$

The approximation always holds for a radial system or a meshed system with a constant r/x ratio, as justified by the theoretical derivation presented in this section. For more complex cases the approximation error is mild, as can be demonstrated by a numerical example.

2.2.4.1 An illustrative example

Let us begin with a 2-bus system (Fig. 2.3) for illustration. In this case, Eq. (2.13) becomes

$$\begin{bmatrix} \boldsymbol{H} & \boldsymbol{N} \\ \boldsymbol{M} & \boldsymbol{L} \end{bmatrix} = \begin{bmatrix} -b_{12} & g_{12} \\ -g_{12} & -b_{12} - b_{22} \end{bmatrix} \tag{2.32}$$

and $\tilde{\boldsymbol{H}}$ and $\tilde{\boldsymbol{L}}$ can be calculated as follows:

$$\begin{aligned} \tilde{\boldsymbol{H}} &= \boldsymbol{H} - \boldsymbol{N}\boldsymbol{L}^{-1}\boldsymbol{M} \\ &= -b_{12} - g_{12}(b_{12} + b_{22})^{-1}g_{12} \\ &\simeq -b_{12} - g_{12}b_{12}^{-1}g_{12} \\ &= \frac{1}{x_{12}} \end{aligned} \tag{2.33}$$

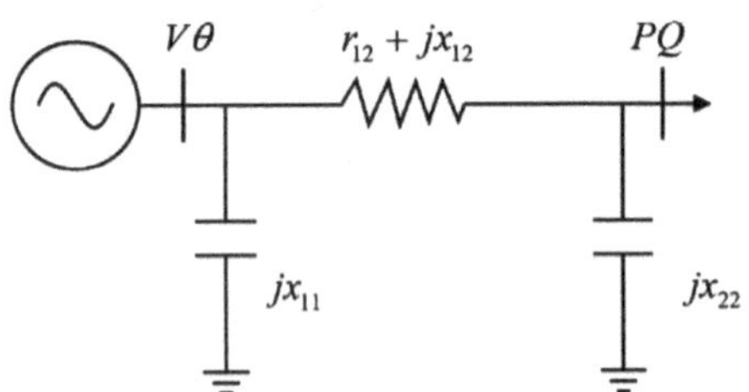

$$g_{12} = \frac{r_{12}}{r_{12}^2 + x_{12}^2} \quad b_{12} = -\frac{x_{12}}{r_{12}^2 + x_{12}^2} \quad b_{11} = -\frac{1}{x_{11}} \quad b_{22} = -\frac{1}{x_{22}}$$

Figure 2.3 A 2-bus system.

$$\begin{aligned}\tilde{\boldsymbol{L}} &= \boldsymbol{L} - \boldsymbol{M}\boldsymbol{H}^{-1}\boldsymbol{N} \\ &= -b_{12} - b_{22} - g_{12}b_{12}^{-1}g_{12} \\ &= \frac{1}{x_{12}} + \frac{1}{x_{22}}\end{aligned} \tag{2.34}$$

It can be seen from (2.33) and (2.34) that the updated elements are exactly identical to the diagonal elements of $\tilde{\boldsymbol{H}}'$ and $\tilde{\boldsymbol{L}}'$. This surprising feature can be extended to more general cases, as demonstrated next.

2.2.4.2 Theoretical derivation

For a general power system, $\tilde{\boldsymbol{H}}$ and $\tilde{\boldsymbol{L}}$ are too complicated to explicitly derive. However, for a radial system or a constant-r/x-ratio system without PV buses, it can be proven that $\tilde{\boldsymbol{L}}'$ is identical to $\tilde{\boldsymbol{L}}$, whereas $\tilde{\boldsymbol{H}}'$ is a good approximation of $\tilde{\boldsymbol{H}}$.

Let $\boldsymbol{y}_{\mathcal{SL}} = \boldsymbol{g}_{\mathcal{SL}} + j\boldsymbol{b}_{\mathcal{SL}}$ denote the primitive admittance matrix of a system in which the $V\,\theta$ buses are grounded and all shunt susceptances are neglected. $\boldsymbol{C}$ represents the node-to-branch incidence matrix. The shunt susceptances are represented by the $(n-1)\times(n-1)$ diagonal matrix $\boldsymbol{diag}(b_0)$.

For a meshed system with a constant r/x ratio, let us assume that for any branch l, $r_l/x_l = \alpha$.

This leads to

$$\begin{aligned}g_l &= -\frac{r_l}{x_l}\frac{-x_l}{r_l^2 + x_l^2} = -\alpha b_l \\ (1+\alpha^2)b_l &= \left(1 + \frac{r_l^2}{x_l^2}\right)\frac{-x_l}{r_l^2 + x_l^2} = -\frac{1}{x_l}\end{aligned} \tag{2.35}$$

and thus

$$\begin{aligned}\tilde{\boldsymbol{H}} &= \boldsymbol{H} - \boldsymbol{N}\boldsymbol{L}^{-1}\boldsymbol{M} \\ &= -\boldsymbol{C}\boldsymbol{b}_{\mathcal{SL}}\boldsymbol{C}T - \left(\boldsymbol{C}\boldsymbol{g}_{\mathcal{SL}}\boldsymbol{C}^T\right)\cdot \\ &\quad \left(\boldsymbol{C}\boldsymbol{b}_{\mathcal{SL}}\boldsymbol{C}^T + \boldsymbol{diag}(b_0)\right)^{-1}\left(\boldsymbol{C}\boldsymbol{g}_{\mathcal{SL}}\boldsymbol{C}^T\right) \\ &\approx -\boldsymbol{C}\boldsymbol{b}_{\mathcal{SL}}\boldsymbol{C}^T - \alpha^2\left(\boldsymbol{C}\boldsymbol{b}_{\mathcal{SL}}\boldsymbol{C}^T\right)\cdot \\ &\quad \left(\boldsymbol{C}\boldsymbol{b}_{\mathcal{SL}}\boldsymbol{C}^T\right)^{-1}\left(\boldsymbol{C}\boldsymbol{b}_{\mathcal{SL}}\boldsymbol{C}^T\right) \\ &= -(1+\alpha^2)\boldsymbol{C}\boldsymbol{b}_{\mathcal{SL}}\boldsymbol{C}^T \\ &= \tilde{\boldsymbol{H}}'\end{aligned} \tag{2.36}$$

$$\begin{aligned}\tilde{\boldsymbol{L}} &= \boldsymbol{L} - \boldsymbol{M}\boldsymbol{H}^{-1}\boldsymbol{N} \\ &= -\boldsymbol{C}\boldsymbol{b}_{\mathcal{SL}}\,\boldsymbol{C}^T - \boldsymbol{diag}(b_0) \\ &\quad - \left(\boldsymbol{C}\boldsymbol{g}_{\mathcal{SL}}\,\boldsymbol{C}^T\right)\left(\boldsymbol{C}\boldsymbol{b}_{\mathcal{SL}}\,\boldsymbol{C}^T\right)^{-1}\left(\boldsymbol{C}\boldsymbol{g}_{\mathcal{SL}}\,\boldsymbol{C}^T\right) \\ &= -\left(1+\alpha^2\right)\boldsymbol{C}\boldsymbol{b}_{\mathcal{SL}}\,\boldsymbol{C}^T - \boldsymbol{diag}(b_0) \\ &= \tilde{\boldsymbol{L}}'\end{aligned} \tag{2.37}$$

For a radial system (usually a distribution system), $\boldsymbol{C}$ is a nonsingular matrix. It can also be easily found that $\tilde{\boldsymbol{L}}' = \tilde{\boldsymbol{L}}$ and $\tilde{\boldsymbol{H}}' \approx \tilde{\boldsymbol{H}}$, in the same manner as for meshed systems.

2.2.4.3 A numerical example

The approximations of $\tilde{\boldsymbol{H}}$ and $\tilde{\boldsymbol{L}}$ also hold for more general meshed systems. Let us demonstrate this claim using a numerical example of a Grainger and Stevenson 4-bus system [14], which is shown in Fig. 2.4.

Using Eq. (2.13), the matrices $\boldsymbol{H}$, $\boldsymbol{N}$, $\boldsymbol{M}$, and $\boldsymbol{L}$ are as follows:

$$\begin{bmatrix}\boldsymbol{H} & \boldsymbol{N} \\ \boldsymbol{M} & \boldsymbol{L}\end{bmatrix} = \begin{bmatrix} 40.97 & -25.85 & -15.12 & -5.17 & -3.02 \\ -25.85 & 44.93 & 0 & 8.99 & 0 \\ -15.12 & 0 & 40.97 & 0 & 8.19 \\ 5.17 & -8.99 & 0 & 44.84 & 0 \\ 3.02 & 0 & -8.19 & 0 & 40.86 \end{bmatrix}$$

The matrices $\tilde{\boldsymbol{H}}$ and $\tilde{\boldsymbol{L}}$ are derived from Eqs. (2.18) and (2.19):

$$\tilde{\boldsymbol{H}} = \begin{bmatrix} 41.79 & -26.88 & -15.72 \\ -26.88 & 46.73 & 0 \\ -15.72 & 0 & 42.61 \end{bmatrix}, \quad \tilde{\boldsymbol{L}} = \begin{bmatrix} 46.63 & 0 \\ 0 & 42.50 \end{bmatrix}$$

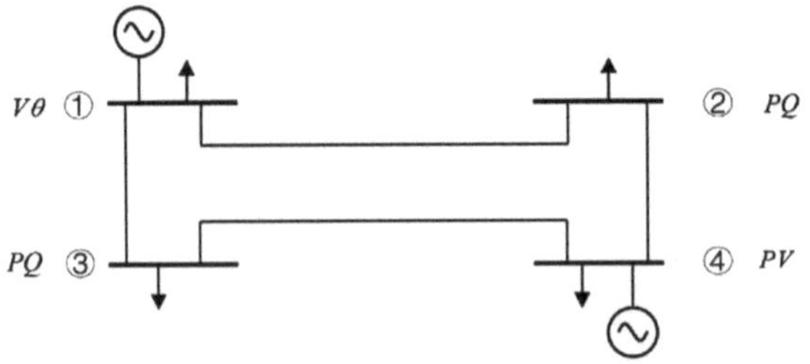

Figure 2.4 Grainger and Stevenson 4-bus system.

The approximate matrices $\tilde{\boldsymbol{H}}'$ and $\tilde{\boldsymbol{L}}'$ are defined by (2.28) and (2.29):

$$\tilde{\boldsymbol{H}}' = \begin{bmatrix} 42.60 & -26.88 & -15.72 \\ -26.88 & 46.72 & 0 \\ -15.72 & 0 & 42.61 \end{bmatrix}, \quad \tilde{\boldsymbol{L}}' = \begin{bmatrix} 46.63 & 0 \\ 0 & 42.50 \end{bmatrix}$$

It can be easily seen that $\tilde{\boldsymbol{L}}'$ and $\tilde{\boldsymbol{L}}$ are identical, whereas $\tilde{\boldsymbol{H}}'$ is extremely close to $\tilde{\boldsymbol{H}}$.

2.3 Linear three-phase power flow models of the unbalanced distribution systems

The basic three-phase power flow equations can be formulated by taking the mutual inductance and interphase capacitance among different phases into consideration. It is shown as follows:

$$P_i^\alpha = V_i^\alpha \sum_{j=1}^{N} \sum_{\beta=a,b,c} V_j^\beta \left(G_{ij}^{\alpha\beta} \cos\theta_{ij}^{\alpha\beta} + B_{ij}^{\alpha\beta} \sin\theta_{ij}^{\alpha\beta} \right) \tag{2.38}$$

$$Q_i^\alpha = V_i^\alpha \sum_{j=1}^{N} \sum_{\beta=a,b,c} V_j^\beta \left(G_{ij}^{\alpha\beta} \sin\theta_{ij}^{\alpha\beta} - B_{ij}^{\alpha\beta} \cos\theta_{ij}^{\alpha\beta} \right) \tag{2.39}$$

where N denotes the number of nodes of active distribution network; i and j are node indexes; α and β are phase indexes; P_i^α and Q_i^α denote the active and reactive injections of phase α and node i; V_i^α denotes the voltage magnitude and phase angle of phase α; and node i; $G_{ij}^{\alpha\beta}$, $B_{ij}^{\alpha\beta}$, and $\theta_{ij}^{\alpha\beta}$ denote the conductance, susceptance, and angle difference between phases α and β, and nodes i and j, respectively.

Three linear approximations are applied to linearize the relationship among active and reactive power, voltage magnitude, and phase angle.

Approximation 1. Since the angle difference $\theta_{ij}^{\alpha\alpha}$ of each branch in active distribution network is very small, the trigonometric sines and cosines of it can be approximated as $\cos\theta_{ij}^{\alpha\alpha} \approx 1$ and $\sin\theta_{ij}^{\alpha\alpha} \approx \theta_{ij}^{\alpha\alpha}$. $\theta_{ij}^{\alpha\beta}$, the angle difference of different phases, is near $2/3\pi$ or $-2/3\pi$, the trigonometric functions of which can also be linearized similarly near $2/3\pi$ or $-2/3\pi$.

Approximation 2. The production of two voltages is linearized as follows by neglecting the second-order small quantities.

$$V_k(V_k - V_j) = (V_k - V_j) + (1 - V_k)^2 - (1 - V_k)(1 - V_j) \approx (V_k - V_j) \tag{2.40}$$

Approximation 3. The ZIP model of the loads describing the load variation with respect to the voltage is (taking active power as example): $P\ (V) = (F_Z V^2 + F_I V + F_P)P_0$ where, F_Z, F_I, and F_P are the constant coefficients corresponding to impedance, current, and power contributions, respectively; P_0 denotes the active power of the load under normal voltage. The second-order small quantities ΔV^2 is neglected to obtain the approximated IP model of the loads.

$$\begin{aligned} P(V) &= (F_Z V^2 + F_I V + F_P)P_0 = (F_Z(1+\Delta V)^2 + F_I(1 + \Delta V) + F_P)P_0 \\ &\approx (F_P - F_Z)P_0 + [(2F_Z + F_I)P_0]V \end{aligned} \tag{2.41}$$

The first two approximations linearize the nonlinear terms and trigonometric functions on the right side of Eq. (2.1) by first-order approximation. The third approximation transforms the ZIP model into PI model of the loads in the left side of Eq. (2.1) by removing the production of two voltages. Thus the relationship between voltages and injections described in Eq. (2.1) is fully linearized. Eq. (2.1) can be viewed as the summation of two parts:

$$A = V_i^{\alpha} \sum_{j=1}^{n} \sum_{\beta=\alpha}^{c} V_j^{\beta} G_{ij}^{\alpha\beta} \cos \theta_{ij}^{\alpha\beta} \tag{2.42}$$

$$B = V_i^{\alpha} \sum_{j=1}^{n} \sum_{\beta=\alpha}^{c} V_j^{\beta} B_{ij}^{\alpha\beta} \sin \theta_{ij}^{\alpha\beta} \tag{2.43}$$

To illustrate out derivation process the first part of Eq. (2.1), which is denoted as A, is provided to demonstrate how the linear relationship is derived. The rest of Eq. (2.1) can be linearized accordingly. According to Approximation 1, $\cos\theta_{ij}^{\alpha\beta}$ can be linearized as follows:

$$
\begin{aligned}
A &= V_i^{\alpha} \sum_{j=1}^{n} \sum_{\beta=\alpha}^{c} V_j^{\beta} G_{ij}^{\alpha\beta} \cos \theta_{ij}^{\alpha\beta} \\
&\approx \sum_{j=1}^{n} V_j^{\alpha} G_{ij}^{\alpha\alpha} + \sum_{j=1}^{n} \sum_{\beta \neq \alpha} V_j^{\beta} G_{ij}^{\alpha\beta} \left(-\frac{1}{2} + \frac{\sqrt{3}}{2} \theta_{ij}^{\alpha\beta} \right)
\end{aligned}
\tag{2.44}
$$

By disassembling the second sum term and according to Approximation 2, Eq. (2.7) can be further approximated:

$$
\begin{aligned}
A \approx & \frac{3}{2} \sum_{j=1}^{n} V_j^{\alpha} G_{ij}^{\alpha\alpha} - \frac{1}{2} \sum_{j=1}^{n} \sum_{\beta=\alpha}^{c} V_j^{\beta} G_{ij}^{\alpha\beta} \\
& + \frac{\sqrt{3}}{2} \sum_{j=1}^{n} G_{ij}^{\alpha\alpha_{lead}} \theta_{ij}^{\alpha\alpha_{lead}} - \frac{\sqrt{3}}{2} \sum_{j=1}^{n} G_{ij}^{\alpha\alpha_{lag}} \theta_{ij}^{\alpha\alpha_{lag}}
\end{aligned}
\tag{2.45}
$$

where α_{lead} and α_{lag} denote the phases that lead or lag to phase α, respectively. Then the original three-phase power equations can be neatly presented in a matrix form:

$$
\begin{bmatrix} P^A \\ P^B \\ P^C \\ Q^A \\ Q^B \\ Q^C \end{bmatrix} =
\begin{bmatrix}
J_{PV}^{AA} & J_{PV}^{AB} & J_{PV}^{AC} & J_{P\theta}^{AA} & J_{P\theta}^{AB} & J_{P\theta}^{AC} \\
J_{PV}^{BA} & J_{PV}^{BB} & J_{PV}^{BC} & J_{P\theta}^{BA} & J_{P\theta}^{BB} & J_{P\theta}^{BC} \\
J_{PV}^{CA} & J_{PV}^{CB} & J_{PV}^{CC} & J_{P\theta}^{CA} & J_{P\theta}^{CB} & J_{P\theta}^{CC} \\
J_{QV}^{AA} & J_{QV}^{AB} & J_{QV}^{AC} & J_{Q\theta}^{AA} & J_{Q\theta}^{AB} & J_{Q\theta}^{AC} \\
J_{QV}^{BA} & J_{QV}^{BB} & J_{QV}^{BC} & J_{Q\theta}^{BA} & J_{Q\theta}^{BB} & J_{Q\theta}^{BC} \\
J_{QV}^{CA} & J_{QV}^{CB} & J_{QV}^{CC} & J_{Q\theta}^{CA} & J_{Q\theta}^{CB} & J_{Q\theta}^{CC}
\end{bmatrix}
\begin{bmatrix} V^A \\ V^B \\ V^C \\ \theta^A \\ \theta^B \\ \theta^C \end{bmatrix}
\tag{2.46}
$$

where P^{α},Q^{α}, V^{α}, and θ^{α} denote the N-dimensional vectors for active power, reactive power, voltage magnitude, and angle of phase α. $J_{(PQ)(V\theta)}^{\alpha\beta}$ is the $N \times N$ dimensional matrix describing the coupling relationship between active (or reactive) power and voltage magnitude (or angle) of phases α and β. This matrix uncovers how the injections of the nodes affect the nodal voltages.

The matrix $J_{(PQ)(V\theta)}^{\alpha\beta}$ can be calculated using three-phase admittance matrix. For example, $J_{PV}^{AB} = -(1/2)G^{AB} + (\sqrt{3}/2)B^{AB}$, $J_{PV}^{AA} = G^{AA}$ where $G^{\alpha\beta}$ and $B^{\alpha\beta}$ denote the conductance and susceptance matrices of phases α and β. If the ZIP model of the loads is considered, the vectors P^{α}, Q^{α} and matrices $J_{PV}^{\alpha\alpha}$, $J_{QV}^{\alpha\alpha}$ should be revised accordingly by Eq. (2.4). Finally, the vectors V^{α} and θ^{α} can be solved by Gauss elimination using Eq. (2.9).

2.4 Case study

2.4.1 Meshed transmission systems

Table 2.1 compares the performances of the different models for various kinds of transmission systems. The model proposed in Ref. [11] is excluded from this comparison because it was designed only for radial networks. Fig. 2.5 visualizes the solutions for the IEEE 118-bus system in terms of voltage magnitude and branch MW flow as an example for illustration. A comparison of the computation times for the large-scale system is also presented in Table 2.2.

From Table 2.1 and Fig. 2.5, it is evident that the results of the DLPF and FDLPF models are the most accurate in terms of voltage magnitude among various cases of large-scale transmission systems. The accuracy of the DLPF model in terms of MW flow is also higher than that of the classical DCPF model. Although the FDLPF model performs a little worse than the DCPF model with respect to MW flow in some cases, the errors are acceptable because the main advantage of FDLPF is its accurate evaluation of bus voltage magnitudes. Moreover, Table 2.2 suggests that the FDLPF model significantly improved computational efficiency by sacrificing only limited precision in terms of MW flow while maintaining the voltage magnitude accuracy of the DLPF model. The FDLPF model is therefore suitable for applications that require a large number of repeated load flow calculations.

A comparison of the computation times for the large-scale system is presented in Table 2.2. Comparing it with Table 2.1 suggests that the

Table 2.1 Errors of different linear power flow models for transmission systems.

Test case	$\varepsilon_V^{DCPF\,a}$	$\varepsilon_V^{md-[12]}$	ε_V^{DLPF}	ε_V^{FDLPF}	ε_P^{DCPF}	$\varepsilon_P^{md-[12]}$	ε_P^{DLPF}	ε_P^{FDLPF}
IEEE 30-bus	0.018	0.011	0.001	0.001	0.004	0.023	0.002	0.007
IEEE 57-bus	0.024	0.012	0.007	0.009	0.015	0.062	0.010	0.017
IEEE 118-bus	0.023	0.003	0.001	0.001	0.035	0.090	0.034	0.042
IEEE 300-bus	0.025	0.016	0.014	0.014	0.105	0.148	0.102	0.105
Polish 2383-bus	0.008	0.005	0.003	0.003	0.026	0.047	0.022	0.038
Polish 3012-bus	0.089	0.005	0.002	0.003	0.022	0.055	0.021	0.034
Pegase 2869-bus	0.034	0.008	0.008	0.008	0.119	0.157	0.115	0.119
Pegase 9241-bus	0.032	0.014	0.007	0.007	0.181	0.208	0.178	0.192

DCPF, DC power flow.
[a]The bus voltage profile is assumed to be flat at 1.0 p.u. in the DCPF model.

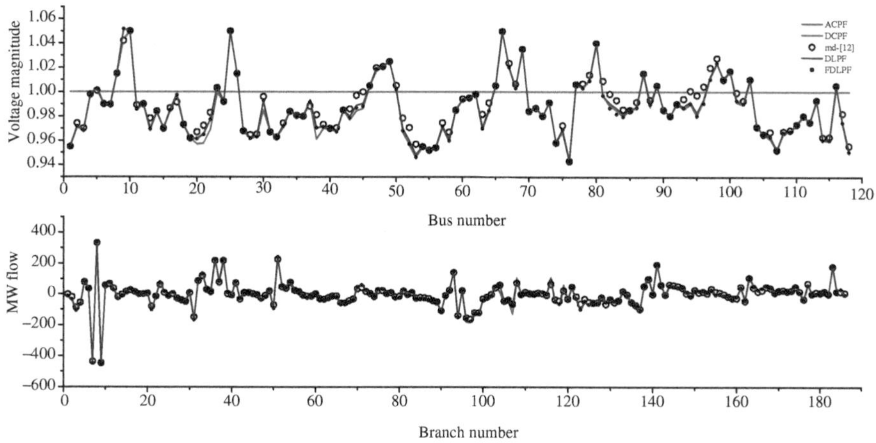

Figure 2.5 Voltage magnitudes and branch MW flows for the IEEE 118-bus system.

Table 2.2 Computational efficiencies of different linear power flow models for large-scale systems.

Test case	t^{ACPF}[a]	t^{DCPF}	$t^{md\text{-}[12]}$	t^{DLPF}	t^{FDLPF}
Polish 2383	0.24	0.01	1.75	1.60	0.01
Polish 3012	0.27	0.02	5.60	9.42	0.02
Pegase 2869	0.30	0.02	1.90	1.71	0.02
Pegase 9241	4.03	0.08	43.0	110	0.09

[a]The unit of computational time is second.

FDLPF model achieves significantly improved computational efficiency by sacrificing only limited precision in terms of MW flow while maintaining the voltage magnitude accuracy of the DLPF model. The FDLPF model is therefore suitable for applications that require a large number of repeated load flow calculations.

2.4.2 Balanced distribution systems

In this section, several cases are analyzed to demonstrate the accuracy and robustness of the DLPF/FDLPF approach and the fast computation speed of the FDLPF model [15]. Various types of power systems are studied.

The analyses were performed in MATLAB with the aid of MATPOWER 5.1 [16]. The simulation platform was a ThinkPad workstation with an Intel i7 CPU@2.60 GHz and 8 GB of RAM. Three important results were obtained for comparison: bus voltage magnitudes, branch MW flows, and computation times. With the results of the AC

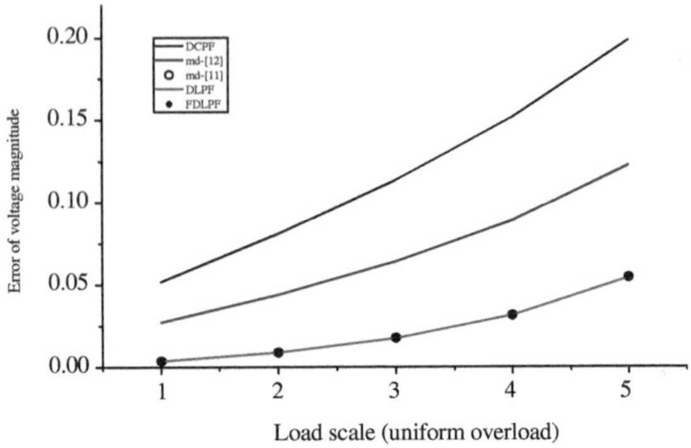

Figure 2.6 33-Node ill-conditioned system (uniform overload).

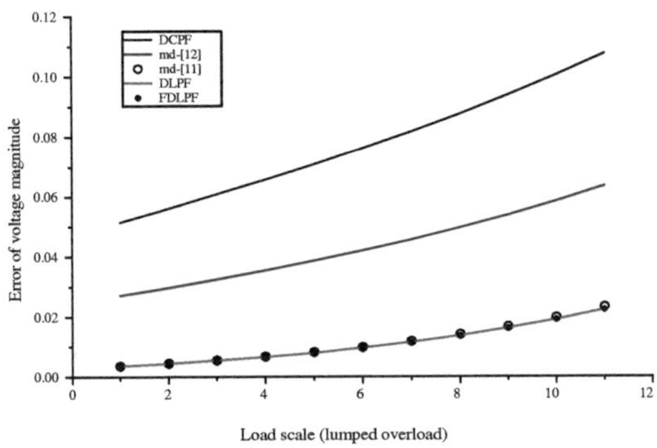

Figure 2.7 33-Node ill-conditioned system (lumped overload).

power flow model as the benchmark, the error for each model (md) was calculated as follows:

$$\varepsilon_V^{md} = \frac{1}{n}\sum_{i=1}^{n}\left|V_i^{md} - V_i^{AC}\right| \quad \text{(p.u.)} \tag{2.47}$$

$$\varepsilon_P^{md} = \frac{1}{m}\sum_{i=1}^{n}\left|P_i^{md} - P_i^{AC}\right| \quad \text{(p.u.)} \tag{2.48}$$

To avoid random timing errors, each MATLAB program was run 10 times, and the average computation time t^{md} was considered.

Table 2.3 compares the performances of different models on the IEEE 33- and 123-node test feeders. The solutions were obtained using the DCPF model, the method proposed in Ref. [12] (md-[12]), the

Table 2.3 Errors of different linear power flow models for distribution systems.

Test case	ε_V^{DCPF}[a]	$\varepsilon_V^{md-[12]}$	$\varepsilon_V^{md-[11]}$	ε_V^{DLPF}	ε_V^{FDLPF}	ε_V^{DCPF}	$\varepsilon_P^{md-[12]}$	$\varepsilon_P^{md-[11]}$[b]	ε_P^{DLPF}	ε_P^{FDLPF}
IEEE 33-node	0.052	0.027	0.004	0.004	0.004	0.026	0.847	0.979	0.026	0.026
IEEE 123-node	0.051	0.027	0.004	0.003	0.003	0.008	0.291	0.173	0.008	0.008

DCPF, DC power flow.

[a]The bus voltage profile is assumed to be flat at 1.0 p.u. in the DCPF model.

[b]The linear expression for branch MW flow is not provided in Ref. [11]. Because the phase angle is accurate, according to the results reported in Ref. [11], we assume the same expression as that for the DCPF model, namely, $\forall$ branches (i, j), $P_{(i,j)} = (\theta_i - \theta_j)/x_{ij}$.

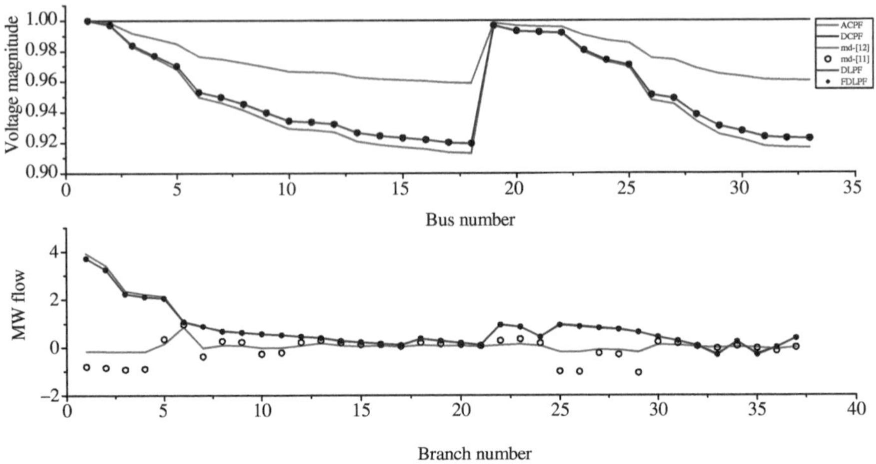

Figure 2.8 Voltage magnitudes and branch MW flows for the IEEE 33-node test feeder.

method proposed in Ref. [11] (md-[11]) and the DLPF/FDLPF approach proposed in this chapter. Fig. 2.8 visualizes the solutions for the 33-node system in terms of voltage magnitudes and branch MW flows. It should be noted that an average error of 0.05 p.u. in voltage magnitude is beyond tolerance, given that the operational voltage limit is typically between 0.95 and 1.05 p.u. From Table 2.3 and Fig. 2.8, it is obvious that the DLPF and the FDLPF models perform the best not only in voltage magnitudes but also in active power flows as well. Note that md-[11] exhibits the same accuracy in bus voltage magnitudes as the DLPF/FDLPF model but a poor performance in branch MW flows. This is not due to the flaws in the model itself but our choice to use DCPF branch flow expression, as no linear expression for branch MW flow is provided in Ref. [11]. In fact, the proposed model is a generalization of md-[11] from radial distribution systems to meshed networks with PV buses. A detailed discussion is provided in the Appendix.

2.4.3 Unbalanced distribution systems

The proposed method is applied to the IEEE 37-node test feeder with unbalanced loads [17]. For simplicity the voltage dependencies of all loads are assumed to be identical ($F_Z = 0.7$, $F_I = 0.2$, and $F_P = 0.1$) for both active and reactive power. Nodes 21 and 27 are specified as PV nodes

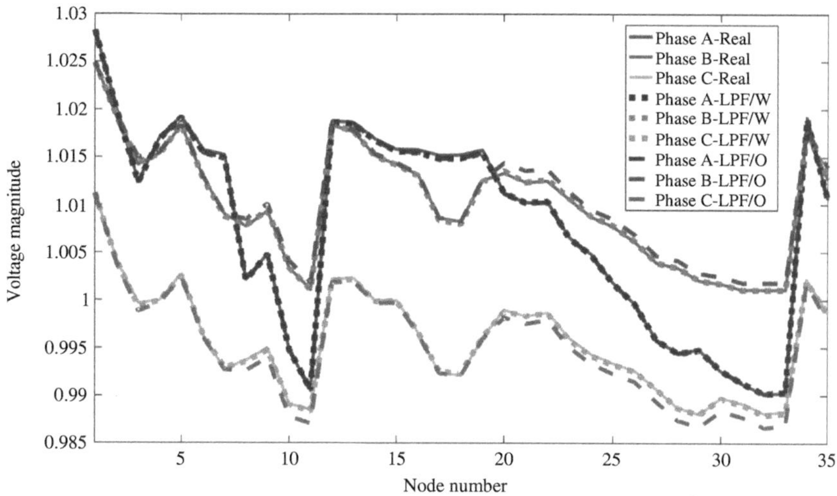

Figure 2.9 Voltage magnitudes of all nodes calculated by AC power flow and the proposed method with (W) and without (O) PV nodes.

with the same voltage magnitudes as the original system in the case with PV nodes. The voltages obtained by the backward–forward sweep algorithm are used as real values. The error of the proposed method is defined as the absolute difference between the calculated value and real value. Fig. 2.9 shows the real and calculated voltage magnitudes with (W) and without (O) PV nodes. The dotted line (calculated) is close to or even overlapped with the solid line (real value). Without PV nodes, the maximum and average errors are 1.49×10^{-3} and 3.70×10^{-4}. Such error meets the needs of many practical applications. Fig. 2.10 presents the errors of the three-phase voltage magnitudes calculated by our proposed method with PV nodes. The maximum and average errors are 2.91×10^{-4} and 8.79×10^{-5}, respectively, which are comparable with the method proposed in Ref. [18]. The introduction of PV nodes can also largely reduce the error. In rural distribution networks, voltage percent variations maybe higher than 5%. To illustrate that our proposed method can also be used indeed to detect lager voltage variations, the original load is expanded by 150%, 200%, and 300%. The average error of the calculated voltage magnitudes are 1.00×10^{-3}, 1.40×10^{-3}, and 2.80×10^{-3}, respectively. The calculated voltage magnitudes of all nodes when the loads are tripled are shown in Fig. 2.11. It can be seen that even though the node voltage can be low as 0.91, the calculated voltages can obtain

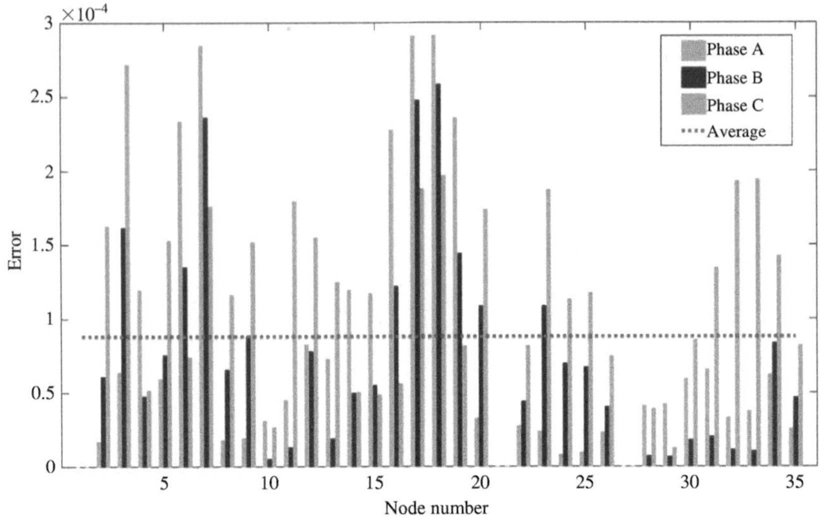

Figure 2.10 Error of the three-phase voltage magnitudes between AC power flow and the proposed method with PV nodes.

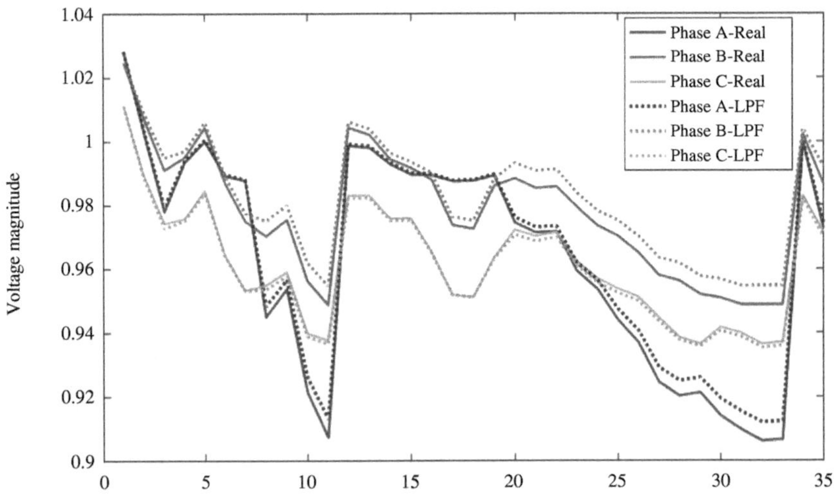

Figure 2.11 Voltage magnitudes of all nodes calculated by AC power flow and the proposed method with loads tripled.

relative higher accuracy. In addition, the reactive power injections of PV nodes (nodes 21 and 27) have been calculated. The relative errors are 0.013% and 0.014%, respectively.

2.5 Conclusion

In this chapter, we propose a state-independent, $V-\theta$ DLPF model that is distinguished by high accuracy in voltage magnitude. An in-depth analysis of the matrices used for calculation is presented, giving rise to a fast version of the DLPF model, namely, the FDLPF model. The approximation that is applied to obtain the FDLPF model from the DLPF model is justified by theoretical derivation and numerical examples. An analysis of various cases, including radial distribution systems with high r/x ratios, large-scale transmission systems, and ill-conditioned systems, proves the accuracy and robustness of the two proposed power flow models. In large-scale cases the FDLPF model is proven to be much faster than the DLPF model, thereby demonstrating the effectiveness of the approximation technique.

We envision that the proposed models can serve as useful tools in power system planning and operation. Specifically, the DLPF model has the potential to be employed in optimization problem in place of the AC power flow equations, which may allow nonlinear, nonconvex problems to be transformed into linear programming problems. In addition, the FDLPF model is characterized by high computational efficiency and has the potential to be applied in contingency analysis and probabilistic load flow problems. By utilizing the FDLPF model, voltage magnitude and reactive power can be considered as in the AC power flow model with a computational efficiency that is as high as that of the classical DCPF model.

References

[1] M.K. Enns, J.J. Quada, B. Sackett, Fast linear contingency analysis, IEEE Trans. Power Apparatus Syst. PAS-101 (4) (1982) 783–791. Available from: https://doi.org/10.1109/TPAS.1982.317142.

[2] R. Billinton, R.N. Allan, Reliability Evaluation of Engineering Systems, 1992.

[3] Z. Hu, X. Wang, A probabilistic load flow method considering branch outages, IEEE Trans. Power Syst. 21 (2) (2006) 507–514. Available from: https://doi.org/10.1109/TPWRS.2006.873118.

[4] F. Dorfler, F. Bullo, Novel insights into lossless ac and dc power flow, Power and Energy Society General Meeting, IEEE, 2013. 1–5. Available from: https://doi.org/10.1109/PESMG.2013.6672260.

[5] F. Li, R. Bo, DCOPF-based LMP simulation: algorithm, comparison with ACOPF, and sensitivity, IEEE Trans. Power Syst. 22 (4) (2007) 1475–1485. Available from: https://doi.org/10.1109/TPWRS.2007.907924.

[6] F. Lee, J. Huang, R. Adapa, Multi-area unit commitment via sequential method and a DC power flow network model, IEEE Trans. Power Syst. 9 (1) (1994) 279–287. Available from: https://doi.org/10.1109/59.317600.

[7] B. Stott, J. Jardim, O. Alsac, DC power flow revisited, IEEE Trans. Power Syst. 24 (3) (2009) 1290–1300. Available from: https://doi.org/10.1109/TPWRS.2009.2021235.

[8] V. Sarkar, S. Khaparde, A comprehensive assessment of the evolution of financial transmission rights, IEEE Trans. Power Syst. 23 (4) (2008) 1783–1795. Available from: https://doi.org/10.1109/TPWRS.2008.2002182.

[9] PJM Training Materials, 2015. <http://pjm.com/>.

[10] A. Leite da Silva, V. Arienti, Probabilistic load flow by a multilinear simulation algorithm, generation, transmission and distribution, IEE Proc. C 137 (4) (1990) 276–282.

[11] S. Bolognani, S. Zampieri, On the existence and linear approximation of the power flow solution in power distribution networks, IEEE Trans. Power Syst. 31 (1) (2016) 163–172. Available from: https://doi.org/10.1109/TPWRS.2015.2395452.

[12] S. Fatemi, S. Abedi, G. Gharehpetian, S. Hosseinian, M. Abedi, Introducing a novel DC power flow method with reactive power considerations, IEEE Trans. Power Syst. 30 (6) (2015) 3012–3023. Available from: https://doi.org/10.1109/TPWRS.2014.2368572.

[13] D. Van Hertem, J. Verboomen, K. Purchala, R. Belmans, W. Kling, Usefulness of DC power flow for active power flow analysis with flow controlling devices, in: The Eighth IEE International Conference on AC and DC Power Transmission, 2006, 2006, pp. 58–62. Available from: https://doi.org/10.1049/cp:20060013.

[14] J.J. Grainger, W.D. Stevenson, Power Systems Analysis, McGraw-Hill, 1994.

[15] Source Code for the Linear Power Flow Models, GitHub, 2017. <https://github.com/Jingwei-THU/linear-power-flow>.

[16] R. Zimmerman, C. Murillo-Sanchez, R. Thomas, MATPOWER: steady-state operations, planning, and analysis tools for power systems research and education, IEEE Trans. Power Syst. 26 (1) (2011) 12–19. Available from: https://doi.org/10.1109/TPWRS.2010.2051168.

[17] W.H. Kersting, Radial distribution test feeders, IEEE Trans. Power Syst. 6 (3) (1991) 975–985.

[18] A. Garces, A linear three-phase load flow for power distribution systems, IEEE Trans. Power Syst. 31 (1) (2015) 827–828.

CHAPTER THREE

Renewable energy integration and system operation challenge: control and optimization of millions of devices

Ying Xu, Wei Sun and Zhihua Qu
Department of Electrical and Computer Engineering, University of Central Florida, FL, United States

Contents

3.1 Introduction 50
3.2 Distribution system model with high penetration of renewables 52
  3.2.1 Distribution network model 52
  3.2.2 An explicit branch model of distribution network 55
  3.2.3 Dynamic distributed generation model 56
3.3 Autonomous distributed voltage control 57
  3.3.1 Distributed subgradient algorithm 58
  3.3.2 Distributed subgradient voltage control 59
  3.3.3 Reactive power control and power factor control 64
3.4 Hierarchical multiagent control of large-scale distribution system 68
  3.4.1 Virtual leader design 69
  3.4.2 Case study 72
3.5 Islanded microgrid with high penetration of distributed generations 74
3.6 Grid-edge situational awareness: enhanced observability by voltage inference 79
  3.6.1 Voltage inference method 79
  3.6.2 Network sensitivity 81
  3.6.3 Implementation 82
3.7 Control-enabled dynamic hosting allowance: *P* and *Q* control capacity and impact analysis 86
  3.7.1 Traditional hosting analysis 86
  3.7.2 Dynamic hosting allowance analysis 87
3.8 Cosimulation of integrated transmission and distribution systems 91
  3.8.1 The framework of cosimulation 91
  3.8.2 Simulation results 92
3.9 Conclusion 95
References 97

New Technologies for Power System Operation and Analysis.
DOI: https://doi.org/10.1016/B978-0-12-820168-8.00003-1

3.1 Introduction

As the fast development of the sustainable, clean and renewable energy, the power industry is believed by most to have the potential to meet the rising challenges of climate change, environment deterioration and resource depletion on the way forward of human beings. The modern grid is expected to integrate diverse generation, energy-efficient and clean resources. While the existing power grid infrastructure was designed and built on large-scale power plants further away from loads, operated under centralized control (Supervisory control and data acquisition system/Energy management system/Distribution Management system-SCADA/EMS/DMS) the integration of distributed energy resources (DERs) is greatly limited. In the scenario of a large-scale distribution system with extremely high penetration of renewables, it is challenging to control thousands to millions of "smaller" distributed generating devices. To address this challenge the distributed control and operation with multiagent implementation are promising approaches, and tremendous efforts have been devoted into the development during the past decade, as reviewed in [1]. The distributed algorithms have several advantages over centralized methods. First, in the distributed architecture, each agent requires only local communication and shares limited information with its neighbors; second, the modular design and robust system enable the secure operation against failure of some agents; and third, the privacy of each agent is well protected.

However, from the control perspective, a distributed design is not always better than the centralized approach, for example, the convergence time is longer especially when the size of system is large. Also due to business practice, the power industry has to move gradually from existing infrastructure mainly relied on centralized control to a modernized grid, which is flexible to integrated large amount of renewables in a cost-effective, secure, resilient, and reliable manner. Therefore, the hierarchical design [2] that combines the advantage of both centralized and distributed methods is more practical.

In this chapter, we follow the layered and divisional design principle for large-scale power system operation and control. While the voltage/reactive power (Volt/VAr) control is mainly treated as a local control, the real power control is a system-level control (frequency) and a supplementary control for local voltage, which will only respond in the case of insufficient reactive power control.

The large-scale distribution system is modeled by algebraic network model and dynamic distributed generation (DG) models. For electrical circuit, both nodal injection and branch power flow models are used to model distribution network. A simplified power control model [3] of DG is used, which is simple but good enough to illustrate the design of system operation and control. It is preferable to design the reactive power control of DGs to respond to voltage events at a local level and in an autonomous way. Nevertheless, as a result of local controls, the reactive power at the feeder head (the slack bus) would change accordingly; hence, the power factor at the substation will fluctuate, and sometimes it might not be acceptable to the transmission side. A detailed discussion is presented in this chapter. Moreover, how the presented control applies to distribution system operated in an islanded mode is also provided.

Considering the limitations of investment and maintenance, in the distribution systems, the ICT system of real-time measurement and control in distribution system is not applicable. Therefore, the observability of distribution network is usually a problem, which is a major obstacle of distribution system state estimation, control, and optimization. However, based on the properties observed from many distribution network case studies, it is highly possible to infer the system situation using limited information from the measured buses [4]. To this end, we present a sensitivity-based grid-edge situational awareness method. Following that, a network sensitivity-based dynamic hosting allowance (DHA) method is presented for system operation. At last, to validate and demonstrate the feasibility and scalability of proposed algorithms, a cosimulation architecture of integrated T&D system is developed.

The main contents of this chapter are listed as follows:

- model of distribution system with high penetration renewables
- distributed cooperative voltage control
- hierarchical multiagent real and reactive power control
- control of islanded distribution system with high penetration of renewables
- grid-edge situational awareness: enhanced observability by voltage inference
- control-enabled DHA: P and Q control capacity and impact analysis
- cosimulation of transmission and distribution

3.2 Distribution system model with high penetration of renewables

This section will focus on the detailed mathematical models of voltage and frequency (VF) control in distribution systems with high penetration of renewables. Both nodal and branch power flows will be applied to model distribution network. A DG model [3] is used, which is simple but good enough to illustrate the design from the system control perspective.

3.2.1 Distribution network model

A distribution grid can be seen as a tree graph $\mathscr{T}:=(\mathscr{N},\mathscr{B})$ with node set $\mathscr{N}$ and branch set $\mathscr{B}$. The notation of each node in the tree graph corresponds with that of each bus in distribution network. With respect to bus i in a distribution network, as shown in Fig. 3.1, it has a unique parent bus Γ_i and a set of children buses, denoted by C_i. We denote the set of neighbors of node i by N_i, which contains its parent Γ_i, and its children set C_i, that is, $N_i := \{\Gamma_i\} \cup C_i$. With root (substation) indexed by 0, it is assumed that each line points to node i from Γ_i and is indexed by a single subscript i for simplicity of denotation, $\mathscr{B}:=\mathscr{N}\ \{0\}$.

To simplify the nomenclature of distribution system, the branch impedance and admittance of line i are denoted by $\tilde{Z}_i = R_i + jX_i$ and $y_i = g_i - jb_i$, respectively, the complex current through line i by $\tilde{I}_i = I_i \angle \theta_{I_i}$, and complex power at the receiving end (as shown in Fig. 3.1) of line i by $\tilde{S}_{s_i} + P_{s_i} + jQ_{s_i}$. Also we denote the complex voltage at bus i by $\tilde{V}_i = V_i \angle \delta_{V_i}$, the angle difference by $\theta_{ik} = \theta_i - \theta_k$ $(k \in N_i)$, and complex power injection by $\tilde{S}_i = P_i + jQ_i$.

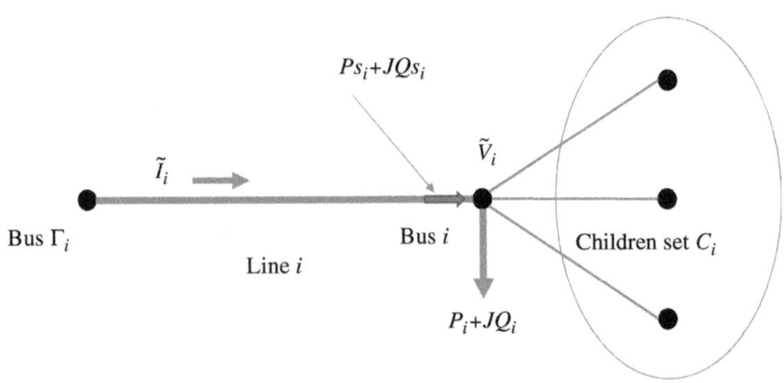

Figure 3.1 Notations of distribution network.

The bus injection power flow and branch power flow equations (Dist-Flow [5]) are two commonly used models in distribution power system analysis. They are essentially equivalent on distribution systems but with different expressions: node injection power flow is a clear expression of power and voltage at all the nodes, while the branch power flow has extra information of current and power flow on each branch. The bus injection power flow equations are written as (3.1a,b) for each node (e.g., the *i*th node):

$$P_i = V_i^2 g_{ii} + \sum_{k\in N_i} V_i V_k (g_k \cos\theta_{ik} + b_k \sin\theta_{ik}) \tag{3.1a}$$

$$Q_i = -V_i^2 b_{ii} + \sum_{k\in N_i} V_i V_k (g_k \sin\theta_{ik} - b_k \cos\theta_{ik}) \tag{3.1b}$$

where $g_{ii} = -\sum_{k\in C_i\cup\{i\}} g_k$, $b_{ii} = -\sum_{k\in C_i\cup\{i\}} b_k$, and the conductance of bus i to ground has been considered as a portion of power injection in $\tilde{S}_i$.

On the other hand, the branch power flow equations are based on each branch of the system (e.g., the *i*th line):

$$\sum_{k\in C_i} \left(P_{s_k} + P_{L_k}\right) = P_{s_i} - P_i, \tag{3.2a}$$

$$\sum_{k\in C_i} \left(Q_{s_k} + Q_{L_k}\right) = Q_{s_i} - Q_i, \tag{3.2b}$$

$$v_{\Gamma_i} - v_i = 2\left(R_i P_{s_i} + X_i Q_{s_i}\right) + |z_i|^2 \ell_i, \tag{3.2c}$$

$$v_i \ell_i = P_{s_i}^2 + Q_{s_i}^2, \tag{3.2d}$$

where $\ell_i = I_i^2$, $v_{\Gamma_i} = V_{\Gamma_i}^2$, $v_i = V_i^2$, P_{L_k}, and Q_{L_k} represent the power flow at the receiving end of the *k*th line:

$$P_{L_i} = R_i \ell_i, \quad Q_{L_i} = X_i \ell_i, \tag{3.3}$$

The physical meaning of branch model is quite straightforward: (3.2a) and (3.2b) represent the power balance of real and reactive power on the *i*th line, respectively; (2) is the RMS value form of power law equation. To explain (3.2c), we use Fig. 3.2. In Fig. 3.2, assuming that the angle of $\tilde{V}_i$ is the reference angle and equals to zero, the angle of $\tilde{V}_{\Gamma_i}$ is θ_{Γ_i}. Define the voltage difference of the *i*th bus and its parent by $\delta\tilde{V}_i = \tilde{V}_{\Gamma_i} - \tilde{V}_i$, and the real and image parts of $\delta\tilde{V}_i$ are δV_{i_x} and ΔV_{i_y}, respectively. It follows Ohm's law on the *i*th line that

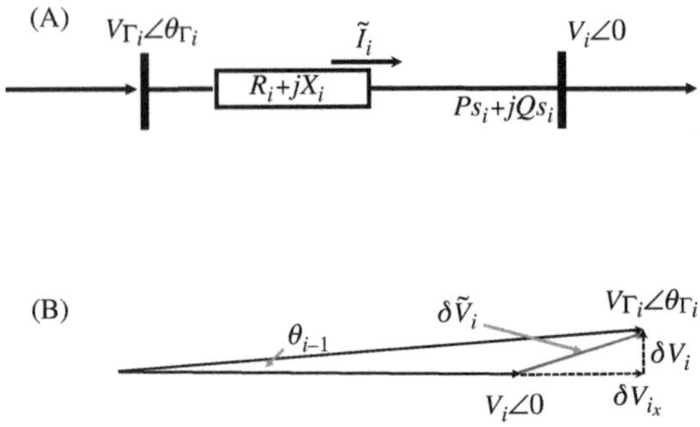

Figure 3.2 The phasor notations on the *i*th line: (A) circuit and (B) phasor diagram.

$$\delta\tilde{V}_i = \tilde{I}_i(R_i + jX_i) \tag{3.4}$$

and

$$\tilde{I}_i = \frac{P_{S_i} - jQ_{S_i}}{V_i}. \tag{3.5}$$

Substituting (3.5) into (3.4) yields

$$\delta V_{i_x} = \frac{P_{S_i}R_i + Q_{S_i}X_i}{V_i}, \quad \delta V_{i_y} = \frac{P_{S_i}X_i - Q_{S_i}R_i}{V_i}.$$

Using triangle formula in Fig. 3.2B, that is,

$$\left(V_i + \delta V_{i_x}\right)^2 + \delta V_{i_x}^2 = V_{\Gamma_i}^2, \tag{3.6}$$

hence (3.2c) can be obtained, which is actually the connection between magnitude deviation of $\tilde{V}_i$ from $\tilde{V}_{\Gamma_i}$ and the power on the *i*th line.

By substituting (3.3) into (3.2a), we have (3.7) that describes the Kirchhoff balance on *i*th line:

$$P_{L_i} = P_{s_i} + P_i - \sum_{k \in C_i} P_{s_k}. \tag{3.7}$$

Remark 3.2.1. One advantage about branch power flow (3.2a,b,c,d) is that it offers more information on the lines, such as current, real, and reactive power through the lines. Another advantage is that no angle variables are involved and hence no terms with triangular functions, which does not need angle measurements and is easier to relax the model as a convex formulation that can be solved by standard mathematical tools.

3.2.2 An explicit branch model of distribution network

The branch model of the distribution network has become popular and been used in many distributed algorithm implementations in power system. For each agent, it has a nonlinear cascade form with four equations of each agent (e.g., agent i as shown in Fig. 3.1, including branch i and bus i), and four variables each: the square of voltage magnitude (v_i), the square of current magnitude (ℓ_i), and the power injection at bus i (P_i and Q_i). From the perspective of power system control, the injection (P_i and Q_i) is the control and system voltage (and/or loss) is the outcome. Hence, a useful transformation of (3.2a,b,c,d), which focuses on the effects of bus injection on system voltage, is written as following:

$$v_k = -\sum_{j\in\mathscr{P}_{1k}\cup\mathscr{T}_k}\left[2\mathscr{R}_{kj}P_j + 2\mathscr{X}_{kj}Q_j + \left(Z'_{kj}\right)P_{L_j}\right] + v_0, \tag{3.8}$$

where the set $\mathscr{P}_{1k}$ is defined as the path from bus 1 to k; $\mathscr{T}_i$ is the node set of the subtree under node i; $\mathscr{R}_{kj}$, $\mathscr{X}_{kj}$, and Z'_{kj} and γ_j are network parameters (constants), that is,

$$\mathscr{R}_{kj} = \sum_{i\in\mathscr{P}_{1j}\cap\mathscr{P}_{1k}} R_i, \quad \mathscr{X}_{kj} = \sum_{i\in\mathscr{P}_{1j}\cap\mathscr{P}_{1k}} X_i, \quad \gamma_j = X_j/R_j,$$

$$Z'_{kj} = \mathscr{R}'_{kj} + \mathscr{X}'_{kj}\gamma_j,$$

and

$$\mathscr{R}'_{kj} = \begin{cases}\mathscr{R}_{kj} - R_j, & \text{if } j <= k\\ \mathscr{R}_{kj}, & \text{otherwise}\end{cases}, \quad \mathscr{X}'_{kj} = \begin{cases}\mathscr{X}_{kj} - X_j, & \text{if } j <= k\\ \mathscr{X}_{kj}, & \text{otherwise}\end{cases}.$$

The explicit form of power flow (3.8) is derived as follows [i.e., from (3.9) to (3.11)]:

It follows (3.2a,b,c,d) that

$$v_{i-1} - v_i = 2P_{S_i}R_i + 2Q_{S_i}X_i + P_{L_i}R_i + Q_{L_i}X_i, \tag{3.9}$$

According to Kirchhoff law, on the subtree $\mathscr{T}_i$, we can write the power balance equations as

$$P_{S_i} = P_i + \sum_{j\in\mathscr{T}_i}\left(P_j + P_{L_j}\right), \quad Q_{S_i} = Q_i + \sum_{j\in\mathscr{T}_i}\left(Q_j + Q_{L_j}\right). \tag{3.10}$$

Plugging (3.10) into (3.9), we can then add (3.9) of all $i\in\mathscr{P}_{1k}$ together which yields

$$
\begin{aligned}
v_0 - v_k = & \sum_{i \in \mathscr{P}_{1k}} \left[2 \sum_{j \in \mathscr{T}_i \cup \{i\}} (R_i P_j + X_i Q_j) \right. \\
& \left. + 2 \sum_{j \in \mathscr{T}_i} (P_{L_j} R_i + Q_{L_j} X_i + (P_{L_i} R_i + Q_{L_i} X_i) \right] \\
= & 2 \sum_{j} \left(P_j \sum_{i \in \mathscr{P}_{1j} \cap \mathscr{P}_{1k}} R_i + Q_j \sum_{i \in \mathscr{P}_{1j} \cap \mathscr{P}_{1k}} X_i \right) \\
& + 2 \sum_{j \in \mathscr{T}_1} \left(P_{L_j} \sum_{i \in \mathscr{P}_{1\min\{j-1,k\}}} R_i + Q_{L_j} \sum_{i \in \mathscr{P}_{1\min\{j-1,k\}}} X_i \right) \\
& + \sum_{i \in \mathscr{P}_{1k}} (R_j P_{L_j} + X_j Q_{L_j}),
\end{aligned}
\tag{3.11}
$$

where Q_{L_i} can be represented by P_{L_i}, that is, $Q_{L_i} = \gamma_i P_{L_i}$. Hence, (3.11) can be wrapped up as the concise form (3.8).

Note 3.2.2. It is important to emphasize that the recursive model (3.8) is the exact model without any approximations.

From (3.8), many practical forms of distribution system equation can be easily derived by some assumptions, for example, by neglecting the system loss, (3.8) becomes

$$
v_k = - \sum_{j \in \mathscr{P}_{1k} \cup \mathscr{T}_k} \left(2\mathscr{R}_{kj} P_j + 2\mathscr{X}_{kj} Q_j \right) + v_0. \tag{3.12}
$$

Given the fact that V_k is always close to 1 pu, (3.8) can be written as the following linearized form

$$
\Delta v_k = - \sum_{j \in \mathscr{P}_{1k} \cup \mathscr{T}_k} \left(\mathscr{R}_{kj} \Delta P_j + \mathscr{X}_{kj} \Delta Q_j \right), \tag{3.13}
$$

which has been used in many distributed optimization algorithms in power system and smart grid analysis.

3.2.3 Dynamic distributed generation model

The previous analysis does not differentiate the power generation and consumption at each bus. To illustrate the system-level design, we define the generation (DGs) and consumption (loads) on each bus i by $P_{g_i} + jQ_{g_i}$ and $P_{d_i} + jQ_{d_i}$, respectively, and define net injection by combining DGs generation and loads by

$$P_i + jQ_i = -\left(P_{g_i} - P_{d_i}\right) - j\left(Q_{g_i} - Q_{d_i}\right). \tag{3.14}$$

The detailed model of DGs will be provided as follows.

For the simplicity of analysis, it is assumed that P_{g_i} and Q_{g_i} are determined by decoupled $d-q$ control method via phase locked loops (PLL), and assume that inner dynamics of inverter-based DG usually diminish much faster compared to the power outputs, their control model [3] can be written as

$$\begin{cases} P_{g_i} = V_i I_{p_i} \\ Q_{g_i} = V_i I_{q_i} \end{cases} \tag{3.15}$$

and

$$\begin{cases} \dot{I}_{p_i} = u_{p_i} \\ \dot{I}_{q_i} = u_{q_i} \end{cases}, \tag{3.16}$$

where I_{p_i} and I_{q_i} are the output current in dq-axis, u_{p_i} and u_{q_i} are the real/reactive power control inputs to be designed, respectively. The active and reactive power of each DG should be well controlled in their operational limits:

$$0 \leq P_{g_i} \leq \overline{P}_{g_i}, \quad |Q_{g_i}| \leq \overline{Q}_{g_i},$$

where $\overline{P}_{g_i}$ is the maxim available real power output of each DG, and

$$\overline{Q}_{g_i} = \sqrt{\overline{S}_i^2 - P_{g_i}^2}, \tag{3.17}$$

$\overline{Q}_{g_i}$ is the maxim available reactive power output, and $\overline{S}_i$ is the apparent power capacity of the inverter, which is usually constant.

Remark 3.2.3. u_{p_i} and u_{q_i} are the real and reactive power of each DG which can be controlled to achieve specific objectives, hence providing immense potential flexibility to improve overall performance of power systems.

3.3 Autonomous distributed voltage control

It is challenging for the existing centralized tools such as EMS/SCADA systems, to control thousands to millions distributed small-sized devices as described by (3.15) in future power grids. Therefore the distributed cooperative control is chosen to handle this operation. This control strategy should be built upon the distributed communication structure, with each node communicating with its neighbors. Using the cooperative control theory [6] and distributed subgradient-based multiagent optimization method [7], the distributed VF control in distribution network is formulated as a multiagent problem.

3.3.1 Distributed subgradient algorithm

We use a general form of vector z_i ($z_i \in R^n$) to represent the local information of agent i, and N_i^c is defined as the neighboring set of the ith agent. Note that the ICT structure does not necessarily have the same topology connection as electrical buses in the power distribution network.

Use a binary matrix to express the communication topology as

$$S = \begin{bmatrix} s_{11}(t) & \cdots & s_{1n}(t); \\ \vdots & \ddots & \vdots \\ s_{n1}(t) & \cdots & s_{nn}(t) \end{bmatrix}, \tag{3.18}$$

where $s_{ij} = 1$ if $j \in N_i^c$ and $s_{ij} = 0$ if otherwise, for all $i, j \in \mathcal{N}$.

We start from a general multiagent optimization problem in which agents cooperatively optimize a common additive objective function. Each agent minimizes its own cost function and communicates with other agents,

$$\min \sum_{i \in \mathcal{N}} f_i(z) \tag{3.19}$$

where f_i is the distributed cost function of agent i with a convex form. Denote the optimal value of (3.19) by f^*, and the optimal solution set by Z^*, which is $Z^* = \left\{ z \in | \sum_{i \in \mathcal{N}} f_i(z) = f^* \right\}$.

In this setting the information state z_i is an estimate of the optimal solution of (3.19). Each agent updates its estimate first, then exchanges information with others, then updates itself again, and iterates until converging. We propose the close-loop cooperative control law based on the conclusion of [7]:

$$z_i(k+1) = \sum_{j \in N_i^c} d_{ij} z_j(k) - \beta_{ik} g_i(k). \tag{3.20}$$

where k denotes the iteration; the vector g_i is a subgradient of control objective function f_i of agent i with respect to z_i; β_i is the step size used by the ith agent; d_{ij} denotes the weights of communication topology:

$$d_{ij} = \frac{w_{ij}s_{ij}}{\sum_{j \in N_i^c} w_{ij}s_{ij}}, \tag{3.21}$$

$w_{ij} > 0$ are the weights and all $w_{ij} = 1$ for symmetric systems which is true for all power systems.

3.3.2 Distributed subgradient voltage control

In this section the distributed control algorithm is implemented through DG inverters to cooperatively control real and reactive power generation of each DG, so that the system voltage performance can be well maintained in order to satisfy the regulation requirement, as follows:

$$|1 - V_i| \leq 0.05. \tag{3.22}$$

Therefore the distributed objective function is designed to minimize the voltage deviation at the ith bus, which is as follows:

$$f_i = \frac{\lambda_i^v}{2}\left(V_i - V_i^{ref}\right)^2 + \frac{\lambda_i^p}{2}\left(P_{g_i} - P_{g_i}^{ref}\right)^2 + \frac{\lambda_i^q}{2} Q_{g_i}^2, \tag{3.23}$$

where the last two terms are designed to minimize the real power curtailment and the reactive power usage; V_i^{ref} and P_i^{ref} are the reference voltage and power, respectively, at the ith bus and can be obtained from OPF or other regulatory level design, for example, $V_i^{ref} = 1$ and $P_{g_i}^{ref} = \overline{P}_{g_i}$; $\lambda_i^v \geq 0$, $\lambda_i^p \geq 0$, and $\lambda_i^q \geq 0$ are the weighting coefficients.

We first justify that (3.23) is convex by the following steps: $\lambda_i^v/2\left(V_i^{ref} - V_i\right)^2$ is proved in Refs. [8,9] to be strictly convex with respect to Q_{g_i} in the context of power system operation; by the same procedure, it is straightforward to prove that it is also strictly convex with respect to P_{g_i}; the last two terms $\lambda_i^p/2(P_{g_i} - P_{g_i}^{ref})^2$ and $\left(\lambda_i^q/2\right) Q_{g_i}^2$ are also convex.

One way to implement the distributed subgradient control (3.20) for voltage control is the fair utilization ratio method [10], which uses the utilization ratios of real and reactive power at the ith bus as the control variables, that is,

$$\alpha_{p_i} \triangleq \frac{P_{g_i}}{\overline{P}_{g_i}}, \tag{3.24a}$$

$$\alpha_{q_i} \triangleq \frac{Q_{g_i}}{\overline{Q}_{g_i}}. \tag{3.24b}$$

Hence, the subgradient of each agent, which is defined by g_i in (3.20), can be calculated by taking the derivative of f_i with respect to α_{p_i} and α_{q_i} as follows:

$$g_i^p = \frac{\partial f_i}{\partial \alpha_{p_i}} = \left(\lambda_i^v \frac{\partial f_i}{\partial V_i}\frac{\partial V_i}{\partial P_{g_i}} + \lambda_i^p (P_{g_i} - P_{g_i}^{ref})\right)\frac{\partial P_{g_i}}{\partial \alpha_{p_i}}, \tag{3.25a}$$

$$g_i^q = \frac{\partial f_i}{\partial \alpha_{q_i}} = \left(\lambda_i^v \frac{\partial f_i}{\partial V_i}\frac{\partial V_i}{\partial Q_{g_i}} + \lambda_i^q Q_{g_i}\right)\frac{\partial Q_{g_i}}{\partial \alpha_{q_i}}. \tag{3.25b}$$

It follows (3.23) and (3.24a,b) that

$$\frac{\partial f_i}{\partial V_i} = V_i - V_i^{ref}. \tag{3.26}$$

and

$$\frac{\partial P_{g_i}}{\partial \alpha_{p_i}} = \overline{P}_{g_i}, \tag{3.27a}$$

$$\frac{\partial Q_i}{\partial \alpha_{q_i}} = \overline{Q}_{g_i}. \tag{3.27b}$$

By taking the derivative of power flow (3.1a) and (3.1b), one can get the following equations for real/reactive power versus bus voltage:

$$\begin{aligned}\frac{\partial Q_i}{\partial V_i} &= -2V_i b_{ii} + \sum_{k\in N_i} V_k (g_k \sin\theta_{ik} - b_k \cos\theta_{ik})\\ &= \frac{Q_i - V_i^2 b_{ii}}{V_i},\end{aligned} \tag{3.28}$$

and

$$\begin{aligned}\frac{\partial P_i}{\partial V_i} &= 2V_i g_{ii} + \sum_{k\in N_i} V_k (g_k \cos\theta_{ik} + b_k \sin\theta_{ik})\\ &= \frac{P_i - V_i^2 g_{ii}}{V_i}.\end{aligned} \tag{3.29}$$

Without loss of generosity, let us assume $\partial P_i/\partial V_i = \partial P_{g_i}/\partial V_i$ and $\partial Q_i/\partial V_i = \partial Q_{g_i}/\partial V_i$. Then following (3.28) and (3.29), the derivatives of bus voltage with respect to real/reactive power generation of DG can be written as

$$\frac{\partial V_i}{\partial P_{g_i}} = \frac{V_i}{P_i + V_i^2 g_{ii}}, \tag{3.30a}$$

$$\frac{\partial V_i}{\partial Q_{g_i}} = \frac{V_i}{Q_i - V_i^2 b_{ii}}. \tag{3.30b}$$

Plugging (3.26), (3.27a,b), and (3.30a,b) into (3.25a,b) yields the subgradient formula:

$$g_i^p = \overline{P}_{g_i}\left[\lambda_i^v\left(V_i - V_i^{ref}\right)\frac{V_i}{P_i + V_i^2 g_{ii}} + \lambda_i^p\left(P_{g_i} - P_{g_i}^{ref}\right)\right] \tag{3.31a}$$

$$g_i^q = \overline{Q}_{g_i}\left[\lambda_i^v\left(V_i - V_i^{ref}\right)\frac{V_i}{Q_i + V_i^2 b_{ii}} + \lambda_i^q Q_{g_i}\right] \tag{3.31b}$$

It can be inferred from (3.31a,b) that the subgradient calculation of voltage control for each agent only requires local information. In summary the continuous form of distributed real power control is written as

$$\dot{\alpha}_{p_i} = \sum_{j \in N_i^c} d_{ij}\left(\alpha_{p_j} - \alpha_{p_i}\right) - \beta_i g_i^p, \tag{3.32}$$

and the reactive power control is

$$\dot{\alpha}_{q_i} = \sum_{j \in N_i^c} d_{ij}\left(\alpha_{q_j} - \alpha_{q_i}\right) - \beta_i g_i^q. \tag{3.33}$$

The overview of the distributed cooperative voltage control is shown by Fig. 3.3. To synthesize the previous control into the feedback loop of each DG as defined by (3.15), we take derivative of (3.15) and (3.24a) and (3.24b), then we get the following equations:

$$\dot{P}_{g_i} = V_i \dot{I}_{p_i} = V_i u_{p_i} \tag{3.34a}$$

$$\dot{Q}_{g_i} = V_i \dot{I}_{q_i} = V_i u_{q_i} \tag{3.34b}$$

and

$$\dot{P}_{g_i} = \overline{P}_{g_i} \dot{\alpha}_{p_i}, \tag{3.35a}$$

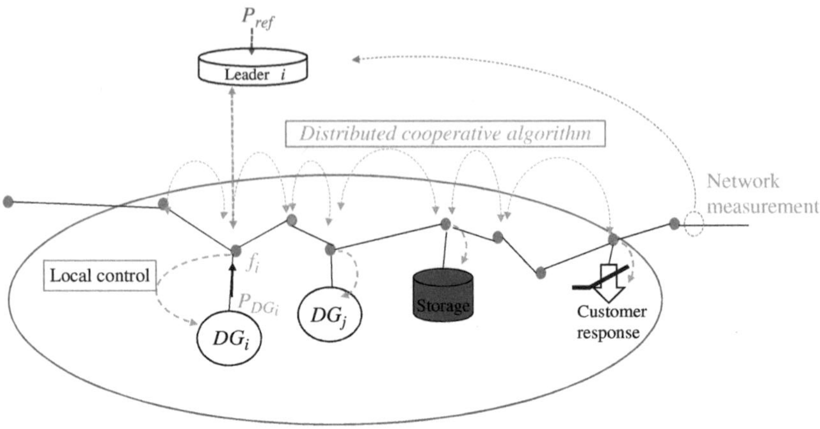

Figure 3.3 Cooperative real power control.

$$\dot{Q}_{g_i} = \overline{Q}_{g_i}\dot{\alpha}_{q_i} + \dot{\overline{Q}}_{g_i}\alpha_{q_i}, \tag{3.35b}$$

where V_i and $\overline{P}_{g_i}$ should be measured from the bus, and $\dot{\overline{Q}}_{g_i}$ can be obtained from (3.17), that is,

$$\dot{\overline{Q}}_{g_i} = -\frac{\overline{P}_{g_i}}{\overline{Q}_{g_i}}\dot{\alpha}_{p_i}. \tag{3.36}$$

Thus by (3.34b), (3.35b), and (3.36), the reactive power control of the ith DG, u_{q_i} is written as

$$\begin{aligned} u_{q_i} &= \frac{\overline{Q}_i}{V_i}\dot{\alpha}_{q_i} - \frac{\alpha_{q_i}\overline{P}_{g_i}}{\overline{Q}_{g_i}V_i}\dot{\alpha}_{p_i} \\ &= \frac{\overline{Q}_i}{V_i}\left[\sum_{j\in N_i^c} d_{ij}(\alpha_{q_j} - \alpha_{q_i}) - \beta_i^q g_i^q\right] \\ &\quad - \frac{\alpha_{q_i}\overline{P}_{g_i}}{Q_{g_i}V_i}\left[\sum_{j\in N_i^c} d_{ij}(\alpha_{p_j} - \alpha_{p_i}) - \beta_i^p g_i^p\right]. \end{aligned} \tag{3.37}$$

Similarly, using (3.34a) and (3.35a), the real power control u_{p_i} is written as

$$u_{p_i} = \frac{\overline{P}_i}{V_i}\dot{\alpha}_{p_i} = \frac{\overline{P}_i}{V_i}\left[\sum_{j\in N_i^c} d_{ij}\left(\alpha_{p_j} - \alpha_{p_i}\right) - \beta_i^p g_i^p\right]. \tag{3.38}$$

Note 3.3.2. According to (3.37), u_{q_i} is dependent on α_{p_i}. The reason is the control strategy of real power control priority is adopted. u_{p_i} and u_{q_i} can have other forms according to different control priorities, but the derivation should be the same.

The performance of the proposed voltage control has been evaluated on multiple test systems such as: IEEE 123-bus, 8500-node, and other standard systems [11]. To show the effectiveness of the proposed approach on even larger scale systems, a 100,000 (100k)-node circuit is built by combining different type of feeders: the circuit is assembled by 12 urban/suburban feeders, as shown in Fig. 3.4.

The total load of the circuit is 122 MW. On top of the physical layer, 268 clusters are predefined for communication and control. Assume all load tap changers in the circuit are fixed at the predefined positions during the fast control of inverters. Two extremely harsh scenarios developed by the greedy search method [11] are studied: Case 1 defines 104 large-scale photovoltaics (PVs) among 12 feeders, which is the large-scale PV Farms scenario; Case 2 defines 2528 PVs distributed across the network, which is the distributed PVs scenario. In both cases, each of the feeders is self-

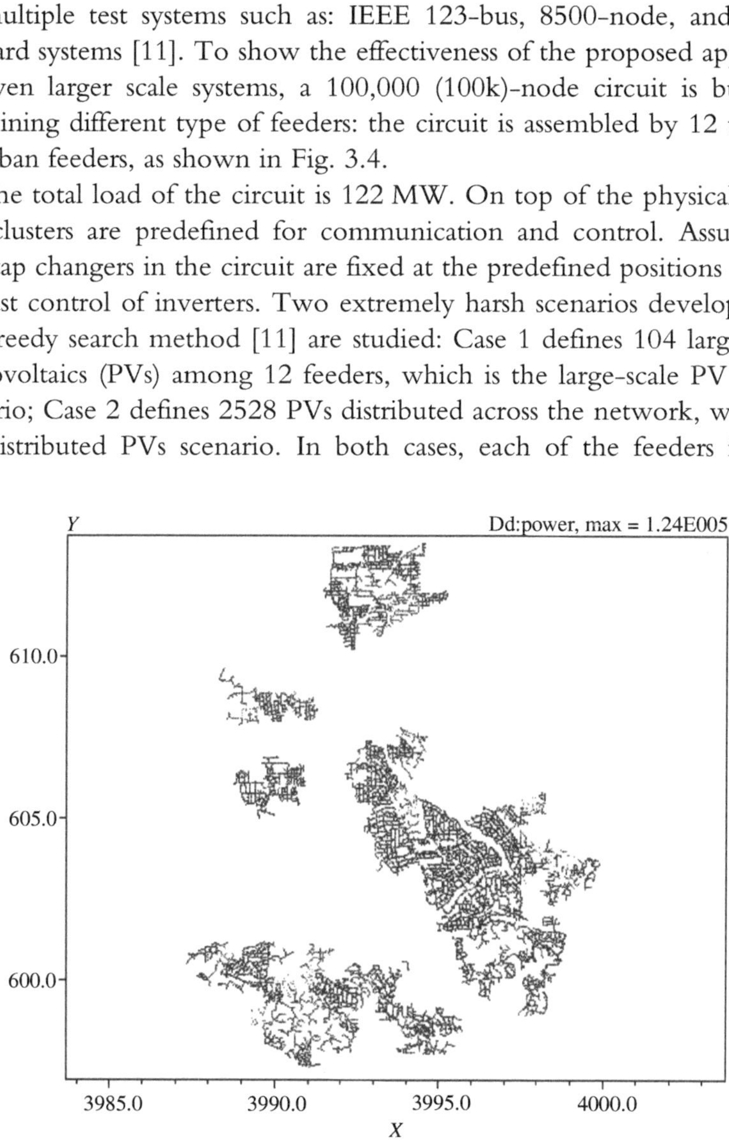

Figure 3.4 100k synthetic circuits.

balanced in terms of its own PV generation and load consumption, and the load in each feeder is evenly distributed among PVs. Simulation results of both cases show that the proposed cooperative reactive power control can effectively regulate the system voltages for distributed PVs, as shown in Fig. 3.5. The spatial voltage profile of all phases is shown in Fig. 3.5, where *Y*-axis is the pu value of bus voltages and *X*-axis denotes the distance from substation. In Scenario 1 a total of 37 MVAr inductive reactive power is generated by PV inverters to suppress the voltage violation; the highest inverter capacity is 108.6%. In Scenario 2 a total of 22 MVAr of inductive reactive power is generated by PV inverters; the highest inverter capacity is 103.3%. Note that it is easier for the distributed scenario to control the voltage.

To better illustrate the control effect, the detailed information about one part of the 100k system in Scenario 1 is shown in Fig. 3.5C. As shown by the results, the voltage is controlled within in the limits, and reactive power utilization ratios of all DGs in each feeder reach consensus.

3.3.3 Reactive power control and power factor control

The voltage control scheme aims to hold the system voltage by the inverter's ability of offering reactive power support at DG buses. Following (3.13), to control system voltage autonomously, the reactive power of DGs must respond to the real power injection of renewables. As a result, the reactive power at the feeder (the slack bus) would change accordingly; hence, the power factor at the substation will fluctuate and sometimes it might not be acceptable to the transmission side. To illustrate the effect of autonomous control on power factor, a simplified system with two concentrated loads and one aggregated DG is used, as shown in Fig. 3.6.

In the system, E, V_l, and V_e are bus voltages; P_0+jQ_0 is the power at the feeder; the total generation is represented by P_g+jQ_g as an aggregated DG; the synthetic electrical line between aggregated DG and substation is assumed to be $R+jX$; the total load is denoted by $P+jQ$. The parameter $\alpha \in [0, 1]$ is used to denote the position of the load bus, and $\beta \in [0, 1]$ is an additional parameter to split the load between two buses. Also define the losses by $P_{L_1}+jQ_{L_1}$ and $P_{L_2}+jQ_{L_2}$ for the two line segments (E, V_l) and (V_l, V_e), respectively.

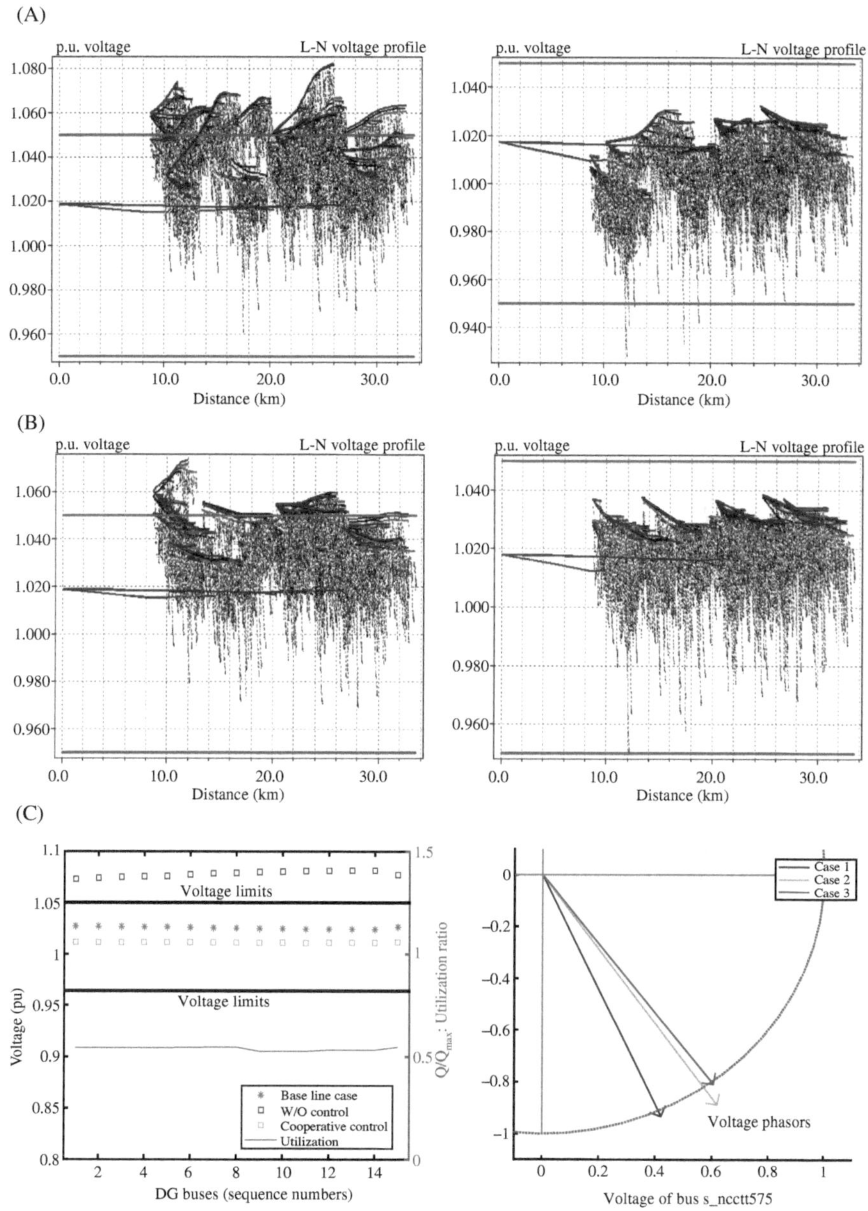

Figure 3.5 Results of voltage control on 100k synthetic circuits: (A) scenario 1, (B) scenario 2, and (C) scenario 1: details.4

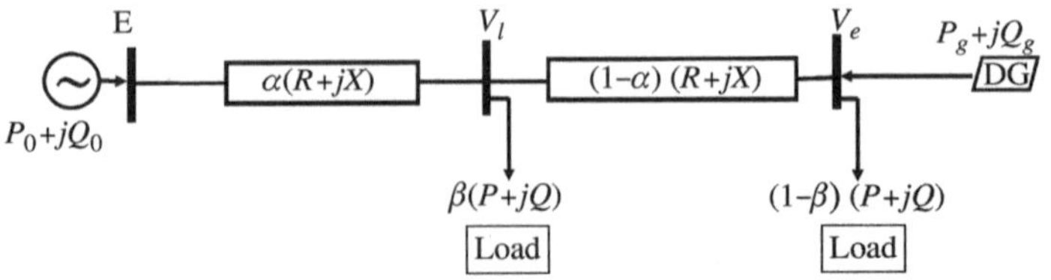

Figure 3.6 Illustration system of distribution network with high penetration renewables.

It follows (3.8) that

$$\begin{aligned}E^2 - V_l^2 &= 2\alpha RP + 2\alpha XQ \\ &\quad + \alpha RP_{L_1} + 2\alpha RP_{L_2} + \alpha XQ_{L_1} + 2\alpha XQ_{L_2}, \\ E^2 - V_e^2 &= 2(\alpha\beta + 1 - \beta)RP + 2(\alpha\beta + 1 - \beta)XQ \\ &\quad + \alpha RP_{L_1} + (1+\alpha)RP_{L_2} + \alpha XQ_{L_1} + (1+\alpha)XQ_{L_2},\end{aligned}$$

and

$$P_{L_1} = \frac{\alpha R}{V_l^2}\left(\left[P - P_g + P_{L_2}\right]^2 + \left[Q - Q_g + Q_{L_2}\right]^2\right),$$

$$P_{L_2} = \frac{(1-\alpha)R}{V_e^2}\left(\left[(1-\beta)P - P_g\right]^2 + \left[(1-\beta)Q - Q_g\right]^2\right),$$

where

$$Q_{L_1} = \frac{P_{L_1}X}{R} \quad \text{and} \quad Q_{L_2} = \frac{P_{L_2}X}{R}.$$

Given the voltage at feeder E, the four variables, V_e, V_l, P_{L_1}, and P_{L_2}, can be solved from the previous four equations. The voltage V_e and the total loss $P_{L_1} + P_{L_2}$ with respect to DG injection are shown in Fig. 3.7A and B, respectively.

Previous results show that system voltage is monotonic to bus power injection [12], and system loss is convex when bus injections are near normal operational points [5]. These two properties are very useful to the control design of large-scale distribution network with high penetration of renewables.

By Kirchhoff law the power at the feeder is then calculated by

$$P_0 = P - P_g + P_{L_1} + P_{L_2},$$

$$Q_0 = Q - Q_g + Q_{L_1} + Q_{L_2}$$

To compensate the power factor at the feeder, the reactive power support required at the substation can be calculated by

$$Q_{bank} \triangleq Q_0 - P_0\sqrt{\left(\frac{1}{pf}\right) - 1} \tag{3.39}$$

The results of Q_{bank} and Q_0 are shown in Fig. 3.8. As shown in Fig. 3.8B, it is clear that the reactive power of DG (Q_g) increases (absorption) along with the increase of its real power (P_g), which is close to a linear relationship, as described in (3.13), given relatively small line loss. This is also true for Q_0 and Q_{bank}. The necessary reactive power support is also

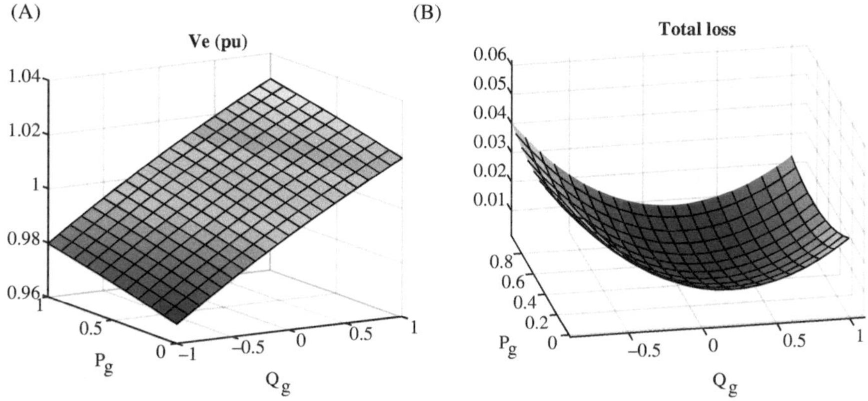

Figure 3.7 Bus voltage and system loss with respect to power injection from DGs: (A) bus voltage and (B) system loss. *DG*, Distributed generation.

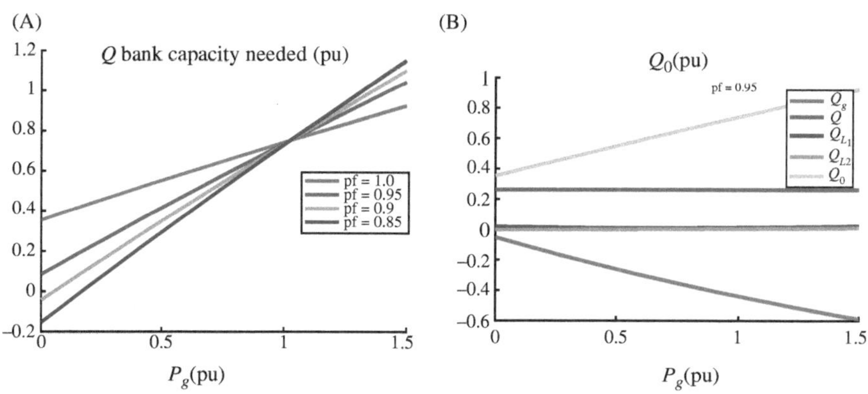

Figure 3.8 Reactive power support at the feeder with respect to power injection from DGs: (A) Q_{bank} and (B) Q_0. *DG*, Distributed generation.

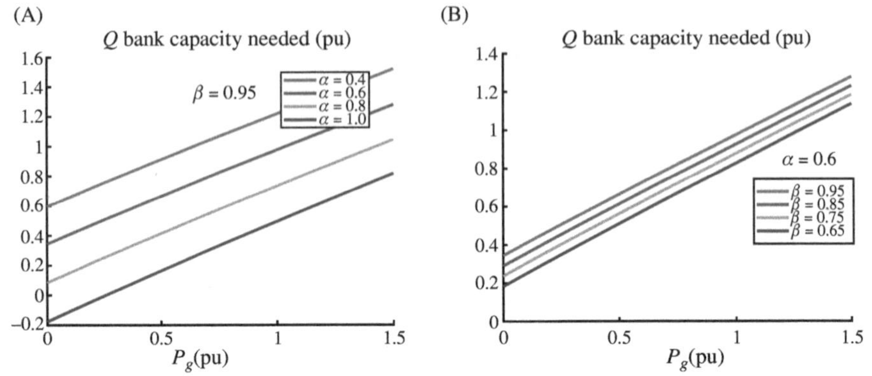

Figure 3.9 Reactive power support at the feeder with respect to power injection from DGs under different load conditions: (A) Q_{bank} under different α and (B) different β. *DG*, Distributed generation.

studied with different α and β (different load conditions), and results are shown in Fig. 3.9A and B.

Note 3.3.2. Previous equivalent model of a feeder is useful to analyze certain properties of distribution network, such as: reactive power schedule on transmission side, and the network planning. Also it can be used as heuristic modeling of feeders with less (or no) measurement, T&D cosimulation and control, etc.

3.4 Hierarchical multiagent control of large-scale distribution system

The layered and divisional *principle* for large-scale power system planning, operation, and control has been practiced for years:

Principal. For large-scale power system operation:
1. Volt/VAr control should be designed as local control, which is within certain electrical or geological area to prevent the transfer of reactive power, and to avoid unnecessary losses;
2. Real power control should respond to system-level objectives such as frequency regulation, and operation optimization.

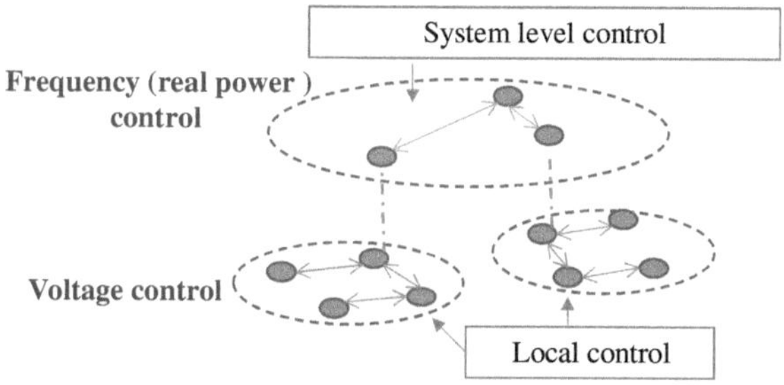

Figure 3.10 Hierarchical control design for voltage and frequency control.

By following the previous operation principal, a hierarchical control is presented, as shown in Fig. 3.10. In the hierarchy, voltage control is considered as a local problem within an electrical or geological area, while real power control (frequency control) is formulated as global problem. Accordingly, the communication network is organized as either local or wide-area network. We define each local area as a cluster and each bus (or each information agent) as a node. We assign a virtual leader (VL) in each cluster which is the connection between the two layers: it works as virtual node in local cluster and an aggregated agent for the upper level control, as shown in Fig. 3.10.

We expand the denotations to fit the hierarchy as follows: $\mathscr{G}^L$ is the set of all VLs in all clusters; $\mathscr{N}^k$ is the set of all nodes in the kth cluster, $k \in \mathscr{G}^L$; 0 is a default member of all $\mathscr{N}^k$s, which means the subscript 0 is specified for variables of VL node within each cluster, for example, P_{g_0} and Q_{g_0} are used to denote the power output of VLs. Then we can redefine the communication matrix (3.18) for each cluster with its augmented form as follows:

$$S = \begin{bmatrix} s_{00}(t) & s_{01}(t) & \cdots & s_{0n}(t); \\ s_{10}(t) & s_{11}(t) & \cdots & s_{1n}(t); \\ \vdots & \ddots & \cdots & \vdots \\ s_{n0}(t) & \cdots & \cdots & s_{nn}(t) \end{bmatrix}, \tag{3.40}$$

where $s_{ij} = 1$ if $j \in N_i^c$, otherwise $s_{ij} = 0$; and d_{ij} is redefined in the same way as in (3.21) for all $i, j \in \mathscr{N}^k$.

3.4.1 Virtual leader design

The properties of VLs are specified by a superscript L and being defined as follows: $\overline{P}_k^L$ and $\overline{Q}_k^L$ are defined to collect the capacities of all DGs within the kth cluster, that is,

$$\overline{P}_{gk}^{L} = \sum_{j\in\mathcal{N}^k} \overline{P}_{gj}, \quad \overline{Q}_{gk}^{L} = \sum_{j\in\mathcal{N}^k} \overline{Q}_{gj}, \quad k\in\mathcal{G}^L; \tag{3.41}$$

$\alpha_{p_k}^L$ and $\alpha_{q_k}^L$ are defined as utilization ratios of the *k*th cluster. For simplicity of expression, in the context of local control within the *k*th cluster, $\alpha_{p_k}^L$ and $\alpha_{q_k}^L$ are represented as α_{p_0} and α_{q_0}, respectively.

The computational sequences in the hierarchy are shown in the information flow in Fig. 3.11: α_q^k is the result of local control; hence, it is calculated from each cluster by standard cooperative law, while real power α_p^k is on the opposite way, from system level to each of the clusters.

The real power control aims to pursue system-level objectives such as driving the downstream or upstream power flow at the feeder to follow the dispatch signals from OPF (or control center), optimizing operation in terms of system loss or power quality, and so on. One practical design of control objective for real power, which focuses on the fast response to load demand and voltage regulation, can be written as following:

$$f_i^L = \frac{\lambda_i^{Lf}}{2}\left(P_f^* - P_f\right)^2 + \frac{\lambda_i^{Lv}}{2}\left(1 - V_i^L\right)^2, \quad i\in\mathcal{G}^L, \tag{3.42}$$

where $\lambda_i^{Lf} \geq 0$ and $\lambda_i^{Lv} \geq 0$ are the coordinate coefficients at the *i*th agent; P_f and P_f^* denote the power at the feeder of distribution systems and its dispatch value, respectively; V_i^L is the representative voltage of the cluster, particularly, the worst measured voltage of the cluster. λ_i^{Lf} is the impact factor of frequency or power dispatch control for main grid (transmission side), which should be zeros when the distribution system is isolated from main grid.

Let us define z_{ik} as the estimation of P_{gk}^* at the *i*th agent. As described by Fig. 3.11, reactive power is not considered as decision variable at system level. We follow [5] to make the assumption that $\lambda_i^{Lf}/2(P_f^* - P_f)^2$ is convex in the context of normal power system operation. Hence, (3.42) is convex and the distributed subgradient control (3.20) is also deployed

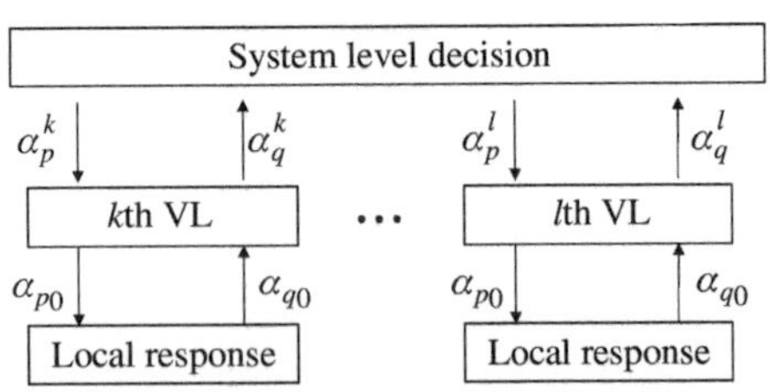

Figure 3.11 Information flow of the hierarchical control.

on system level. Then distributed subgradient law (3.20) can be used, and we have

$$\dot{z}_{ik} = \sum_{j \in N_i^c} d_{ij}\left(z_{jk} - z_{ik}\right) - \beta_i g_{ik}, \tag{3.43}$$

where g_{ik} is the gradient of f_i^L with respect to z_i which can be written as

$$g_{ik} = \lambda_i^f\left(P_f - P_f^*\right)\frac{\partial p_f}{\partial p_{g_k}} + \lambda_i^{Lf}(V_i - 1)\frac{\partial V_i}{\partial P_{g_k}}. \tag{3.44}$$

Note 3.4.1. Different from local subgradient calculation at lower level, the standard subgradient calculation requires information across the system to pursue the optimal operation point on the system level $(\partial V_i/\partial P_{g_k})$. However, the computation should not be too large for reasonably planned systems, because only the clusters that cannot control the voltage by themselves need the subgradients calculation.

According to the hierarchy design, the voltage control term in (3.44) is a supplementary control and will only be performed when the local reactive power control is not enough. So $\partial V_i/\partial P_{g_k} = 0$ when $i = k$ because the local voltage control has already been considered at the lower level. When $i \neq k$, the supplementary voltage control is needed and hence the calculation of $\partial V_i/\partial P_{g_k}$ is performed only under the following conditions:

C1: α_q^i (or α_{q_0} in the ith cluster) has reached its limit;

C2: V_i violates the regulation limits;

C3: V_k is well controlled within the regulation limits.

The condition C3 implies that the reactive power at the kth cluster is sufficient to maintain the local voltage within that cluster, and the reactive power control is a fast response to change of P_k given V_k being well controlled. So we can estimate the reactive power response at VLs and use (3.13) to approximate the derivative of Q_k to P_k, that is, $-\mathcal{R}_{kk}/\mathcal{X}_{kk}$. We refer to branch power flow model (3.8), condition C3, and through derivation we can get the following results:

$$\frac{\partial V_i}{\partial P_{g_k}} = \frac{1}{2V_i}\left(\mathcal{R}_{ik} - \frac{\mathcal{R}_{kk}\mathcal{X}_{ik}}{\mathcal{X}_{kk}}\right) - \sum_{j \in \mathcal{P}_{1k} \cap (\mathcal{P}_{1i} \cup \mathcal{T}_i)} Z'_{kj}\frac{\partial P_{L_j}}{\partial P_{g_k}}, \quad k \in \mathcal{G}^L, \tag{3.45}$$

where the derivative of P_{L_j} with respect to P_{g_k} can be calculated from (3.2a,b,c,d) and (3.10), that is,

$$\frac{\partial P_{L_j}}{\partial P_{g_k}} = \begin{cases} -\frac{2R_j}{v_j}\left[P_{S_j} - Q_{S_j}\frac{\mathscr{R}_{kk}}{\mathscr{X}_{kk}}\right] & \text{if } k \in \mathscr{T}_j \\ 0 & \text{if otherwise} \end{cases}, \tag{3.46}$$

where P_{S_j} and Q_{S_j} can be measured. Note that (3.46) is an approximation made by assuming that the impact of both losses in the subtree of bus j on P_{L_j} and $\partial P_{L_j}/\partial v_j$ is negligible.

From the previous analysis, z_{ik} are consensus variables and the decision variable P_{g_i} should be its estimation at the ith agent, that is,

$$P_{g_i} = z_{ii}. \tag{3.47}$$

3.4.2 Case study

IEEE 8500-node system is used to test the proposed hierarchical control on both the system- and cluster-level real/reactive power control. First, we use a case with low voltage problem when only reactive power control is implemented, as shown in Fig. 3.12. The spatial power unbalance in the circuit causes huge voltage deviations between power sources and loads, so it is necessary to implement a system-level control that can work as supplementary control to fix these problems, especially for large-scale systems. In this case the area FM4 is experiencing low voltage problem due to the voltage deviations between power sources and loads in the large-scale system. The supplementary control is to compensate the shortfall that the local power control is only for the local voltage.

The control response is shown in Fig. 3.13. It is shown that the system voltages increase due to the power injection at $t = 2$ second, then the local reactive power control starts to push the voltage down. At around $t = 11$ second the voltage at the lowest bus starts to violate the lower limit of voltage regulation. This is because the local reactive power control is only for the local voltage, which is shown in the second plot of the figure. About the same time the system-level control starts to kick in and the real power control (curtailment) responds to the system-level objectives. Eventually, the control settles down and all the voltages throughout the network are well controlled.

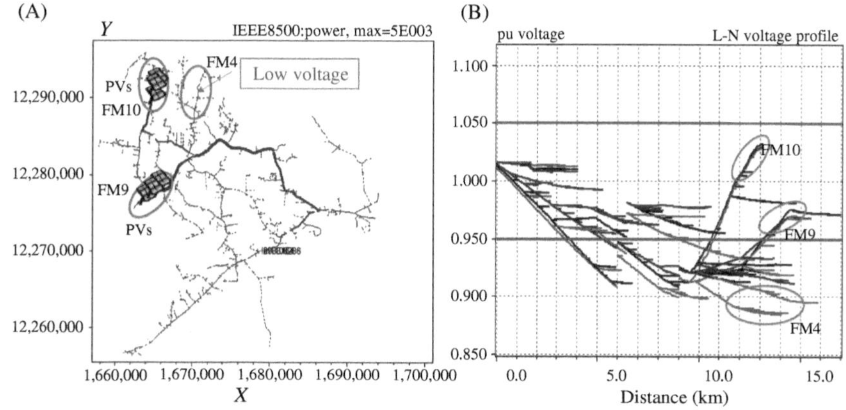

Figure 3.12 Low voltage problem due to the spatial unbalance of power flow in IEEE 8500-node system: (A) circuit and (B) voltage profile.

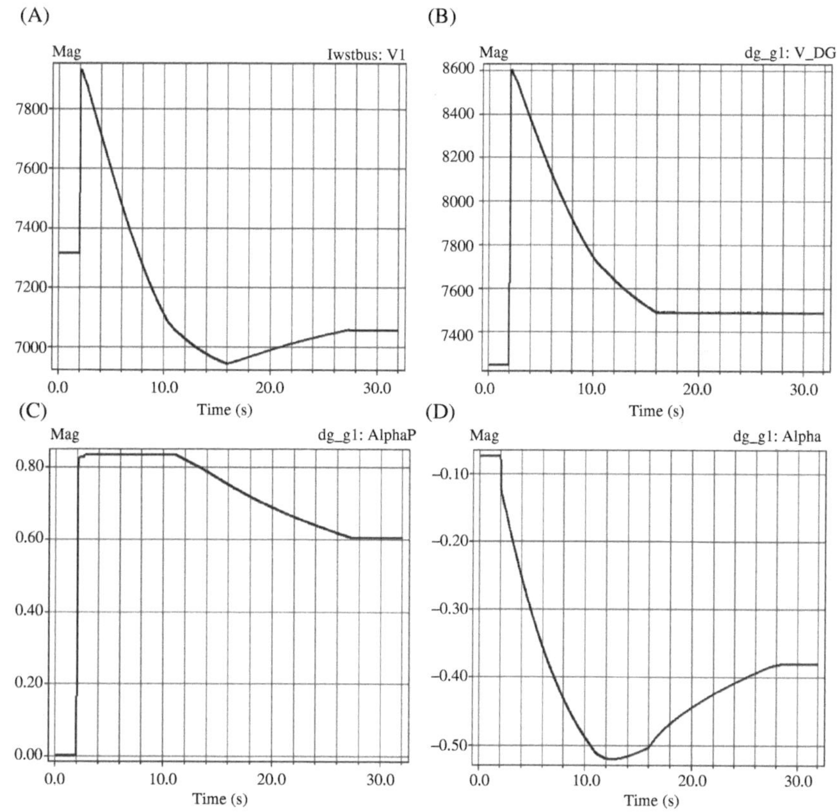

Figure 3.13 The control response of DGs to disturbance in the system: (A) the lowest bus voltage, (B) bus voltage of DG1, (C) real power utilization ratio of DG1, and (D) reactive power utilization ratio of DG1. *DG*, Distributed generation.

3.5 Islanded microgrid with high penetration of distributed generations

In the previous control design, all inverters can be operated in PQ-mode (the grid-following control), because there are VF references when the distribution network is connected to the main power grid through mid-voltage network. However in islanded mode, the reference for frequency and voltage control will not be available. In some design a VF-mode voltage source inverter [13] is used to provide the reference for VF as shown in Fig. 3.14; thus it is possible to operate the MG in islanded mode.

In islanded mode of microgrids the output power of VF inverter (e.g., the energy storage systems) will balance out the load changes. The VF-mode inverter control design can be categorized into three types [14]: (1) droop control, (2) virtual synchronous machine (VSM), and (3) nonlinear oscillator synchronization. Take VSM method, for example, the bus angle, δ_{ref}, and frequency, ω_{ref}, can be modeled by the following strategy:

$$\begin{cases} \dot{\delta}_{ref} = \omega_{VF} - \omega_0 \\ M\dot{\omega}_{VF} = P_{VF} - P_{V\ F0} + D(\omega_{VF} - \omega_0) \end{cases}, \tag{3.48}$$

where ω_0 is the nominal value of frequency, P_{VF} and P_0 are the instantaneous and nominal power of VF-mode ESS, respectively, M and D are coefficiencies of VSM design.

The PQ-mode DGs are assumed to follow the control command. In this scenario, let us assume the droop ratio m_i for each of the PQ inverters, then the power desired in the system (which is the output power of the

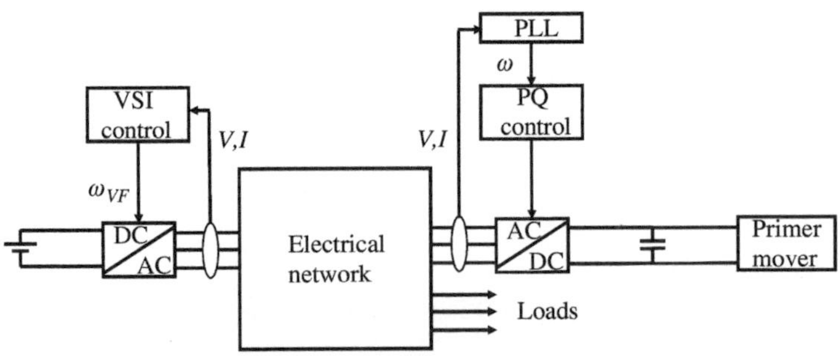

Figure 3.14 The reference for the islanded system operation.

VF inverters) can be distributed among all PQ inverters by the following droop strategy (as shown in Fig. 3.15):

$$P_{VF} - P_{VF0} = \sum m_i(\omega_0 - \omega_i), \tag{3.49}$$

where ω_i is the instantaneous frequency measured by PLL at the connection bus, which is generated by VF inverters (ω_{VF}) and spread through the electrical network. Hence, we have

$$\omega_i = \omega_{VF}, \tag{3.50}$$

given the dynamic of PLL is neglected.

It is clear that the VF inverters in the islanded mode work the same as the slack bus. So in the islanded mode, we can split all VF inverters as the slack bus and apply the same control strategy (3.43) to the rest DGs. The power dispatch in (3.42) is equivalent to the output of VF inverters, that is,

$$P_f^* - P_f = P_{VF} - P_{VF0}, \tag{3.51}$$

then we have

$$P_f^* - P_f = \sum m_i(\omega_0 - \omega_{VF}). \tag{3.52}$$

Hence, similar to (3.42), the control objective for the real power control in islanded mode can be written as follows:

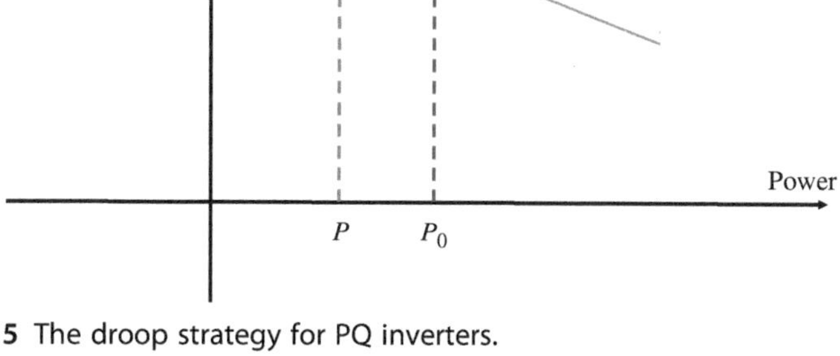

Figure 3.15 The droop strategy for PQ inverters.

$$f_i = \frac{\lambda_i^f}{2}(\omega_0 - \omega_{VF})^2 + \frac{\lambda_i^v}{2}\left(V_i^* - V_i\right)^2 + \frac{\lambda_i^p}{2}\left(P_{g_i}^{ref} - P_{g_i}\right)^2 \tag{3.53}$$

where $\lambda_i^f \geq 0$, $\lambda_i^v \geq 0$, and $\lambda_i^p \geq 0$ are the coordinate coefficients at the ith agent; P_{g_i} stands for the power injection of the ith agent; V_i is the representative voltage of the ith agent, particularly, the worst measured voltage at that agent. We use λ_i^f to include the information of the droop gains (m_is) for the simplicity of expression.

The distributed subgradient algorithm (3.20) is used again to control the real power in order to pursue the previous goal. The subgradient of objective function defined by (3.53) can be calculated in the same way as in Section 3.4: the subgradients of $\lambda_i^v/2(V_i^* - V_i)^2$ and $\lambda_i^p/2(P_{g_i}^{ref} - P_{g_i})^2$ are the same as in (3.44). The derivative of $\lambda_i^f/2(\omega_0 - \omega_{V\ F})^2$ with respect to P_{g_k} can be calculated as follows:

$$\frac{\partial \lambda_i^f/2(\omega_0 - \omega_{V\ F})^2}{\partial P_{g_k}} = \lambda_i^f m_k(\omega_0 - \omega_{V\ F}). \tag{3.54}$$

Using (3.54) and (3.45), the subgradient of (3.53) can be obtained and hence the real power control for islanded mode is complete. It is necessary to emphasize in the islanded mode that the frequency and power balance is of much more importance than other objectives. This will be reflected in the design of coordinate coefficients, and an extreme case is to set λ_i^v and λ_i^p to zero and the objective becomes

$$f_i = \frac{1}{2}(\omega_0 - \omega_i)^2. \tag{3.55}$$

In this case the algorithm regresses to droop control. The fair utilization ratio method (3.32) is also applicable:

$$\dot{\alpha}_{p_i} = \sum_{j \in N_i^c} d_{ij}\left(\alpha_{p_j} - \alpha_{p_i}\right) - \beta_i g_i^p, \tag{3.56}$$

where g_i^p is the subgradient of f_i which can be calculated from the previously mentioned procedure. For reactive power control in the islanded microgrid, the same control (3.33) is applied according to the previous local voltage control strategy (principal 4).

Using OpenDSS, we evaluate the performance of the previous control scheme by standard IEEE system: 8500-node circuit. A virtual synchronous generator model is used to generate the frequency reference in the system, as shown in Fig. 3.16. The stiff source bus is simulated by the

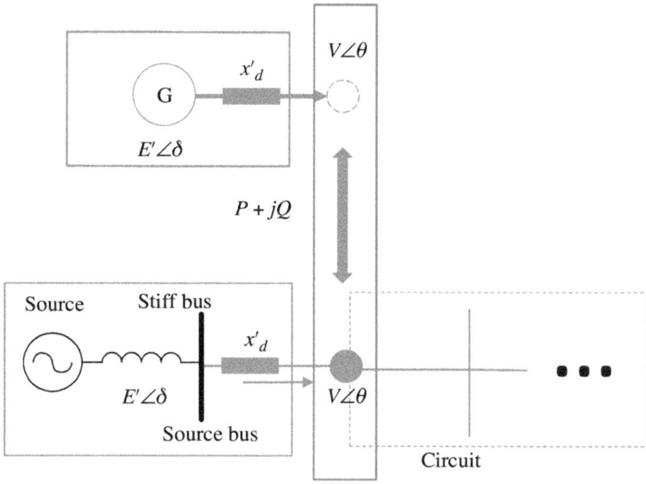

Figure 3.16 The stiff source bus is simulated as the inner bus behind transient admittance of a virtual synchronous generator.

classic synchronous machine model with the constant inner bus voltage behind transient admittance of the generator. Then the frequency of islanded system can be generated by the frequency of the VSM. A worst scenario found by the greedy search method [11] of the system with 4 large-scale PV farms and 100% penetration is shown in Fig. 3.17.

First, we built up an islanded system by opening the breaker at the feeder of IEEE 8500-node circuit, and using VSM to supply 300 kW at connection bus, assuming the reactive power supply is sufficient to maintain the inner bus voltage. Second, all regulators of the circuit are fixed at the predefined positions during the control. To test the active power and frequency control, we use dynamic simulation mode in OpenDSS. The simulation is set as simulation time $T = 30$ second with time step $h = 0.005$ second. A system disturbance of 2 MW load drop happens at $t_0 = 0.7$ second. As shown in Fig. 3.18, the proposed algorithm is effective for system frequency control in the case of islanded operation of distribution system. When the demand of one load decreases at 0.7 second, the system frequency starts to increase correspondingly. Then the active power control of PVs responds to the change of frequency. As a result, both the frequency is well maintained by the proposed control. At the same time, the reactive power control is also effective to maintain the system voltages.

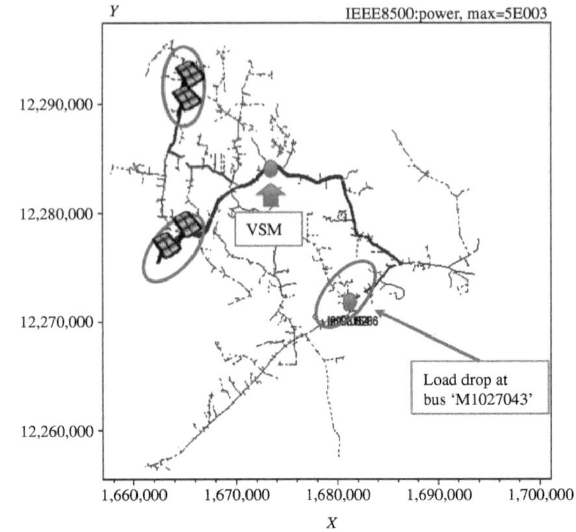

Figure 3.17 The worst scenario of IEEE 8500 system with four large-scale PVs.

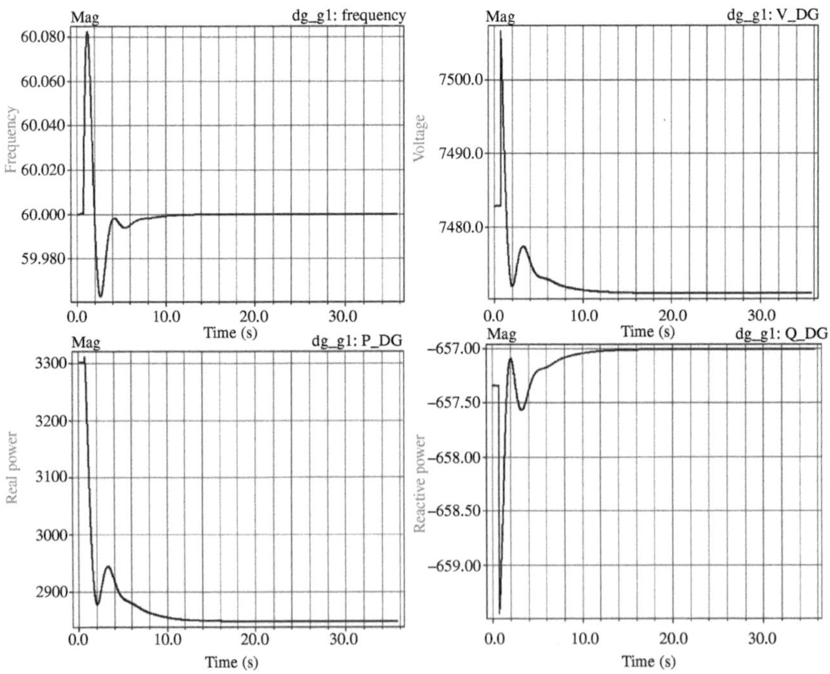

Figure 3.18 The frequency control of IEEE 8500-node system on islanded mode.

3.6 Grid-edge situational awareness: enhanced observability by voltage inference

There are synchronous measurement and control units (SMCUs) and asynchronous measurement units (AMUs) in distribution systems. While SMCUs are probably limited, AMUs should be deployed on all the load buses. The insufficiency of SMCUs is the reason of the low observability of distribution network. However, it is not difficult and less expensive to upgrade AMUs for a smaller time interval of information gathering. It is possible to infer the system situation using information from both SMCUs and AMUs. To this end, we present a sensitivity-based grid-edge situational awareness method in this section.

3.6.1 Voltage inference method

Assume some of the buses have SMCUs, while the rest buses do not. Instead, the rest buses have AMUs which can update their information for a particular period of time, T, which can be 5 minutes or several hours.

We summarize two types of distribution network voltage inference scenarios in Fig. 3.19. In this figure, there are prosumers under some of the buses i, j, k, m, and n. The net power injection (direction defined by arrow) of each prosumer is denoted by $P_* + jQ_*$, where the superscript * represents the bus number. The blue block represents the real-time measurement and control unit, while the yellow block denotes off-line measurement. The two scenarios of operational situation awareness for prosumer-dominated distribution systems are (I) subtree type (Fig. 3.19A), and (II) the two-end type (Fig. 3.19B). Type (I) represents a typical scenario in distribution network that bus i is an SMCU, but in the subtree of bus i, all buses are AMUs; type (II) is for a portion of circuit that the two ends are SMCUs; and most of other cases can be considered as a kind of the mix of these two types. Hence, in the following analysis, we will focus on the voltage inference of the two types of distribution network.

Let us assume bus i as the relative root bus in both circuits, type (I) and (II), and also assume that it is an SMCU. So Eq. (3.8) is applicable to all buses in both circuits considering bus i as the slack bus:

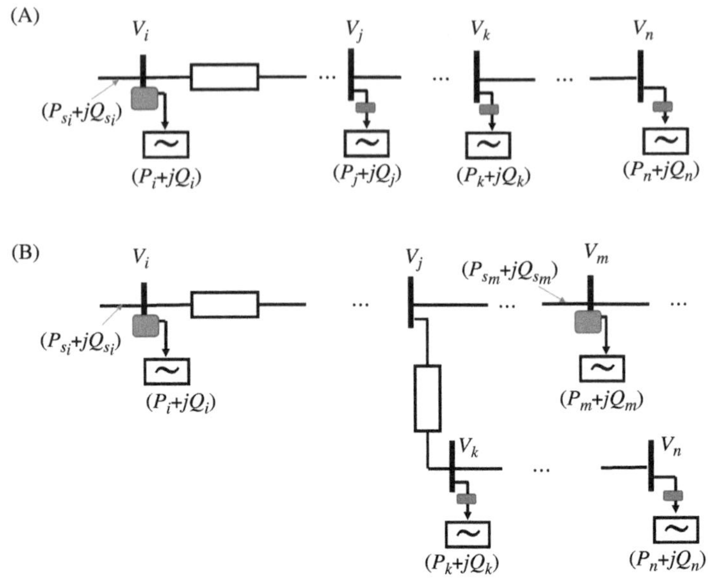

Figure 3.19 Example of distribution network with measurements: (A) subtree type and (B) mix type.

$$V_k^2 - \sum_{m\in \mathscr{P}_{ik} \cup \mathscr{T}_k} \left(2\mathscr{R}_{km}P_m + 2\mathscr{X}_{km}Q_m + Z'_{km}P_{L_m}\right) + V_i^2, \quad k\in \mathscr{T}_i, \tag{3.57}$$

where V_i can be measured in real-time.

Define the sensitivities of voltage at bus k to real and reactive power injection at bus j by ζ_{kj} and ξ_{kj}, respectively:

$$\zeta_{kj} \triangleq \frac{\partial V_k}{\partial P_j} \text{ and } \xi_{kj} \triangleq \frac{\partial V_k}{\partial Q_j}, \tag{3.58}$$

then we can infer all AMU bus voltages by

$$V_k = V_k^0 + \Delta V_k, \tag{3.59}$$

where V_k^0 is the initial value and ΔV_k can be estimated by

$$\Delta V_k = - \sum_{m\in \mathscr{P}_{ik} \cup \mathscr{T}_k} \left(\zeta_{km}\Delta P_m + \xi_{km}\Delta Q_m\right). \tag{3.60}$$

Therefore, the key of voltage inference is the network sensitivity calculation, which will be discussed in the next subsection.

3.6.2 Network sensitivity

It follows (3.57) that the sensitivities can be written as

$$\zeta_{kj} \triangleq -\frac{\mathscr{R}_{kj}}{V_k} - \frac{1}{2V_k} \sum_{m \in \mathscr{P}_{ik} \cup \mathscr{T}_k} Z'_{km} \frac{\partial P_{L_m}}{\partial P_j}, \tag{3.61}$$

and

$$\zeta_{kj} \triangleq -\frac{\mathscr{X}_{kj}}{V_k} - \frac{1}{2V_k} \sum_{m \in \mathscr{P}_{ik} \cup \mathscr{T}_k} Z'_{km} \frac{\partial P_{L_m}}{\partial Q_j}. \tag{3.62}$$

From the previous analysis the system loss should always be relatively small compared to the power injection, for example, the loss of the illustration example (Fig. 3.6) is shown in Fig. 3.20. As in the example, within the studied region of (P_g, Q_g), the derivatives are close to zero and monotonic along with the injection: loss is decreasing at the negative part and increasing after certain point (zero).

By neglecting the system loss the voltage sensitivities of real and reactive power can be written as follows:

$$\zeta_{kj} \approx -\frac{\mathscr{R}_{kj}}{V_k} \text{ and } \xi_{kj} \approx \frac{\mathscr{X}_{kj}}{V_k}. \tag{3.63}$$

Using the illustration system, the previously defined sensitivities of bus voltage with respect to real and reactive power injections are shown in Fig. 3.21. As the plots show, V_i increases when the power P_g and Q_g increase, so that the derivatives slightly decrease. Also it is clear that

$$\zeta_{kj} \approx -\mathscr{R}_{kj} \text{ and } \xi_{kj} \approx -\mathscr{X}_{kj}, \tag{3.64}$$

as shown in the illustration example ($\mathscr{R}_{kj} = 0.025$ and $\mathscr{X}_{kj} = 0.05$), and bus voltages are always close to 1 pu

Hence, the voltage inference Eq. (3.60) is modified as the following application:

$$V_k = V_k^0 - \sum_{m \in \mathscr{P}_{ik} \cup \mathscr{T}_k} \left(\zeta_{km}\Delta P_m + \xi_{km}\Delta Q_m\right), \tag{3.65}$$

where V_k^0 is the base value; ΔP_m and ΔQ_m represent the power fluctuation at AMU buses and will be discussed in the next subsection.

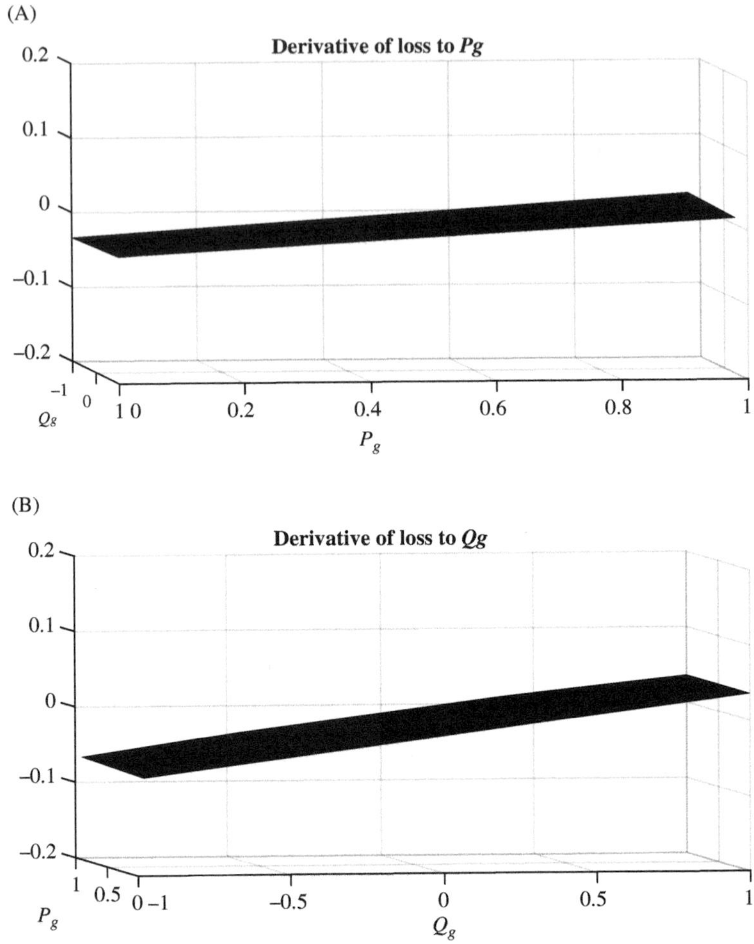

Figure 3.20 The derivative of loss with respect to power injection: (A) loss to P_g and (B) loss to Q_g.

3.6.3 Implementation

From (3.65) the last question for voltage inference is to determine the power injection of AMU buses. Use the proposed model (3.13); we use standard estimation methods to tackle this problem. Load estimation is not the focus of this chapter; hence, we will briefly describe the problem and will not expand the detailed process here. According to the setup, at time T_t, all the measurements from both SMCUs and AMUs are known; in the time between T_t and T_{t+1} the information of SMCUs is available,

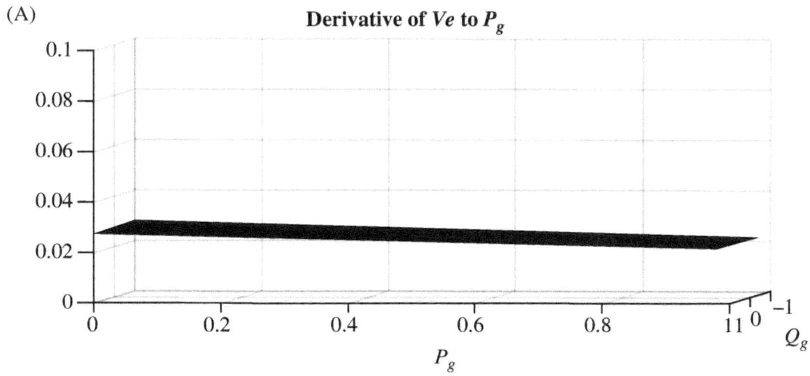

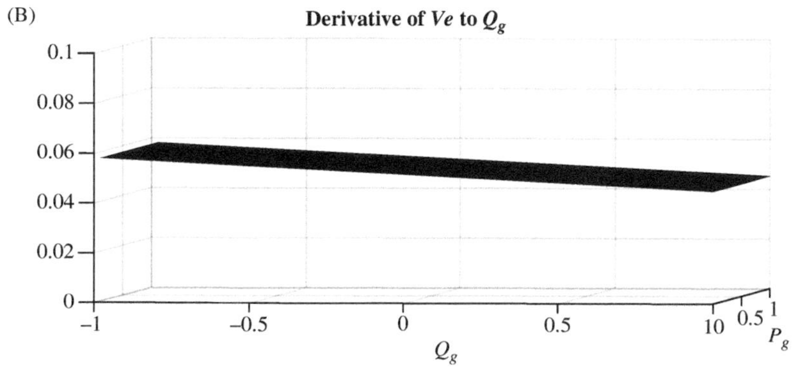

Figure 3.21 The derivative of loss with respect to power injection: (A) loss to P_g and (B) loss to Q_g.

while that of AMUs is not until T_{i+1}. Under the same assumption of neglecting system loss, the total power injection of all AMU buses (defined by $P_t + jQ_t$) can be calculated by applying Kirchhoff law in Fig. 3.19: in type (I)

$$P_t = P_{S_i} - P_i \text{ and } Q_t = Q_{S_i} - Q_i, \tag{3.66}$$

and in type (II)

$$P_t = P_{S_i} - P_i - P_{S_m} \text{ and } Q_t = Q_{S_i} - Q_i - Q_{S_m}. \tag{3.67}$$

There are mainly four steps to implement the grid-edge situational awareness method: first, the total injection $P_t + jQ_t$ and voltage deviation ΔV_i (and/or ΔV_m) can be obtained from measurement; second, using historical data and measurement at time T_t, we can apply one of the short-term prediction methods such as [15,16] to form a participation factor of each bus, then $P_t + jQ_t$ can be distributed among these buses; third, using the real-time measurement at SMCUs (voltage, power flow), standard estimation method, for example, least square method, can be employed to improve the results of ΔP_m and ΔQ_m; and last, plugging them into (3.65) to solve V_k, which completes the inference process.

We use IEEE-123 system to test the performance of the proposed method. As shown in Fig. 3.22, four PVs are installed in the subtree of bus 72, two SMCUs (*black stars*) at bus 72 and 77, and 16 AMUs (*emptys-tars*) at some other buses. Two stars are at bus 77, one for each case: case 1 with bus 77 as an AMU and case 2 with bus 77 as an SMCU.

The sensitivities of bus voltages with respect to different real and reactive power injections at bus 89 are shown in Fig. 3.23. The results are PU values with the base power as 3.6 MVAr and base voltage as 2.4 kV. The lines are the results from simulation of changing the power injection at bus 89, while the stars are the approximated sensitivities at each bus using (3.64) by the network parameters. It is clear that the numerical results are close to the results approximated by system parameters. So in the case of fast calculation, network sensitivity

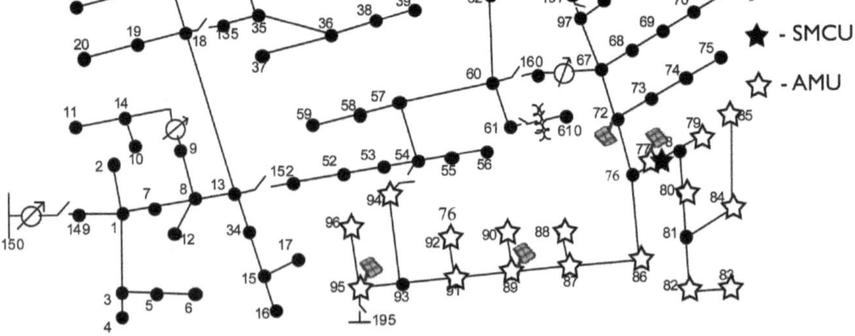

Figure 3.22 IEEE 123 system with SMCUs (*black stars*), AMUs (*empty stars*), and PVs on bus 72, 77, and 89. *AMU*, Asynchronous measurement unit.

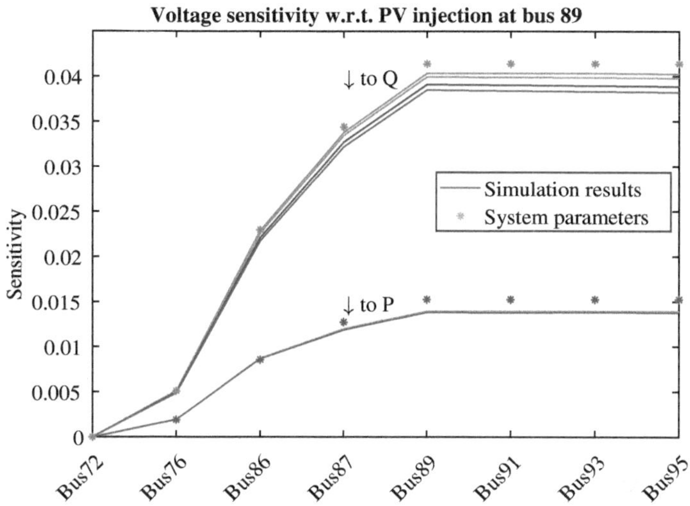

Figure 3.23 Voltage sensitivity with respect to different real and reactive power injections.

can be roughly approximated by network parameters, that is, $\zeta_{kj} \approx -\mathcal{R}_{kj}$ and $\xi_{kj} \simeq -\mathcal{X}_{kj}$.

Case 1 is designed as type (I) voltage inference scenario, where no SMCU is placed in the subtree of bus 72 (SMCU). Assume that, at time T_k, all the measurements are available at bus 72, but all measurements from AMUs are not reachable at time $T_k + \Delta t$, $\Delta t \in (T_k, T_{k+1})$. At time $T_k + \Delta t$ the voltage and the downstream (upstream) power at SMCU can be measured; given a forecast load, the output of PVs can be estimated. Then bus voltages can be inferred by the proposed method. Simulation results in Fig. 3.24A show that the voltages can be accurately inferred. The red and blue lines are the system voltage lines of two scenarios with different PV allocations. The red one is worse because the load prediction is worse than the blue. Case 2 is designed as type (II) voltage inference scenario, where another SMCU is placed at bus 77. The extra information help improve the result, especially when the load change is larger in the subtree of bus 77. In this case, we set a load disturbance of PV 77. The results are shown in Fig. 3.24B, where the stars are the results with real-time measurement at bus 77 and the diamonds are without SMCU there. It is obvious that the voltage inference results with real-time measurement are better than without real-time measurement.

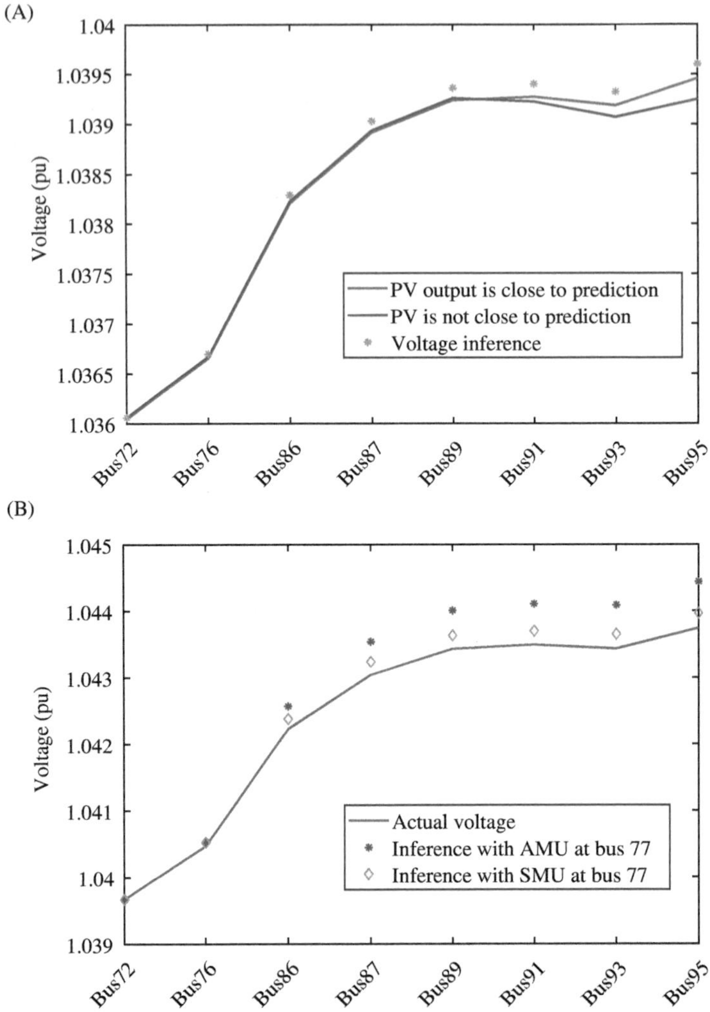

Figure 3.24 Simulation results of voltage inference on IEEE 123 system: (A) Case 1 and (B) case 2.

3.7 Control-enabled dynamic hosting allowance: *P* and *Q* control capacity and impact analysis

3.7.1 Traditional hosting analysis

Hosting capacity (HC) is an important planning tool for both distribution system operators and DG investors to assess how much DG generation can be integrated into the distribution network [17]. It is a limit of the quantity of renewables allowed to be integrated without imposing any changes to the existing infrastructure and without violating operational limits. Up until HC, DGs can be easily interconnected and could be approved through fast-track process. There are numerous unspecified factors for HC calculation, such as DER locations and capacities, control settings of feeder equipment such as voltage regulators, and the changing operational conditions (load level). The exhaustive detailed HC calculation is usually not preferred in practice, so instead the stochastic methods are widely used. Based on the trends observed from many detailed study cases, EPRI presented a streamline method to speed up HC calculation [18]. However, all existing hosting analyses are generally conservative and require further development in the following aspects:

- to incorporate time-serious analysis using prediction and trend data other than worst-case snapshot data
- to involve advanced inverter control and complimentary DER technologies
- to explore rapid approaches for HC calculation and scanning
- to further study the interactivities between DERs and the impact on HC

It is very possible and feasible, along with the development of renewable technologies, to install more DER than HC either by upgrading or installing additional equipment, or developing advanced operational and control strategies. Furthermore, HC is a system-oriented terminology and specific for planning. To the extent of system operation, it is required to go beyond the HC which needs more extensive and much faster analysis and control strategy. To this end a network sensitivity–based DHA method is presented in this section, which includes both the operating condition and the control impact of DER inverters.

3.7.2 Dynamic hosting allowance analysis

Voltage problems, specially voltage rise, are considered as the most significant problem for high penetration of DG integration [19], so the system-wide max voltage is a major factor in hosting analysis. As shown in Fig. 3.25, the trend of maximum feeder-wide primary voltage in a particular deployment of PV with respect to penetration level appears roughly linear [18], but the physical nature behind this observation has not been

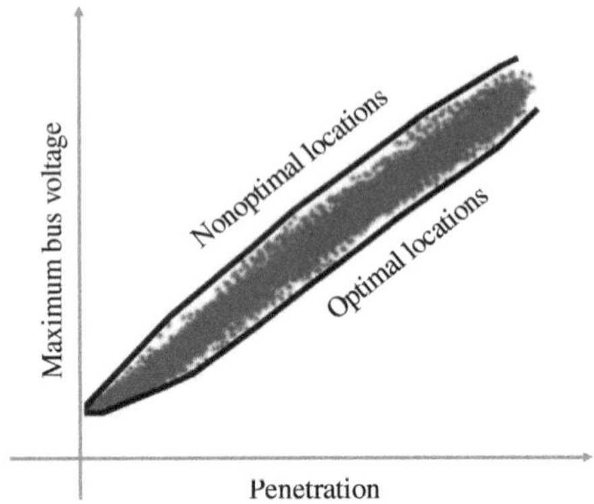

Figure 3.25 System voltage shows a linear trend with respect to renewable penetration [18].

discussed yet. In the previous section, we built the network sensitivity, and through analysis, we found that the network sensitivity is a property of distribution network. Here we rewrite (3.60) by the following matrix form:

$$\Delta V = \Pi \Delta P + \Xi \Delta Q, \tag{3.68}$$

where

$$\Pi = \left\{ \zeta_{kj} \right\} \text{ and } \Xi = \left\{ \xi_{kj} \right\}, \tag{3.69}$$

which can be approximated as constant (the system parameter of distribution network, $\mathscr{R}_{kj}$ and $\mathscr{X}_{kj}$). The properties of network sensitivity, that is, Π and Ξ, can explain the linear trend shown in the next figure.

Using sensitivity matrix, we can improve the HC maps [17] by revealing the additive effect of system-wide DERs to critical buses—DHA. As shown in Fig. 3.26, the traditional HC map did not answer the question of how $P_i + jQ_i$ and $P_k + jQ_k$ contribute to the critical voltage V_n; on the other hand, with certain head room of V_n (and/or other critical bus voltages), how to strategically deploy $P_i + jQ_i$ and $P_k + jQ_k$, as well as the injection of other DERs.

As shown in Fig. 3.27, the network sensitivity together with synchronous and asynchronous measurement can greatly improve the functionality of HC maps. This can be further extended into real-time DG operation and control (if the DG inverters are able to offer extra control ability).

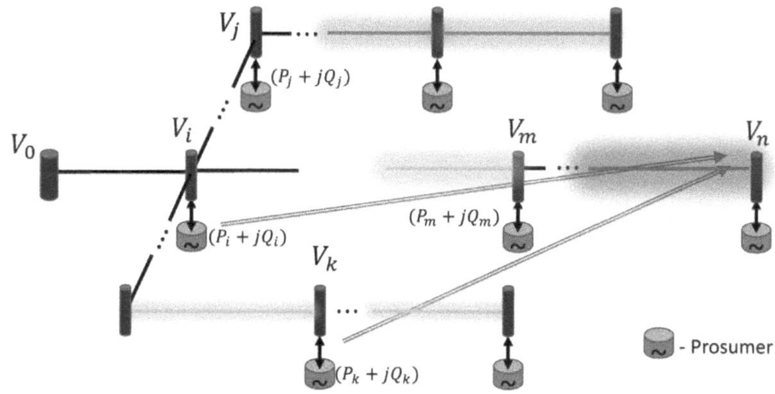

Figure 3.26 HC map can indicate how HC varies along feeder segments using appropriate color coding in regards to the current percentage of each segment, but additive effect of different DGs to different bus voltages is not analyzed. *DG*, Distributed generation; *HC*, hosting capacity.

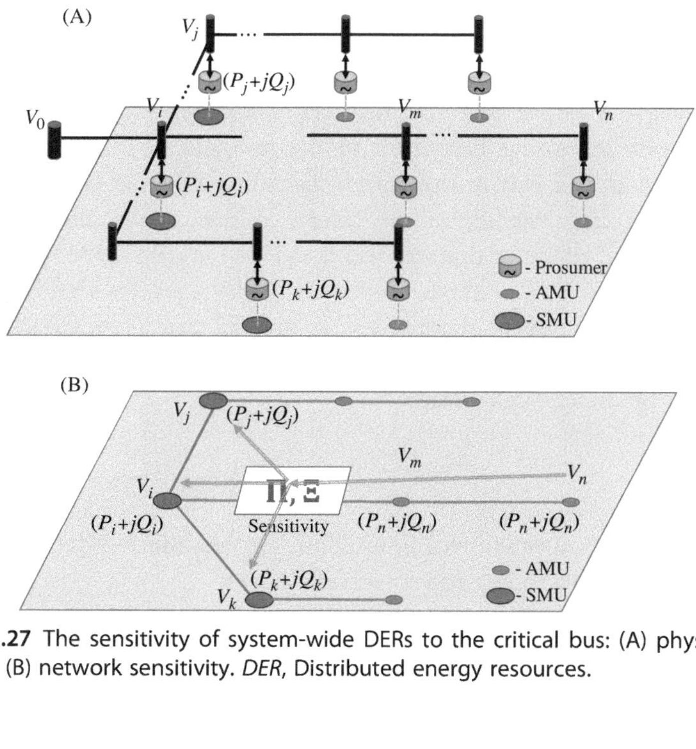

Figure 3.27 The sensitivity of system-wide DERs to the critical bus: (A) physical system and (B) network sensitivity. *DER*, Distributed energy resources.

In the context of operation control, the critical bus voltage is the key factor of DHA. We can first identify the head room of critical voltage, which can be assessed through the grid-edge situational inference process. Assume that the voltage head room of critical bus n is ΔV_n, then we have

$$\Delta V_n = -\sum_{m \in \mathscr{P}_{in} \cup \mathscr{T}_n} \left(\zeta_{nm} \Delta P_m + \xi_{nm} \Delta Q_m \right). \tag{3.70}$$

Then, with specific control strategy in the system, we can quickly calculate the DHA for the system. For example, define a parameter as σ to describe capability of inverter, that $S = (1+\sigma)\overline{P}$. We can calculate ΔQ_m by

$$\Delta Q_m = -\overline{Q}_m = \sqrt{\sigma^2 + 2\sigma}\overline{P}_m = -\sqrt{\sigma^2 + 2\sigma}(P_m + \Delta P_m), \tag{3.71}$$

and finally solve (3.70) for ΔP, which is the result of DHA. Note that as reactive power control is enabled, the system current increase; hence, the thermal limits should be checked.

We use IEEE 123-bus system to illustrate the DHA process. We use case 1 in the previous section as the example to calculate DHA at bus 72 in IEEE 123-bus system. From the simulation results (Fig. 3.23) in the previous section, we know that the network sensitivity has the property of monotonically increasing from root to the leaf side of the circuit. So the highest DHA of this part of the system should be DHA at bus 72 with the concern of voltage rise. From the results of case 1, we also know that the voltage head room of this scenario is $\Delta V = 1.05 - 1.0396 = 0.0104$ pu (base power is 3.6 MW and base voltage is 2.4 kV). Also we know that the real and reactive power sensitivities at bus 72 are $\zeta_{in} = 0.048701$ and $\xi_{in} = 0.0936$ pu. If reactive power control is not considered, the DHA at bus 72 can be directly calculated:

$$\Delta P = \frac{\Delta V}{\zeta_{in}} = 0.2135 \text{ pu}. \tag{3.72}$$

When reactive power control is enabled, we assume $\sigma = 0.02$, then we can plug (3.71) into (3.70) and solve ΔP as

$$\Delta P = \frac{\Delta V + \sqrt{\sigma^2 + 2\sigma}\xi_{in} P}{\zeta_{in} - \sqrt{\sigma^2 + 2\sigma}\xi_{in}} = 0.52 \text{ pu}. \tag{3.73}$$

3.8 Cosimulation of integrated transmission and distribution systems

3.8.1 The framework of cosimulation

3.8.1.1 Integrated T&D system simulation

The cosimulation architecture of integrated T&D system is developed as follows. The snapshot of transmission system power flow is first solved, and the voltage at each bus connected with the distribution systems is saved. Each distribution system is assigned a process with parameters of voltage levels, circuit name, and process number. Once each process is established, they start and execute simultaneously. The various sizes and complexity of distribution systems cause the power flow to solve in different time duration. Thus the algorithm waits for the convergence of all distribution systems before sending the real and reactive power values at the substation back to the transmission bus. A text file is generated containing the real and reactive power at the substation after each power flow solution with respect to the transmission voltage. The transmission system power flow is then solved again with the updated parameters. The algorithm iterates through this process until the $\Delta V \leq 0.0001$. In summary the cosimulation architecture follows the following steps:

S-1. Load and solve transmission system.

S-2. Assign a distribution system to 10 of the load buses.

S-3. Extract V_{pu} at each load bus.

S-4. Define a process for each distribution system with its corresponding voltage value.

S-5. Solve each distribution system in parallel.

S-6. Send P and Q from substation node to the transmission system.

S-7. Update transmission load bus with P and Q obtained from distribution.

S-8. Repeat steps 1–7 until ΔV_{pu} (of each load bus) ≤ 0.001.

The integrated T&D system cosimulation architecture was modeled and simulated using cosimulation of Python and MA-OpenDSS. Loads in the buses of transmission system were analyzed to determine real and reactive power values which resembled the total real and reactive power of various distribution systems. Once the buses were identified, the distribution systems are allocated to each individually. Python was used to interface with MA-OpenDSS and facilitate the usage of parallel processing when solving the power flow of each of the distribution systems.

3.8.1.2 Parallel cosimulation of integrated T&D systems

The parallel simulation was performed on the distribution systems, while the transmission system waited to receive the real and reactive power values. The same type of simulation was performed but in a sequential manner. The distribution systems ran one at a time, while the transmission system waited for the results. This was done to compare and verify the results. When running the distribution systems in sequence, the simulation time was increased, but real and reactive power values remained the same. Simulation results show that even though the sequential simulation of each distribution system tends to be slightly faster than parallel simulation (less processing power required to run one distribution system at a time), it takes longer to complete the overall simulation. While the parallel simulation is bottlenecked by the simulation of the largest distribution system, the sequential simulation waits for one distribution circuit to converge before running the next, thus taking much longer time to perform the integrated T&D simulation.

The simulation framework has been implemented on the latest version of MA-OPenDSS with the parallel capability, which allows different distribution networks to be simulated in parallel by assigning them to different cores in a multicore computer. This makes it scalable to handle multiple large-scale distribution networks in integrated T/D studies.

3.8.2 Simulation results

3.8.2.1 Power flow of integrated T&D systems

The topology of T&D cosimulation is shown in Fig. 3.28. The T&D test system consists of an IEEE 14-bus transmission system with three detailed

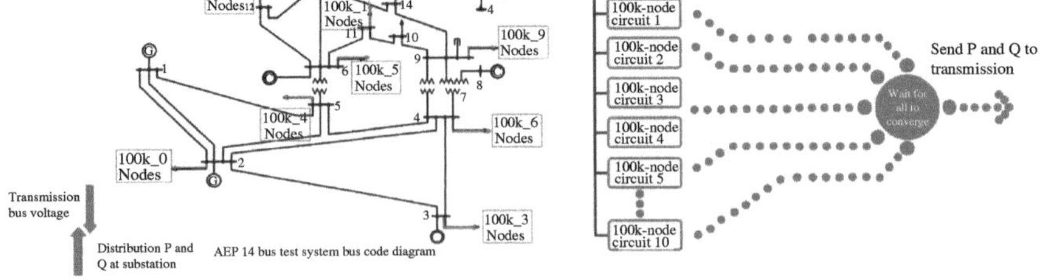

Figure 3.28 Integrated T&D simulation topology.

distribution networks: IEEE 8500-node system, 11,000-node (IEEE 8500 + EPRI Circuit 7) system, and 100,000-node (12 of 8500-node circuits) system. The system can easily be expanded by replacing lumped transmission-level loads with distribution networks.

Table 3.1 shows the simulation results of four iterations with total 27 seconds of simulation time. One can observe the response of distribution systems to various voltage changes in each iteration.

The 14-bus transmission system can be scaled up to accommodate for ten 100k-node distribution circuits located at each of its load buses. Preliminary simulation of ten 100k-node distribution circuits was executed in both parallel and sequential manners without any transmission connection, in order to compare the computation time. For sequential simulation, each of the 10 circuits took approximately 4.7 seconds to solve, and the overall simulation time was about 53 seconds (exporting results, setting up the systems counts are part of the total simulation time). On the other hand, the parallel simulation of 10 circuits resulted in a total of 35 seconds, which counts for one iteration of the integrated T&D without the transmission power flow simulation time. Thus it can be projected that there will be a tremendous benefit in reducing computation time. Furthermore, a faster machine with more processing power and core availability can improve the parallel processing computation time.

3.8.2.2 Integrated T&D systems with PVs and control

PVs and control are added to the integrated T&D systems. A synthetic 100,000-node NREL system with 100% penetration of small-scale PVs and distributed control was selected as the distribution system. Table 3.2 displays the real and reactive power for the base case, the case adding PVs without control, and the case adding PVs with control.

When PVs are present, the reactive power demand increases significantly (approximately 1.4 times the base case) in order to address overvoltage problems. On the other hand, the distribution system is supplying the transmission with about 1.05 MW. In order to address high reactive power demand, capacitor banks are placed on the secondary side (12.47 kV) of 14 feeders of NREL-100k system. Two T&D cosimulations were carried out with an aggregated capacitance of 110.8 and 153.5 MVAr, as shown in Table 3.3. Two capacitor banks were used in order to address the fluctuations of the substation voltage.

Table 3.4 displays T&D cosimulation results of three different simulation scenarios: (3), (4), and (5). The distribution system was connected to

Table 3.1 Simulation results of integrated T&D study.

	8.5k Nodes			**11k Nodes**			**100k Nodes**		
Ite	**V bus 13**	***P***	***Q***	**V bus 11**	***P***	***Q***	**V bus 14**	***P***	***Q***
1	0.9495	10,092.358	2707.363	0.9495	77,552.727	35,871.226	0.9493	144,610.512	21,991.715
2	0.9666	10,422.447	2760.587	0.9665	79,997.143	36,923.693	0.9664	144,026.118	18,664.282
3	0.9660	10,410.870	2758.983	0.9659	79,907.645	36,887.208	0.9658	144,055.285	18,825.498
4	0.9659	10,409.064	2758.731	0.9658	79,894.028	36,881.875	0.9657	144,059.901	18,850.989

Table 3.2 T&D cosimulation cases.

Simulation type	*P* (kW)	*Q* (kW)
1. Base case	123,992.6	48,282.3
2. PVs, no control	− 870.58	38,468.4
3. PVs, with control	− 1052.5	66,138.5

Table 3.3 T&D cosimulation cases with capacitor banks.

Simulation type	Aggregated capacitance (MVAr)	*P* (kW)	*Q* (kVAr)
4. DG, with control	110.80 (12.47 kV)	− 1147.8	− 390.2
5. DG, with control	153.50 (12.47 kV)	− 1192.26	− 34,922.2

DG, Distributed generation.

bus 11 of the IEEE 14-bus system, considering the smaller load at this bus (3.5 MW, 1.8 MVAr). It can be observed that it takes nine iterations to converge without the capacitor banks. On the order hand, it takes five and four iterations to converge with aggregated capacitor banks of 110.8 and 153.50 MVAr, respectively.

Another observation can be made with respect to the difference in total reactive power supplied by the capacitors. Although the power factor at the substation when solving distribution system power flow is much better with the total Q of 110.80 MVAr provided by capacitor banks, this is not the case when the voltage changes at the transmission side. The reactive power shifts from supplying 390.2 kVAr to transmission to demanding 8370.72 kVAr from transmission. Meanwhile, the feeder supplies reactive power when there is approximately 43 MVAr increase in the total capacitor banks. Further studies can be performed in order to find the optimal capacitor placement and capacity in order to maintain a relatively reasonable power factor.

3.9 Conclusion

This chapter presented a hierarchical distributed control framework to model, analyze, optimize, and control large-scale distribution system with extremely high penetration of renewables. Following the layered and divisional principle for large-scale power system planning, the Volt/VAr control is mainly treated as a local control, while the real power

Table 3.4 Comparison of T&D cosimulation cases with capacitor banks.

(3) No Q_{bank}			(4) $Q_{bank} = 110.80$ MVAr			(5) $Q_{bank} = 153.50$ MVAr		
V (pu)	*P* (kW)	*Q* (kVAr)	Voltage	*P* (kW)	*Q* (kVAr)	*V* (pu)	*P* (kW)	*Q* (kVAr)
1.0179	− 1038	62,519	1.04536	− 1070.58	8527.31	1.0179	− 1022	− 35,725
0.9341	− 2824	21,250	1.04418	− 1101.21	8354.14	1.06945	− 1009	− 30,867
1.0154	− 1167	60,371	1.04439	− 1084.81	8364.2	1.06955	− 1070	− 30,698
0.9398	− 2847	26,773	1.04446	− 1079.17	8366.38	1.06956	− 1069	− 30,699
1.0068	− 1566	51,582	1.04453	− 1074.32	8370.72			
0.9602	− 2458	45,637						
0.9736	− 2344	47,560						
0.9697	− 2396	47,585						
0.9698	− 2363	47592						

control is a system-level control (frequency) and a supplementary control for local voltage, which will respond only when reactive power control is insufficient. For electrical circuit, both nodal injection and branch power flow models are used to model distribution network; and a simplified power control model of DG is used, which is simple but good enough to illustrate the design of system operation and control. A detailed analysis of the system-level effect of local autonomous controls is presented in this chapter. Also how the presented control runs in islanded mode of the distribution system is provided. To tackle the problem of low observability in distribution network, a sensitivity-based grid-edge situational awareness method is presented for distribution system state estimation, control, and optimization. Following that a network sensitivity–based DHA method is presented for system operation, as an extensive and much faster analysis compared to the traditional HC. At last, to validate the feasibility and demonstrate the scalability of developed models and algorithms, a cosimulation architecture of integrated T&D system is developed.

References

[1] D.K. Molzahn, F. Dörfler, H. Sandberg, S.H. Low, S. Chakrabarti, R. Baldick, et al., A survey of distributed optimization and control algorithms for electric power systems, IEEE Trans. Smart Grid 8 (6) (2017) 2941–2962.

[2] S. Xia, S. Bu, C. Wan, X. Lu, K.W. Chan, B. Zhou, A fully distributed hierarchical control framework for coordinated operation of DERs in active distribution power networks, IEEE Trans. Power Syst. 34 (2018) 5184–5197.

[3] H. Xin, Z. Lu, Z. Qu, D. Gan, D. Qi, Cooperative control strategy for multiple photovoltaic generators in distribution networks, IET Control Theory Appl. 5 (14) (2011) 1617–1629.

[4] Y. Xu, Z. Qu, R. Harvey, T. Namerikawa, Data-driven wide-area control design of power system using the passivity shortage framework, IEEE Trans. Power Syst., 2019.

[5] L. Gan, S.H. Low, An online gradient algorithm for optimal power flow on radial networks, IEEE J. Sel. Areas Commun. 34 (3) (2016) 625–638.

[6] Z. Qu, Cooperative Control of Dynamical Systems: Applications to Autonomous Vehicles, Springer Science & Business Media, 2009.

[7] A. Nedic, A. Ozdaglar, Distributed subgradient methods for multi-agent optimization, IEEE Trans. Autom. Control. 54 (1) (2009) 48–61.

[8] R. Harvey, Y. Xu, Z. Qu and T. Namerikawa, "Dissipativity-based design of local and wide-area DER controls for large-scale power systems with high penetration of renewables," *2017 IEEE Conference on Control Technology and Applications (CCTA)*, Mauna Lani, HI, 2017, pp. 2180–2187.

[9] A. Gusrialdi, Y. Xu, Z. Qu, M.A. Simaan, A real-time big data control-theoretical framework for cyber-physical-human systems, in: M. Blondin, P.M. Pardalos, J.S. Sáez (Eds.), Computational Intelligence and Optimization Methods for Control Engineering, Chapter 7, Springer Nature Switzerland AG, Switzerland, 2019, pp. 1–25.

[10] A. Maknouninejad, Z. Qu, Realizing unified microgrid voltage profile and loss minimization: A cooperative distributed optimization and control approach, IEEE Trans. Smart Grid 5 (4) (2014) 1621–1630.

[11] M. Rathbun, Y. Xu, Z. Qu, W. Sun, et al., Impact studies and cooperative voltage control for high pv penetration, IFAC-PapersOnLine 51 (28) (2018) 684–689.

[12] K.N. Miu, H.-D. Chiang, Existence, uniqueness, and monotonic properties of the feasible power flow solution for radial three-phase distribution networks, IEEE Trans. Circuits Syst. I: Fundam. Theory Appl. 47 (10) (2000) 1502–1514.

[13] J.A.P. Lopes, C.L. Moreira, A.G. Madureira, Defining control strategies for microgrids islanded operation, IEEE Trans. Power Syst. 21 (2) (2006) 916–924. Available from: https://doi.org/10.1109/TP-WRS.2006.873018. ISSN 0885-8950.

[14] B. Kroposki, B. Johnson, Y. Zhang, V. Gevorgian, P. Denholm, B. Hodge, et al., Achieving a 100% renewable grid: operating electric power systems with extremely high levels of variable renewable energy, IEEE Power Energ. Mag. 15 (2) (2017) 61–73.

[15] C. Guan, P.B. Luh, L.D. Michel, Z. Chi, Hybrid Kalman filters for very short-term load forecasting and prediction interval estimation, IEEE Trans. Power Syst. 28 (4) (2013) 3806–3817.

[16] Y. Yu, T. Choi, C. Hui, An intelligent quick prediction algorithm with applications in industrial control and loading problems, IEEE Trans. Autom. Sci. Eng. 9 (2) (2012) 276–287.

[17] S.M. Ismael, S.H.E. Abdel Aleem, A.Y. Abdelaziz, A.F. Zobaa, State-of-the-art of hosting capacity in modern power systems with distributed generation, Renew. Energy 130 (2019) 1002–1020.

[18] M. Rylander, J. Smith, W. Sunderman, Streamlined method for determining distribution system hosting capacity, 2015 IEEE Rural Electric Power Conference, IEEE, 2015, 3–9.

[19] T. Walla, J. Widén, J. Johansson, C. Bergerland, Determining and increasing the hosting capacity for photovoltaics in Swedish distribution grids, in: 27th European Photovoltaic Energy Conference (EU-PVSEC), 24–28 September 2012, Frankfurt, Germany, 2012.

CHAPTER FOUR

Advances of wholesale and retail electricity market development in the context of distributed energy resources

Qin Wang
Department of Power Delivery & Utilization, Electric Power Research Institute, Palo Alto, CA, United States

Contents

4.1 Introduction 99
4.2 Modern wholesale electricity market 100
 4.2.1 Driving forces of the electricity market development 100
 4.2.2 Operation process of wholesale electricity markets 102
 4.2.3 New products and designs in wholesale electricity markets 104
 4.2.4 Running the electricity market with 100% renewables 117
4.3 Modern retail electricity market 130
 4.3.1 State-of-the-art for retail electricity market development 130
 4.3.2 Design of retail electricity market operation framework 133
 4.3.3 Implementation of blockchain technology in electric power systems 138
4.4 Conclusion 140
References 140

4.1 Introduction

The supply of variable wind and solar generation with zero marginal operating cost is increasing rapidly around the world. Emerging customer behavior and public policy is driving the growth of smaller scale, behind-the-meter (BTM) distributed energy resources (DERs) such as solar photovoltaic (PV) and battery storage in many systems. The increasing penetration of utility-scale renewables and BTM DERs is changing the way how power system is operated in both transmission and distribution grids. In the wholesale electricity markets, many profound questions have

New Technologies for Power System Operation and Analysis.
DOI: https://doi.org/10.1016/B978-0-12-820168-8.00004-3

been raised about whether the current market design can provide both long-term and short-term price signals to support efficient investment and maintain reliable operations. For example, when most of the resources in the system have zero marginal costs, the resource owners will have to recover their cost for participating into the market either from the capacity market or by operational scarcity price. The short-term energy and ancillary service market, which was built upon middle 20th-century models of optimal pricing and investment, should be modified to consider the variability and uncertainty of intermittent generation at scale. From the retail side the current distribution grid operators are not designed to integrate and coordinate large volume of flexible customer, merchant DER, and microgrids capable of seamless islanding. There is an increasing need to develop retail electricity markets for DER products and services. The retail market operators are responsible for three major functions, including integrated system planning, distribution grid operation, and retail market operation. They balance the supply and demand variations at the distribution level and link the wholesale and retail market agents.

This chapter introduces the recent advances of wholesale and retail electricity market development in the context of DERs. From the wholesale perspective, the new market design and models to address the challenges faced by the bulk grid is discussed. In particular, the changes of power system operation paradigm at 100% renewable energy penetration level are discussed. From the retail perspective, a framework to construct retail market platform is proposed and the use of blockchain technology to facilitate retail transactions is introduced.

4.2 Modern wholesale electricity market

4.2.1 Driving forces of the electricity market development

During the past half century the power industry has experienced several transformations. Prior to 1960s, almost all the electricity was supplied from vertical integrated power utilities. In the United States, it was generally accepted that the first time the power industry introduced competition was in 1978, when the Public Utility Regulatory Policies Act (PURPA) was passed [1]. The most significant progress of this law was creating a market for power from nonutility power producers (which is called independent power producer or IPP today). Before PURPA, only utilities are authorized to own and operate electricity generation plants because energy companies were classified as natural monopolies. The PURPA enabled utilities to buy power from nonutility generators, which could produce power at a lower cost to server customers attached to the utilities' grids [2]. The second wave of the electric power industry reform started in the middle of 1990s. On April 24, 1996 the Federal Energy Regulation Commission (FERC) issued two landmark orders—Order 888 and Order 889 [3]. The Order 888's primary goal was to establish and promote generation competitions in the wholesale side by ensuring equal access and market treatment of transmission customers. The Order 889 conceived the establishment of the Open Access Same-Time Information System and set standards for how utilities and customers would come to share information in the transmission system. On December 20, 1999 FERC issued another landmark order—Order 2000 [4], to encourage the establishment of regional transmission organizations (RTOs) and further prompt the goals in previous orders by bringing all transmission control under independent entity's management. These orders led to the creation of three independent system operators (ISOs), including California ISO (CAISO), New York ISO, and the Electric Reliability Council of Texas and four RTOs, including PJM Interconnection, ISO New England, Midwest ISO, and the Southwestern Power Pool.

In the past decade the driving forces for the new development of electricity markets are the increasing penetration of renewable and DERs, along with the implementation of smart grid technologies such as the smart meters and phasor measurement units, and the participation of electricity consumers into the market. From the market design perspective, the fast growing of renewables and DERs could lead to several significant challenges. First, renewable and DERs are variable and uncertain by nature. As a result, additional actions are required to balance the system, including higher operating reserves, manual operations from system operators, and commitment of fast start units. Second, the system will see higher frequency of extreme events on net load ramping, so there is greater need of flexibility in the system. Insufficient flexibility may lead to area balance violations, price volatility, and significant renewable energy curtailments. Third, many DERs are located behind the meter, so there are increasing needs to perform transmission–distribution coordination in system operations. In New York State the utilities are encouraged to form distributed system platforms (DSPs) that will serve as neutral gatekeeper between the energy producers and energy consumers [5].

4.2.2 Operation process of wholesale electricity markets

The objective of establishing an electricity market is to facilitate an economical operation while ensuring the security of the system. Two major components are included for the operation of today's ISO-based electricity markets: day-ahead market clearing process and real-time market operation. Their structures are indicated in Figs. 4.1 and 4.2, respectively.

Fig. 4.1 shows the clearing process of day-ahead electricity market. The market participants submit their supply and demand bids to the ISO. At first, a pure unit commitment problem is solved in the resource commitment (RSC) application procedure. This generally involves solving a mixed integer programing problem. The result of RSC (on and off status) is sent to the market clearing engine (MCE) to identify the optimal dispatch (MW and PAR angle) of resources. Then the dispatch result is sent to the network security analysis procedure to check if there is overload violation on the circuits for both normal and postcontingency states. This is called the simultaneous feasibility test (SFT) [6] process. If there are no violations, the market clearing result is obtained. Otherwise, the SFT will

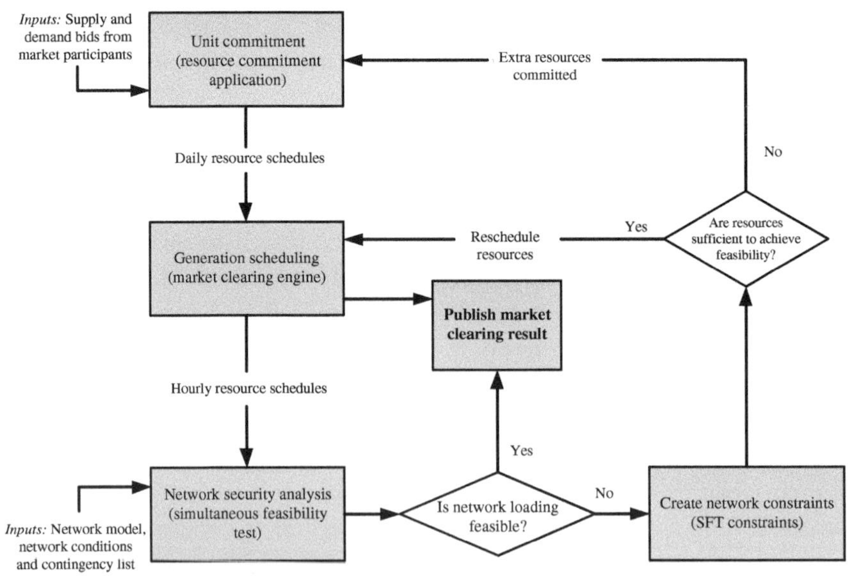

Figure 4.1 The clearing process of day-ahead electricity market.

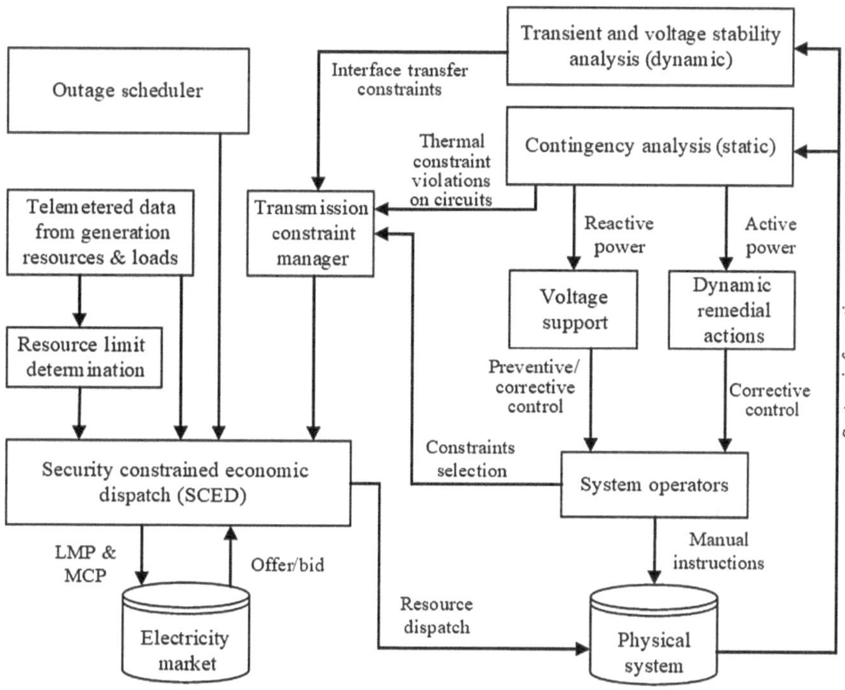

Figure 4.2 Real-time operation of power system and wholesale electricity market.

generate generic constraints that will eliminate the violations. The generic constraints are fed back to the MCE to redispatch the resources. If the MCE finds that the current RSCs are not sufficient to support the secure dispatch, the unit commitment problem is solved again to recommit the available resources. Otherwise, the MCE will find the optimal result thus the market clearing results could be obtained.

Fig. 4.2 demonstrates how the real-time power system and wholesale electricity market are operated [7]. The system and the market are two interconnected components in the operation of ISOs. Historically, ensuring secure operation of the system has always been a critical task. Thus the system conditions are sent to a so-called security assessment (SA) procedure to check if the current system is secure. The SA includes static SA, which includes contingency analysis (CA), and dynamic SA (DSA), which includes transient stability analysis and voltage stability analysis. The static SA will generate thermal constraints, which are able to protect the transmission facilities from thermal overload. The DSA will generate generic constraints, which are able to protect the transmission system from transient instability and voltage collapse. On the other hand, the CA result will be sent to a security enhancement procedure to generate control recommendations to operators. The purpose of security enhancement is to implement control actions to enhance the system's security level, which is realized through the security-constrained optimal power flow (SCOPF). The SCOPF formula is usually divided into two parts: the active power subproblem and the reactive power subproblem. The SCOPF is a nonlinear optimization problem, with the objective function of minimizing the control changes from the base-case. The constraints generated in security enhancement procedure are selected by the system operators and then sent to the transmission constraints management (TCM), along with the thermal and generic constraints generated in SA. The TCM will determine activated constraints for the security-constrained economic dispatch (SCED). Based on market participants' offers, SCED produces a least cost dispatch of resources to meet the system requirements, including the transmission constraints and resource limit constraints. The SCED generates locational marginal price (LMP) for the market and base points for system operation.

4.2.3 New products and designs in wholesale electricity markets

4.2.3.1 Considering flexibility in resource adequacy

In the past the electric power system was considered to have sufficient resources if the generator capacity was sufficient to meet the expected peak demand with a certain margin or risk tolerance. This was typically based on a capacity margin such as 115%—120% of annual peak demand which in turn was a heuristic representation of a loss-of-load expectation level equivalent to, for example, an expected reliability loss during no greater than one peak load day in 10 years. In recent years, there is an increasing trend to include the ability to supply load during challenging ramping conditions. This is rooted from the fact that with more variability and uncertainty in the system, mainly driven by wind and solar penetration, the grid may still face reliability issues in term of insufficient ramping capability, even though there is enough generation capacity it the system. While variability and uncertainty have typically been managed in the operational time frame, planning decisions may need to be made to ensure that there is sufficient flexibility available from a system's resources that can be called on in operation, in the same way that capacity must be planned for to be called upon when needed in operations. There may be a need for resource adequacy analysis, which has traditionally been focused on ensuring sufficient capacity to meet peak demand, to also consider flexibility issues.

The CAISO, under mandate from the California Public Utilities Commission, developed a new resource adequacy—related procurement target. Load Serving Entities (LSEs) are required not only to procure sufficient capacity to meet forecasted peak load as they currently do but also to meet additional flexibility requirements with their capacity. The requirement is split into three categories: base flexibility, peak flexibility, and super peak flexibility, which are determined based on the time of the day and as a percentage of largest and second-largest 3-hour net load ramps, contingency reserves, and a peak demand factor.

Once the total procurement requirement has been calculated, the next step is to distribute the total requirements among each LSE based on the contribution of the LSE to the underlying components. The effective flexible capacity (EFC) is calculated for each resource based on their capability to ramp in 3 hours. The contribution from quick start units is their maximum ramp from cold start in 3 hours, whereas for longer start units, the EFC is calculated as the 3-hour ramping capability when the resources are online and at minimum generation (even if these are not online when ramps happen, it is assumed they would be if the ramps were extreme cases).

Each LSE must procure sufficient EFC to meet their seasonal requirement and also ensure that the capacity is dispatchable in the day-ahead and intraday markets. The FRAC-MOO process, originally useful to identify flexibility needs, is currently under revision to ensure it procures the right type of flexibility. This may include more detailed consideration of minimum generation levels, overgeneration issues when solar power is high and understanding what a generator is likely to be doing when its flexibility is needed.

In CAISO the EFC for each resource is calculated. The EFC quantifies the effective MW a resource contributes toward avoiding reliability events caused by incapability to meet the ramping needs. For conventional generation units the EFC is calculated with the following formula [8]:

$$EFC = \begin{cases} \min\{NQC - P_{\min},\ \ T \times RR_{average}\}, \\ \quad \text{for } SUT > 90 \text{ minutes or } P_{\min} \leq 0 \\ \min\{NQC,\ \ P_{\min} + (T - SUT) \times RR_{average}\}, \\ \quad \text{for } SUT \leq 90 \text{ minutes and } P_{\min} > 0 \end{cases} \tag{4.1}$$

where NQC is the unit's net qualifying capacity, which equals to the maximum output the unit can operate for 4 consecutive hours; $P_{\min}$ is the minimum capacity of the unit, which is energy storage or demand response (DR) resource (DRR) when its value is less than 0; T is the ramping period, which is set to 3-hour in CAISO; $RR_{average}$ is the average ramp rate; and SUT is the start-up time.

$$\begin{aligned} FlexibilityNeed_{MTH_y} = {} & \text{Max}\left\{ \left(3RR_{HR_x}\right)_{MTH_y} \right\} \\ & + \text{Max}\left(MSSC, 3.5\%E\left(PL_{MTH_y}\right)\right) + \varepsilon \end{aligned} \tag{4.2}$$

where MTH_y is the yth *month*, *FlexibilityNeed* is the flexibility capacity need, $\text{Max}\left\{ \left(3RR_{HR_x}\right)_{MTH_y} \right\}$ is the maximum 3-hour ramp rate in month y, $MSSC$ is the most severe single contingency, $E(PL)$ is the expected peak demand, and ε is the error adjustment item and represents the load forecasting errors.

4.2.3.2 Flexible ramping products

To address the operational challenges of maintaining system balance, especially the increasing need of generation ramp capacity, two major RTOs/ISOs in the United States—CAISO and Midcontinent ISO (MISO)—have proposed a new market design called flexible ramp products (FRPs) [9,10]. The FRPs include both flexible ramp-up (FRU) and the flexible ramp-down (FRD) capabilities. The FRPs are a new market design to address the operational challenges of maintaining balance. It is expected that the implementation of FRPs in the electricity markets can reduce the chance for ramp shortages and price spikes and maintain reliable system operations. Generally, the FRPs are provided from conventional generation such as coal and natural gas units, because their generation outputs can be more easily dispatched. Renewable energy resources such as wind and solar are deemed not suitable for providing ramping service because their output is more uncertain and reducing this uncertainty relies on accurate forecasting. This feature makes it difficult to be relied upon by the system operator. However, in recent years, with the development of advanced forecasting [11–13] and control [14] technologies, the generation output and ramping capability of wind turbines or PV panels can be better controlled to meet the system dispatch requirements. Thus it is possible to allow renewables to provide FRPs in modern electricity markets.

In order to integrate the FRPs into the electricity market, there is a need to set up the system-wide ramp capacity requirement (RCR), which is calculated before the resource dispatch process. It represents the expected system net load movement between the current dispatch interval and the posttime (e.g., 10 minutes) beyond the dispatch interval. The RCR includes both ramp-up and ramp-down capacity requirements. They are composed of two parts: the one captures the forecasted net load variability and the other captures the forecasted net load uncertainty, as shown in the following equation:

$$URR_t^{System} = \max\left\{ \underbrace{ND_{t+m} - ND_t}_{Variability} + \underbrace{UT_{t+m}}_{Uncertainty},\ 0 \right\} \tag{4.3}$$

$$DRR_t^{System} = \max\left\{ \underbrace{ND_t - ND_{t+m}}_{Variability} + \underbrace{DT_{t+m}}_{Uncertainty},\ 0 \right\} \tag{4.4}$$

where URR_t^{System} (DRR_t^{System}) is the system-wide up (down) ramping capability requirement at t, m is the time beyond the current dispatch interval, ND_t is the net demand at t, and UT_{t+m} (DT_{t+m}) is the uncertainty of the net load ramp-up (down) at $t+m$. The variability part of the RCR is calculated by the difference in net load forecasting between time $t+m$ and t. The uncertainty part of the RCR can be calculated based on historical data using statistical approaches. In MISO the uncertainty part is calculated from the historical net load probability distribution function using the Gaussian-sigma rule, that is, 2.5 standard deviations for 99% confidence level [15].

An alternative method to integrate FRPs is adopting the ramp demand curves (RDCs), which are defined to represent the value for the ramp capability service. They provide a mechanism for limiting the clearing of ramp capacity and determining the clearing price when ramp capacity is in short supply. The RDCs include both the upward and downward flexible ramping demand curves. They generally have a stair-step shape with several MW-price segments. In general, the amount of unexpected ramp procurement will be less if the supply price is lower, and vice versa. A sample RDC is shown in Fig. 4.3. Two slack variables s_u and s_d, which represent the FRU and ramp-down surplus respectively, are constructed as a ramp DR (i.e., ramp requirement reduction) in response to the price.

The objective function of the economic dispatch problem with FRPs model is:

$$\begin{aligned} \text{Min}\Bigg\{ & \sum_{i=1}^{NG} C_i(P_i) + \sum_{i\in I_{RU}} C_i^{RU}(RU_i) + \sum_{i\in I_{RD}} C_i^{RD}(RD_i) \\ & + \sum_{i\in I_{SP}} C_i^{SP}(SP_i) + \sum_{i\in I_{NS}} C_i^{NS}(NS_i) + \sum_{s} \left(UP_s\cdot s_s^u + DP_s\cdot s_s^d\right) \Bigg\} \end{aligned} \tag{4.5}$$

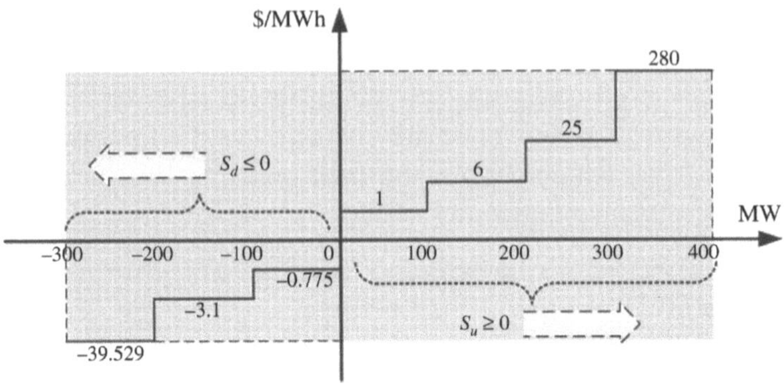

Figure 4.3 Ramp demand curves.

where the symbols are defined as follows: NG is the number of generation resources; NL is the number of transmission lines; I_{RU} is the set of resources providing regulation up service; I_{RD} is the set of resources providing regulation down service; I_{SP} is the set of resources providing spinning reserve service; I_{NS} is the set of resources providing nonspinning reserve service; s is the segment of the ramping capacity demand curve; P_i is the generation MW of resource i; RU_i is the cleared regulation up capacity of resource i; RD_i is the cleared regulation down capacity of resource i; SP_i is the cleared spinning reserve capacity of resource i; NS_i is cleared nonspinning reserve capacity of resource i; $C_i(P_i)$ is the energy cost function for resource i; $C_i^{RU}(RU_i)$ is the regulation up cost function for resource i; $C_i^{RD}(RD_i)$ is the regulation down cost function for resource i; $C_i^{SP}(SP_i)$ is the spinning reserve cost function for resource i; $C_i^{NS}(NS_i)$ is the nonspinning reserve cost function of resource i; UP_s is the price at s_s^u in the ramp capacity demand curve; DP_s is the price at s_s^d in the ramp capacity demand curve; s_s^u is the FRU surplus; and s_s^d is the FRD surplus.

The objective function in (4.5) contains seven terms, including energy cost, regulation up cost, regulation down cost, spinning reserve cost, nonspinning reserve cost, ramp capability demand price, and transmission line violation penalty cost. Since no bids/offers for FRC are submitted, the FRU and FRD are priced with the opportunity costs. The ramp capability demand price term in (4.5) is formulated with the RDC in Fig. 4.3, wherein the surplus variables are the decision variables of the model.

The constraints of the model are shown as follows:

System-wide power balance constraint:

$$\sum_{i\in I_{Gen}} P_i - \sum_{i\in I_{Load}} D_i - Loss = 0 \tag{4.6}$$

where I_{Gen} is the set of generation resources, I_{Load} is the set of load resources, D_i is the load at bus i, and $Loss$ is the system loss.

Transmission constraints:

$$\sum_{i=1}^{N} PTDF_{l-i}^{0} \times (P_i - D_i) \le P_l^0, \quad \text{for } l \in \{all \ \ lines\} \tag{4.7}$$

where N is the number of buses, $PTDF_{l-i}^{0}$ is the power factor shift factor to line l from bus i at normal state, and P_l^0 is the normal limit of transmission constraint l.

Market-wide regulating up reserve requirement:

$$\sum_{i\in I_{RU}} RU_i \ge REQ^{RU} \tag{4.8}$$

where REQ^{RU} is the market-wide regulating up reserve requirement in the system.

Maximum and minimum requirements for market-wide regulating down reserve requirements:

$$REQ_{MAX}^{RD} \ge \sum_{i\in I_{RD}} RD_i \ge REQ_{MIN}^{RD} \tag{4.9}$$

where REQ_{MAX}^{RD} (REQ_{MIN}^{RD}) is the market-wide maximum (minimum) regulating down reserve requirement in the system.

Market-wide regulating up plus spinning reserve requirement:

$$\sum_{i\in I_{RU}} RU_i + \sum_{i\in I_{SP}} SP_i \ge REQ^{RU} + REQ^{SP} \tag{4.10}$$

where REQ^{SP} is the market-wide spinning reserve requirement in the system. According to ancillary service cascading, regulation up can be used as spinning reserve after the regulation up requirement is met.

Market-wide regulating up plus spinning plus nonspinning reserve requirement:

$$\sum_{i\in I_{RU}} RU_i + \sum_{i\in I_{SP}} SP_i + \sum_{i\in I_{NS}} NS_i \ge REQ^{RU} + REQ^{SP} + REQ^{NS} \tag{4.11}$$

where REQ^{NS} is the market-wide nonspinning reserve requirement in the system.

System-wide requirements for FRU and FRD procurements:

$$\sum_{i \in I_{FRU}} R_i^{UP} + s_s^u \geq URR^{System} \tag{4.12}$$

$$\sum_{i \in I_{FRD}} R_i^{DN} + s_s^d \leq DRR^{System} \tag{4.13}$$

$$DRR^{System} \leq s_s^d \leq 0 \leq s_s^u \leq URR^{System} \tag{4.14}$$

where I_{FRU} (I_{FRD}) is the set of resources providing flexible ramping up (down) capacity, R_i^{UP} (R_i^{DN}) is the cleared up (down) ramping capacity at resource i, s_s^u (s_s^d) is the FRU (FRD) surplus at segment s of the demand curve, URR^{System} (DRR^{System}) is the system-wide up (down) ramping capability requirement, and $R_{i,t}^{DN}$ is nonpositive.

Capacity limit constraints for eligible resources:

$$P_i + RU_i + SP_i + R_i^{UP} \leq PG_i^{\mathrm{MAX}} \tag{4.15}$$

$$P_i - RU_i + R_i^{DN} \geq PG_i^{\mathrm{MIN}} \tag{4.16}$$

where PG_i^{MAX} and PG_i^{MIN} is the maximum and minimum capacity of resource i, respectively. Note that RU_i is greater than or equal to 0 and R_i^{DN} is less than or equal to 0.

Resource ramping capability constraints:

$$P_{i,t} - P_{i,t-1} \leq L_t \cdot UPRATE_{i,t} \tag{4.17}$$

$$P_{i,t} - P_{i,t-1} \geq -L_t \cdot DNRATE_{i,t} \tag{4.18}$$

where L_t is the length of time interval t, $UPRATE_{i,\ t}$ is the ramp-up limit of resource i at time t, and $DNRATE_{i,\ t}$ is the ramp-down limit of resource i at time t.

4.2.3.3 Energy storage market modeling

A number of energy storage techniques currently exist in the wholesale electricity market. The most common one is the pumped storage hydro (PSH), which is over than 18 GW capacity in the United States. The PSH can participate into the market in either generation participation mode or load participation mode. When in generation model, it is usually allowed to provide regulation and spinning reserves. When in load model, it is generally not allowed to provide regulation and/or spinning reserves, because PSH is generally installed with fixed speed pumps [16]. Some other energy storage techniques include battery (particularly lithium-ion), flywheels, and compressed air. In recent years, the dominant gird-connected energy storage technique is the lithium-ion battery. In the United States the Federal Energy Regulatory Commission (FERC) issued the Order 841 to direct all the RTOs/ISOs to make changes to their market design and market clearing software to enhance the participation of electric storage into the market. According to the requirement of this order, the ISOs must develop a unique participation model for energy storage resources (ESRs) so that they can act as either a generator or a load, and they can participate in energy, ancillary service, and capacity markets. Although the order requires the ISOs to account for the physical parameters of ESRs, it does not make a hard requirement on whom should be responsible for the state-of-charge (SOC) management. As a result, different markets will develop their own SOC management strategies. Some ISOs would require the ESR asset owners to provide the cost/quantity offer curves and self-manage the SOC of the ESRs. Other ISOs would prefer to manage the SOC through the system operator, that is, the ISO makes scheduling decisions and optimizes the SOC levels to minimize the system's operating costs. In what follows, we will give the detailed formulations for the two SOC management strategies on the ISO' perspective.

When the ESR asset owner self-manages the SOC, the day-ahead security-constrained unit commitment (SCUC) formulation in the ISO will incur additional equations that correspond to other resources and system constraints. For the sake of simplicity the following formulation only includes equations that are related to the subset of ESRs.

The objective function is to minimize the system operation costs

$$\mathrm{Min}\left\{\sum_i \left(c_i^G PG_i - c_i^L PL_i\right) + \sum_i \left(c_i^{REG} R_i^{REG} + c_i^{SPIN} R_i^{SPIN}\right)\right\} \tag{4.19}$$

Subject to the following constraints:

Power balance constraints:

$$\sum_i (PG_i - PL_i) + G^{other} = Load \tag{4.20}$$

Cleared energy and reserve constraint:

$$\sum_i (PG_i - PL_i) + R_i \leq \overline{u}_i^G PG_i^{\max} \tag{4.21}$$

$$\sum_i (PG_i - PL_i) - R_i \geq -\overline{u}_i^L PL_i^{\max} \tag{4.22}$$

Ramp constraints for reserve deployment:

$$R_i \leq T \times RAMP_i \tag{4.23}$$

Resource limit constraints:

$$PG_i \geq 0, PL_i \geq 0, R_i \geq 0 \tag{4.24}$$

When the SOC of ESRs is managed by the ISOs instead of the ESR owners, the SCUC model is more complicated. We assume that ESR can provide energy, regulation, and spinning reserves in the market. For simplicity the following formulations only include equations related to the subset of ESRs.

The objective function of SCUC when ISO manages the SOC is:

Minimize

$$\begin{aligned}\sum_{t\in T}\Bigg[&\sum_{i\in ESR}\left(CG_{i,t}^{ESR}P_{i,t}^{ESR} - CL_{i,t}^{ESR}L_{i,t}^{ESR}\right)\\ &+\sum_{i\in ESR}\left(CREG_{i,t}^{ESR}REG_{i,t}^{ESR} + CSPIN_{i,t}^{ESR}SPIN_{i,t}^{ESR}\right)\Bigg]\\ &+ \upsilon(GD_i - LD_i)\end{aligned} \tag{4.25}$$

Subject to:

Power balance constraints:

$$\sum_{i\in G, i\notin ESR} Pg_{i,t} + \sum_{i\in ESR}\left(P_{i,t}^{ESR} - L_{i,t}^{ESR}\right) = Load_t \tag{4.26}$$

Define four variables to represent the reserves provided by ESR: $r_{i,t}^{G,+}$ is the upward reserves provided by ESR in generation mode, $r_{i,t}^{G,-}$ is the downward reserves provided by ESR in generation mode, $r_{i,t}^{L,+}$ is the upward reserves provided by ESR in load mode, $r_{i,t}^{L,-}$ is the downward reserves provided by ESR in load mode, and $r_{i,t}^{G,+}$, $r_{i,t}^{G,-}$, $r_{i,t}^{L,+}$, $r_{i,t}^{L,-} \geq 0$.

The capacity limits constraints with reserves:

$$r_{i,t}^{G,+} + P_{i,t}^{ESR} \leq u_{i,t}^{ESR} MD_{i,t} \tag{4.27}$$

$$r_{i,t}^{G,-} \leq P_{i,t}^{ESR} \tag{4.28}$$

$$r_{i,t}^{L,+} \leq L_{i,t}^{ESR} \tag{4.29}$$

$$r_{i,t}^{L,-} + L_{i,t}^{ESR} \leq (1 - u_{i,t}^{ESR}) MD_{i,t} \tag{4.30}$$

Note that Eqs. (4.27)−(4.30) ensure that the ESR could be either in generation mode or in load mode. When it is in generation mode ($u_{i,t}^{ESR} = 1$), the right-hand side in (4.30) is zero. Since all the variables are greater than or equal to 0, we have $L_{i,t}^{ESR} = r_{i,t}^{L,-} = r_{i,t}^{L,+} = 0$. When it is in generation mode ($u_{i,t}^{ESR} = 0$), the right-hand side in (4.27) is zero, and there exists $P_{i,t}^{ESR} = r_{i,t}^{G,+} = r_{i,t}^{G,-} = 0$.

Upward reserve equation:

$$REG_{i,t}^{ESR} + SPIN_{i,t}^{ESR} = r_{i,t}^{G,+} + r_{i,t}^{L,+} \tag{4.31}$$

Downward reserve equation:

$$REG_{i,t}^{ESR} = r_{i,t}^{G,-} + r_{i,t}^{L,-} \tag{4.32}$$

Energy ramp-down constraints:

$$\left(P_{i,t-1}^{ESR} - L_{i,t-1}^{ESR}\right) - \left(P_{i,t}^{ESR} - L_{i,t}^{ESR}\right) \le RR_i \times I_{DA} \tag{4.33}$$

Energy ramp-up constraints:

$$\left(P_{i,t+1}^{ESR} - L_{i,t+1}^{ESR}\right) - \left(P_{i,t}^{ESR} - L_{i,t}^{ESR}\right) \le RR_i \times I_{DA} \tag{4.34}$$

Reserve ramp constraints:

$$REG_{i,t}^{ESR} \le RR_i \times T_{REG} \tag{4.35}$$

$$SPIN_{i,t}^{ESR} \le RR_i \times T_{SPIN} \tag{4.36}$$

SOC constraints:

$$SOC_{i,1} = SOC^{Initial} \tag{4.37}$$

$$SOC_{i,24} = SOC^{End} + GD_i - LD_i \tag{4.38}$$

$$SOC_{i,t+1} = SOC_{i,t} + I_{DA}\left(-\frac{1}{\eta}P_{i,t}^{ESR} + \eta L_{i,t}^{ESR}\right) \tag{4.39}$$

Reserve activation constraints:

$$SOC_{i,t} + \kappa I_{DA}\sum_{t'=1}^{t}\left(\frac{1}{\eta}r_{i,t}^{G,-} + \eta r_{i,t}^{L,-}\right) \le SOC_i^{\max} \tag{4.40}$$

$$SOC_{i,t} - \kappa I_{DA}\sum_{t'=1}^{t}\left(\frac{1}{\eta}r_{i,t}^{G,+} + \eta r_{i,t}^{L,+}\right) \ge SOC_i^{\min} \tag{4.41}$$

The constraints (4.40) and (4.41) ensure that in the worst scenario when the activation of all reserves is in one direction, the ESR still has sufficient capability to provide reserves.where the meaning of the symbols is shown as follows: T is the time horizon; $CG_{i,t}^{ESR}$ is the cost when the ESR is on generation mode; $CL_{i,t}^{ESR}$ is the cost when the ERS is on load mode; $P_{i,t}^{ESR}$ is the cleared quantity of generation when the ESR is on generation mode; $L_{i,t}^{ESR}$ is the cleared quantity of load when the ESR is on generation mode; $CREG_{i,t}^{ESR}$ is the offer for providing regulation reserve; $CSPIN_{i,t}^{ESR}$ is the offer for providing spinning reserve; $REG_{i,t}^{ESR}$ is the cleared quantity of regulation reserve; $CSPIN_{i,t}^{ESR}$ is the cleared quantity of spinning reserve; v is the penalty price for SOC violations; GD_i is the positive deviations of SOC at the last time interval (MWh); LD_i is the negative deviations of SOC at the last time interval (MWh); $Pg_{i,t}$ is the generation output from non-ESR resources; $Load_t$ is the system demand at hour t; $u_{i,t}^{ESR}$ is the binary variable to indicate the ESR mode. 1 refers to generation mode, and 0 refers to load mode; $MD_{i,t}$ is the maximum discharge limit (MW); $MC_{i,t}$ is the maximum charge limit (MW); RR_i is the ramp rate of ESR i (MW/min); I_{DA} is the day-ahead resolution (usually 1 hour); T_{REG} is the regulation reserve activation time (hour); T_{SPIN} is the spinning reserve activation time (hour); $SOC^{Initial}$ is the initial SOC; SOC^{End} is the desired SOC at the last hour; $SOC_{i,t}$ is the SOC for ESR i at time t (decision variable; MWh); η is the charging/discharging efficiency; κ is the assumed percentage of reserve deployment in the SOC throughput constraint. It equals to 1 by default; $SOC_i^{\max}$ is the maximum SOC; and $SOC_i^{\min}$ is the minimum SOC.

4.2.3.4 Demand response resources market modeling

From the system operator's perspective, there are two types of DRRs:

- Interruptible load (Type I DR). The Type I DRRs are capable of supplying a specific quantity of energy through physical load interruption. These generally include curtailable or interruptible loads. In the wholesale electricity market, they are modeled as generation resources with block loaded for a specific quantity of energy.
- Controllable load (Type II DR). The Type II DRRs can provide energy and/or operating reserves over a dispatchable range through controllable load or BTM generation. In contrast to the Type I DRRs, they have more flexibility in that they can adjust the load consumption based on system needs. In the wholesale electricity markets, the Type II DRRs are modeled as generations with (negative) dispatchable ranges.

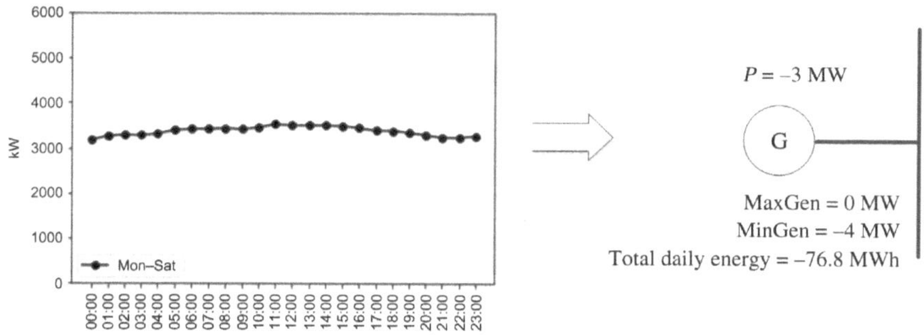

Figure 4.4 Type II DR is modeled as a generator. *DR*, Demand response.

The Type II DRRs are modeled as generators in the electricity markets. Different from conventional generators, they have negative output ranges between the minus of the maximum capacity and zero. For instance, the daily load curve of a Type II DRR is shown on the left half of Fig. 4.4. The maximum power usage of the resource is 4 MW. In most hours of the day the energy usage is between 3 and 4 MWh. The daily energy used is 76.8 MWh. The corresponding generator model is shown on the right half in Fig. 4.4. The maximum and minimum generation capacity is 0 and −4 MW, respectively. The total daily energy generation is −76.8 MWh. When the net output of the generator is −3 MW, it means the DRR is supplying 1 MW of DR, which is the difference between the net output (−3 MW) and the minimum limit (−4 MW).

The DRRs will incur additional items to the objective function and additional constraints to the SCUC in the electricity market. In what follows, we only include the equations related to the subset of Type II DR.

The objective function is:

Minimize

$$\sum_{t\in T}\left[\sum_{i\in DRR2}\left(CG_{i,t}^{DRR2}P_{i,t}^{DRR2}+CSU_{i,t}^{DRR2}u1_{i,t}^{DRR2}\right)\right] + \alpha\sum_{i\in DRR2}(DailyEnergyViolation_i) \tag{4.42}$$

Subjective to:

Generation capacity limits for Type II DRRs:

$$-P_i^{DRR2,\text{Max}}u_{i,t}^{DRR2}\le P_{i,t}^{DRR2}\le 0 \tag{4.43}$$

Ramp-up and ramp-down constraints:

Many utilities and independent system operations are trying to prepare the grid for rapid growth of renewable energy, and some have successfully demonstrated how large amounts of DERs can be integrated and sustain system reliability as successfully as conventional generation plants. The CAISO is a leading example. On May 16, 2017 the CAISO hit a new renewable energy record with renewables serving nearly 42% of electricity demand [17]. Similar is for the Southern Power Pool (SPP) system, at 3:45 a.m. On March 16, 2015, wind accounted for 13,928.94 MW of the system's total load of 22,998.71 MW, a penetration level of 60.56% [18]. In EriGRID an Irish leading utility, the percentage of wind power could account for 60% of electricity usage over a typical day [19]. In Colorado, Senator Matt Jones and Representative Mike Foote submitted a bill to Colorado General Assembly in January 2018 to require 100% renewable energy by 2035 [20]. In July 2017 Hawaii Public Utilities Commission approved Hawaiian Electric Companies' Power Supply Improvement Plan, which contains the recipe for hitting the state's 100% renewables mandate by 2040 [21].

In order to remove barriers to the participation of electric storage resources and DER aggregations in the capacity, energy, and ancillary service markets, the Federal Energy Regulatory Commission opened a rule-making for the nation's six grid operators in 2016 [22] and issued the final rule on February 28, 2018 [23]. In PJM, it has been recognized that inverter-based resources (IBRs) can participate into the wholesale and capacity markets and provide synchronized reserves, but their dispatch parameters are different from synchronized resources and the corresponding rules need to be updated. In CAISO, new market designs such as the flexible ramping products and the flexible resource adequacy and must offer obligation (FRAC-MOO) have been implemented to address the challenges from increased penetration of renewables. In July 2017 the Hawaiian Electric Company secured regulatory approval for a plan to power multiple island grids with 100% green energy by 2045 [24]. Though many ambitious renewable energy targets have been set up, none of today's utilities and system operators are designed to operate their girds under all (100%) or extremely high (greater than 80%) IBRs.

Traditional rules and concepts for operating thermal resource—based system do not apply to systems with all IBRs. This section sought to close the gaps between the existing grid and market design and the new operation paradigm in the all IBRs systems, as demonstrated in Fig. 4.5. In addition, we will propose corresponding models to illustrate how those

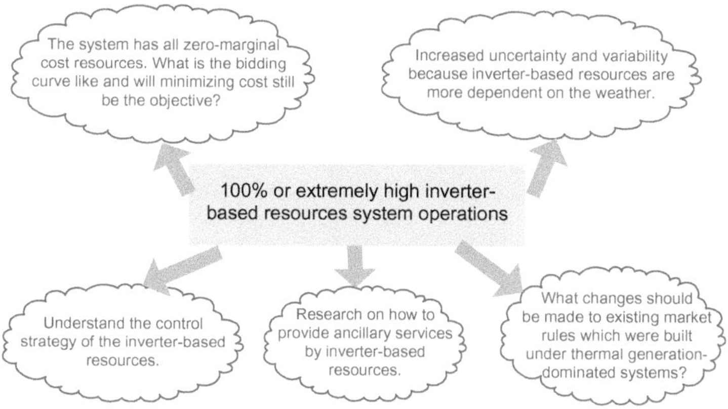

Figure 4.5 Bridging the gaps in perspectives.

Table 4.1 Features of gird-following and grid-forming inverters.

Grid-following inverters	Grid-forming inverters
• Current source (PQ bus) • Follow the grid • Inject active and reactive power • Fast response to the intermittent irradiation levels with no buffers	• Voltage source (PV bus) • Set grid voltage and frequency • Provide active and reactive power to the load via voltage control • Slow response due to large inertia

resources are dispatched to meet the generation and demand balancing requirement.

4.2.4.2 Bulk electric power grid dispatch in the all inverter-based resources systems

Two popular types of inverters connected to the transmission and distribution systems are the grid-following and grid-forming inverters, and their control schemes are different. A grid-following inverter can follow the desirable set points for its real power (e.g., MPP for renewables) under normal conditions, and within the maximum power limit its real power can be changed for power sharing purposes upon a disturbance. A grid-forming inverter should have bidirectional power-transfer capability (e.g., an inverter connects to a battery or microgrid system). A comparison of the features of the two inverters is shown in Table 4.1.

From an ISO or the utility's perspective, an inverter-based facility can be either front-of-the-meter (FTM), which normally refers to the utility-scale plants (such as PV, wind and storage) that generate power and feed

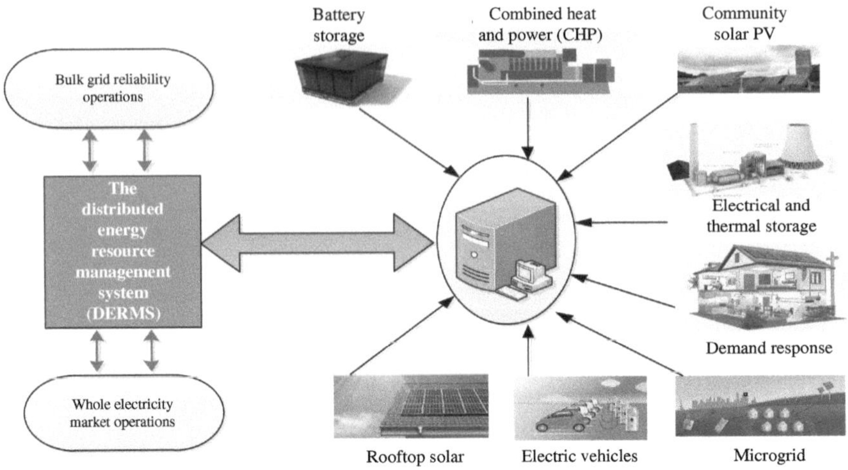

Figure 4.6 Inverter-based resources that are located behind the meter.

into the grid to supply a utility with energy, or BTM, which refers to the inverter-based systems that produce power intended for on-site use in a house, office building, or other commercial facility. The BTM inverter–based systems are usually on the owner's property, not on the side of the electric power grid/utility. The strategies to dispatch the FTM- and BTM inverter–based resources are different. The FTM resources can participate into the wholesale electricity market directly and provide energy and ancillary services. The BTM resources, however, are treated as negative loads on zonal levels based on the generation output forecasting.

IBRs that are located behind the meters may include rooftop solar, electric vehicles, microgrid, DR, electrical and thermal storage, community solar PV, combined heat and power (CHP) plants, and battery storage, as shown in Fig. 4.6. These resources can be controlled and operated by the DER management system. In addition, the BTM resources can interact with the bulk transmission systems in two different layers. First, they can provide reliability services to the grid through active and reactive power dispatch and scheduling, phase balancing, and loss minimization. Second, they can provide electricity market services through supply of capacity, energy, ancillary services, and frequency regulation.

4.2.4.3 Increased uncertainty and variability

Since the generation of renewable energy highly depends on the weather, the challenge to operate a 100% IBRs (mostly are renewables like wind and solar) system is much greater than to operate today's gird. Forecasting will play a significant role in the operation of the 100% IBRs system. In order to participate into the wholesale market, the utility-scale IBRs should either provide the generation forecasting to the ISO or provide related meteorological data to let the ISO do the forecasting. For example, the following information are required to be provided by the New York ISO:

- Provide the plane of array irradiance and back panel temperature to the ISO.
- Employ a sufficient quantity of measuring stations such that all panels are within a certain number of kilometers (e.g., 5 km) of a measuring station.
- The frequency of data transmitting must be sufficient (e.g., at least once every 30 seconds).

4.2.4.4 Increased complexity for the controls of utility-scale inverter-based resources

The control strategy of a utility-scale PV or wind plant, which includes a combination of inverters as well as many other devices, may be quite different from that of a synchronized generator. This session does not aim at designing the control model of the plant-level inverters. Instead, we will discuss the control functions of the plant and how they can be adapted to the operation procedure of the grid. According to Ref. [25], a typical IBR plant has the following plant-level control functions:

- load-following control;
- real power output curtailment of the solar power plant when required so that it does not exceed an operator-specified limit;
- ramp-rate controls to ensure that the plant output does not ramp up or down faster than a specified ramp-rate limit, to the extent possible; and
- frequency control (governor-type response) to lower plant output in the case of an overfrequency situation or increase plant output (if possible) in the case of an underfrequency situation.

These control functions can be transferred to corresponding dispatch functions in the operation procedure of the all or extremely high IBRs system. Table 4.2 shows the control functions and dispatch functions for utility-scale IBRs. The fact that utility-scale inverters can provide load-following control, real power output curtailment, and ramp-rate control makes it possible to run unit commitment and economic dispatch models in the all IBRs systems. Thus lots of concepts in running today's

Table 4.2 Control functions versus dispatch functions for utility-scale inverter-based resources.

Plant-level control functions	Dispatch functions
Real power output curtailment	Control the real power output of the plant for ED
Ramp-rate controls	Ramping limit constraints in the UC, ED, and AGC models
Frequency control	Different from synchronized generators, the frequency of inverters can be recovered very quickly after a disturbance or contingency. New AGC design is required, as is detailed in Chapter 3, Renewable energy integration and system operation challenge: control and optimization of millions of devices

AGC, Automatic generation control; *ED*, economic dispatch.

synchronized electric power grid can be used for running the all IBRs systems—for example, dividing the operating procedure into day-ahead unit commitment and real-time economic dispatch processes, conducting energy and ancillary services cooptimization during the dispatch, and adopting new market design such as flexibility ramping products to increase the system's operation flexibility.

The frequency control and automatic generation control (AGC) operation of inverter-based plants are significantly different from those of synchronized machines. The voltage and reactive power control of inverters is important for maintaining the grid operation reliability in the 100% inverter system. In today's electric power grid, voltage and reactive power are coordinated by three different parties: (1) reliability coordination, which monitors pre- and postcontingency voltage and reactive power and manages them by adjustment of transfers, generation redispatch, and unit commitment; (2) transmission operators, which monitor, analyze, and control voltage and reactive power flows as stated in NERC Standard VAR-001; and (3) generator operators, the primary responsibility of which is to improve voltage and reactive power flow controls as required by NERC Standard VAR-002. We will discuss how frequency control and AGC operation will be shifted in a later section in this chapter.

4.2.4.5 All zero marginal cost resources

The pricing mechanism in today's wholesale electricity market such as LMP and market clearing pricing reflects the marginal cost to supply energy or provide ancillary services. Typical marginal price for energy in the CAISO is about $50/MWh. In the all inverter system, energy and ancillary services are provided mostly by renewable energy such as wind, solar and energy storage, the marginal cost of which is close to zero. While many publications have discussed this issue from the market design or the utility's revenue sufficiency perspectives [25,26], this session mainly focuses on the operation side for a system with all zero marginal cost resources.

SCUC and SCED are the main models to commit and dispatch resources in today's electric power grids, in which the way a synchronous generator is committed and dispatched depends on many factors. For example, a coal unit is usually much cheaper but has longer minimum run time and minimum shut-down time than an oil unit. Thus the coal unit is more suitable to serve baseload, and the oil unit is usually treated as fast starting resource to serve peak load. In the 100% inverter system, all the resources are with close to zero marginal cost, and the physical operation requirements such as minimum on time, minimum off time, start-up time, and shut-down time are neglectable. The SCUC and SCED models can still be applied to commit and dispatch the IBRs to balance load and meet transmission line requirements. Since the marginal costs of the resources are close to 0, the magnitude of the LMPs should be trivial. However, the resource owners can recover the variable operation cost through capacity market and/or scarcity price. In addition, it is expected that ancillary services will play a significant role in the 100% inverter system, and the price to provide ancillary services might be higher than the price to provide energy.

4.2.4.6 Unit commitment and economic dispatch for up to 100% renewables

A 100% renewable energy power system has been proved to be feasible, and has been realized in some places. While lots of previous work was focused on how individual inverters impact the operation of the grid, less was focused on how the balancing paradigm will be changed for the whole system. This section does not intent to propose a totally new operating diagram for the 100% renewables system[1]. Instead, we focus on how the existing unit commitment and economic dispatch framework might be changed to run a 100% renewables power system. In addition, we will also discuss the new method to determine the reserve requirements and

[1] From the system balancing perspective, it is more general to use "100% renewables" rather than "100% inverter-based resources". But they refer to the same thing in this report.

the new design of frequency response mechanism in the following sections.

The SCUC and SCED models are the key models in today's electricity markers to manage the resources and balance the demand. The SCUC is used in the day-ahead and intra-day market to meet the forecasted demand and ancillary service requirements subject to transmission constraints. The SCED is used to achieve real-time reliable grid operation at the lowest costs.

With more IBRs being integrated into the grid, the operation paradigm will have significant changes. The IBRs have almost zero start-up and shut-down time, zero minimum generation level, zero minimum on and off time, and almost zero start-up and shut-down costs. These features are significantly different from those of conventional thermal generators. When a system has 100% IBRs, the UC is no longer required for the system operations. In today's market operations, fast start units can be called up during the intraday to meet the load imbalance caused by uncertainty. In the 100% IBR system the logic to call up quick start units should be different because all the resources can be turned on and off very quickly.

The SCED model should also have significant changes. During intervals where only the zero-cost resources need to provide energy, there is no cost for SCED to minimize. The optimization engine may randomly select resources to be curtailed. As a result, there may be multiple solutions to the SCED model. It is important to identify new rules for determining which units are selected for curtailments.

4.2.4.7 Determination of reserve requirements

In a power system where the demand is 100% supplied by renewables, solar and wind resources form the baseload. In the case of the scenario when the sun is mostly blocked and the wind is weak, the system must build sufficient extra capacity so that there is enough capacity to power the grid under all operating scenarios. This requires building variable energy resource (VER) capacity well beyond the reserves required from thermal plants. With so much extra solar and wind on the grid, the system operators have to deal with overgeneration when the weather conditions are optimal, and VER curtailment is unavoidable. Besides overbuilding capacity, other measures to ensure sufficient reserves in the grid are to build a whole lot of storage and to improve the regional import and export of electricity.

Operating reserves refer to the generating capacity available to meet demand in case the scheduled or forecasted supply of power is disrupted.

A shortage of deployable active power reserve may cause frequency response deterioration in the system with all inverters. This may not only lead to underfrequency load shed but also cause transient stability issue in the system. In the steady state, VREs increase the variability and uncertainty of power supply and hence impact the amount of operating reserves that need to be held to maintain system reliability. It is a challenge for system operators to determine how much each reserve product needs to be increased with high VRE penetration levels. In particular, when the system has 100% renewable energy, the grid operators should find out the answers to the following key questions:

- How to determine the requirement for each reserve product?
- Who provides those reserves?
- Can IBRs provide reserves? If yes, how much can be counted toward reserve?

Traditionally, the operating reserves have been provided by conventional generators such as thermal and hydro units. In the 100% renewable energy systems, those reserves have to be provided by DR, energy storage, and other IBRs (such as wind and solar units).

According to the FERC definition, the types of ancillary services and the associated timescales are shown in Table 4.3. In the US ISOs the system operating reserves mainly include the following categories:

- Regulation reserves, which are used for the balancing of fast second-to-second and minute-to-minute random variations in load or generation. They generally include regulation up and regulation down reserves.
- Regulation mileage reserve, which is a reserve product to measure regulation movement required by FERC Order 755.
- Spinning reserve, also referred to synchronized reserve, is intended to help the system respond quickly to forced outages or other contingency events.

Table 4.3 Type of market products and the associated timescales.

Market products	Timescale
Reactive supply and voltage control	Seconds
Scheduling, system control, and dispatch	Seconds to hours
Regulation and frequency response	Minutes
Synchronized operation reserve	Seconds to less than 10 min
Supplemental operation reserve	Greater than 10 min
Energy imbalance	Hours

- Nonspinning reserve, also referred to supplemental reserve, is intended to help the system recover from unplanned contingencies and can be provided by offline generators.
- Ramping product, or flexibility ramping capacity in MISO and CAISO electricity market, which allows the ISO to procure sufficient ramping capability via economic bids.

In this section, we are going to discuss the reserve requirements in the system under two scenarios: (1) the system has sufficient thermal generators and has 100% renewable generation only in some hours of a day. (2) The system has only few (or none) thermal generators, and the demand is supplied by renewables almost all the time.

The first scenario is an extreme case of today's market operation paradigm. Four types of reserves are generally considered, including regulation up reserve, regulation down reserve, spinning reserve, and supplemental reserve. The requirement for each type of reserve is calculated as follows:

- Regulation up reserve:

  We adopted the method used in SPP market to calculate the regulation up reserve. The regulation reserve is determined by four components: load magnitude, load variability, renewable energy resource magnitude, and renewable energy resource variability. The total regulation reserve requirement is calculated by the summation of each component times a coefficient, as shown in the following equation:

$$RegReq^{up} = a^{up}LF(t) + b^{up}[LF(t+1) - LF(t)] + c^{up}RF(t) + d^{up}[RF(t+1) - RF(t)] \tag{4.50}$$

where $LF(t)$is the load forecasting at time t; $RF(t)$ is the renewable energy resource forecasting at time t; and a^{up}, b^{up}, c^{up}, and d^{up} are coefficients to determine the weight of each component. One typical value of a^{up}, b^{up}, c^{up}, and d^{up} is 1%, 1%, 5%, and 10%, respectively.

- Regulation down reserve:

  The regulation down reserve requirement can be calculated in the following equation:

$$RegReq^{down} = a^{down}LF(t) - b^{down}[LF(t+1) - LF(t)] + c^{down}RF(t) - d^{down}[RF(t+1) - RF(t)] \tag{4.51}$$

where a^{up}, b^{up}, c^{up}, and d^{up} are coefficients to weight each component.

- Spinning reserve:

  The spinning reserve is a 10-minute product in the electricity market. We set the spinning reserve requirement equal to the largest single contingency (in MW) in the system.
- Supplemental reserve:

  Supplemental reserves must be able to become synchronized with the grid and ramp to a specified output level within 30 minutes. In the electricity markets, its requirement can be calculated based on historical data. For example, in ERCOT it is calculated by using the 70th–95th percentile (depending on the risk of net load ramping) of hourly net load uncertainty (load minus the estimated uncurtailed total output from intermittent renewable resource) from the same month of the previous 3 years. In ISO New England, it is set as 50% of the second-largest system contingency.

In the second scenario, when there are few or no thermal generators in the system, all the reserves should be provided by IBRs or DRRs. One significant difference from the first scenario is that there is no need to have various types of reserve products. In conventional power systems, different types of reserves need to be defined because the thermal generators have different response time. For example, those units equated with AGC devices are qualified to provide regulation reserve, because they can respond to ISO's set point signal in a few seconds. Units with longer start-up (or ramp-up) time may be qualified to provide spinning or supplemental reserves. However, the inverters have total different features than the conventional generators. They can respond to the set points quickly in a few seconds. In addition, they have very short start-up or ramp-up time. Considering these features of IBRs, we only propose two types of operating reserves to the system:

- Load-following reserve: It refers to the capacity available during *normal conditions* for assistance in active power balance to correct future anticipated imbalance in the system.
- Contingency reserve: It refers to the capacity available for assistance in active power balance during *infrequent events* that are more severe than balancing needed during normal conditions and are used to correct instantaneous imbalances.

The load-following reserve should consider the variability and uncertainty of the load and renewable energy generations. There are three central needs for operating reserves:

1. hold capacity now to meet the variability that occurs within the scheduling time interval (intrainterval variability need);

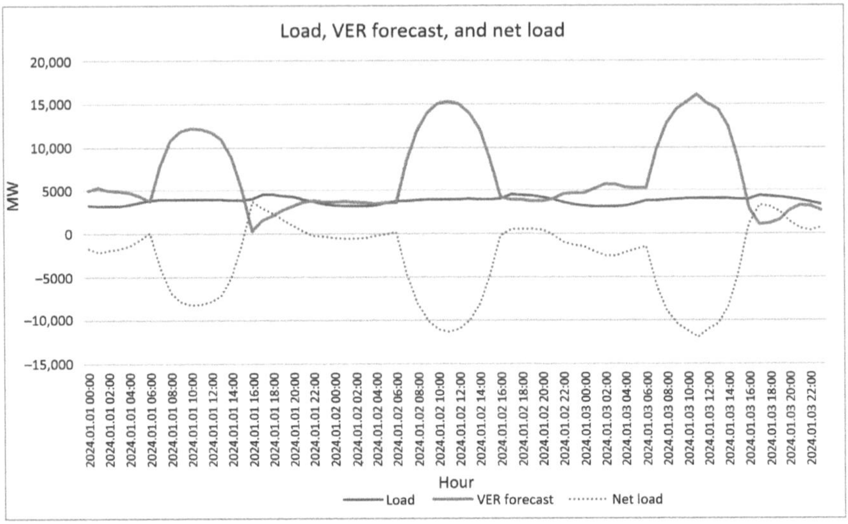

Figure 4.7 Net load is negative when VER penetration level is high.

2. hold capacity now to meet the variability that occurs beyond the scheduling time interval (interinterval variability need); and
3. hold capacity now to meet the uncertainty that occurs within and beyond the scheduling time interval (uncertainty need).

Normally the requirement of load-following reserve is determined based on the system net load. In a system with very high VER penetration, the net load can be negative in most hours, as shown in the example in Fig. 4.7. Thus it would be better to use the VER forecasting data to calculate the reserve capacity need for intrainterval variability, interinterval variability and uncertainty.

The contingency reserve requirement can be defined as the capacity of the largest inverter in the system, in case it is lost under an event. Since the respond time of inverters is very fast, both online and offline inverters can provide the contingency reserve.

4.2.4.8 New frequency response mechanism design

In the current industry practice the balancing and frequency control occur over a continuum of time using different resources, as shown in Fig. 4.8 [27]. The primary control is also called frequency response, which occurs within the first few seconds following a change in system frequency (disturbance) to stabilize the interconnection. The frequency response can be provided by the governor action of synchronized generators or system

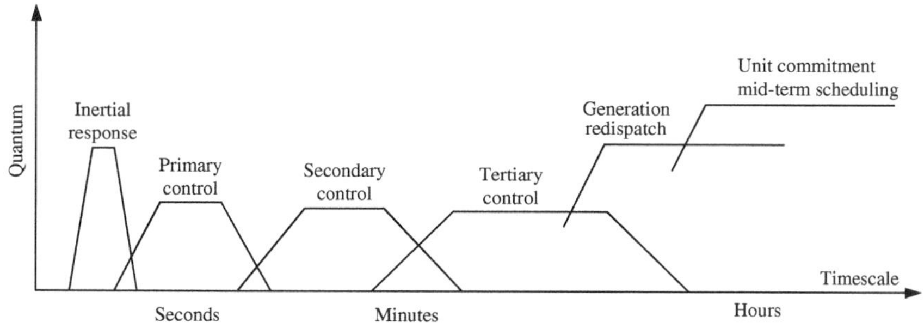

Figure 4.8 Control continuum in the power industry.

load (such as motors). The secondary control maintains the minute-to-minute balance of the system and is used to restore frequency to its scheduled value following a disturbance. The AGC is the major means to exercise the secondary control.

By setting proper control schemes for inverters, the system's frequency can be back to normal in just a few seconds after a disturbance. As a result, instead of control the governors in the primal frequency response block in Fig. 4.8, the primary control in 100% IBR systems will control the power electronic devices so that the system frequency can recover to the target value quickly. Traditionally, secondary control relies on the AGC, logic of which contains the calculation of Balancing Authority's area control error (ACE). The equation of ACE is shown in the following equation:

$$ACE = (NI_A - NI_S) - 10B(F_A - F_S) - I_{ME} \tag{4.52}$$

where NI_A is the actual net interchange, NI_S is the scheduled interchange, B is Balancing Authority bias, F_A is actual frequency, F_S is scheduled frequency, and I_{ME} is interchange metering error. When the system has 100% IBRs, the frequency-related term in (4.52) can be dropped because the system frequency can be quickly adjusted from inverters. The new ACE formulation is shown in the following equation:

$$ACE^{new} = (NI_A - NI_S) - I_{ME} \tag{4.53}$$

The calculated ACE is then allocated to each inverter based on their capacity (MW). For each inverter the set point has three components: the load-following set point, which is calculated by allocating the ACE from (4.53) to each inverter; the energy set point, which is from the economic dispatch result (5 minutes); and the contingency reserve set point (5 minutes). A new AGC design framework for 100% IBR is shown in Fig. 4.9.

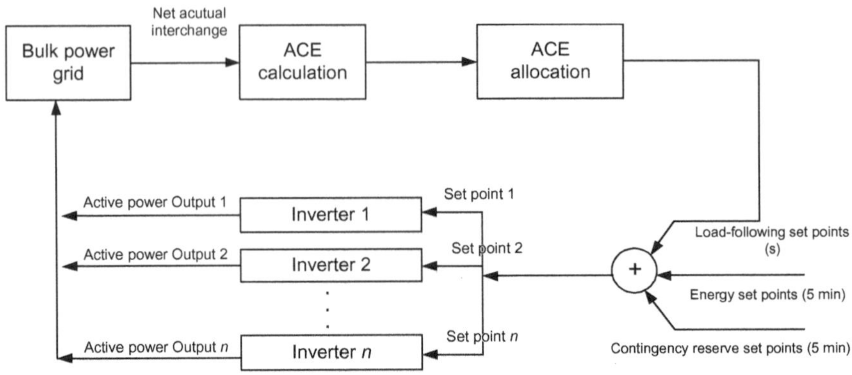

Figure 4.9 New AGC design for 100% inverter-based resource system.

4.3 Modern retail electricity market

4.3.1 State-of-the-art for retail electricity market development

Historically, retail customers were subjected to electricity rates that are determined by electric utilities with governmental regulations. With the deregulation of the power industry, many states in the United States have enabled the retail electricity choice that allows end use consumers (including residential, commercial, and industrial customers) to buy electricity from competitive retail suppliers. The purpose of the retail electricity choice is to increase retail competition so that customers would enjoy lower prices, improved service, and innovative product offerings. By the end of 2017, there are 13 states in the United States that have fully restricted retail electricity market [28]. Some of them have full retail electricity choice, such as Texas, New York, and Illinois; some only have partial retail electricity market choice, such as California, Oregon, and Virginia. In order to attract customers and differentiate themselves from the competitors, some retail electricity suppliers would turn to renewable energy offerings. In six states of United States, community choice aggregation (CCA), also known as municipal aggregation, is formed to allow local governments to procure power on behalf of their residents, businesses, and municipal accounts from an alternative supplier while still receiving transmission and distribution service from their existing utility [29]. Many CCAs have chosen renewable energy for their supply. The retail electricity can exist regardless of the existence of wholesale electricity market in many states of the United States. For example, some states have both wholesale market and retail customer choice, such as Texas, New York, and Pennsylvania. A number of states that are part of wholesale markets do not have retail market, such as Iowa, Minnesota, and Indiana. A few states have retail electricity choice but do not participate into the wholesale market, such as Georgia and Oregon (although Oregon utilities partially participate into the wholesale market through CAISO's Energy Imbalance Market program).

There is a long debate if competition should be restricted to the wholesale side or be extended fully to the retail side [30]. However, with the massive integration of BTM DERs and increased need of customers' participation, more states have realized the importance of establishing the full retail competition markets. The policymakers would pursuit a reliable, efficient, resilient, and clean energy supply within the state to increase the social welfare. The utilities are changing from only providing conventional transmission and distribution services to evolving roles of integrated energy service suppliers with innovations in financing infrastructure, management, and technologies. The end users are expecting new market rules, system monitoring, and legal framework to ensure a fair, transparent, and competitive retail market. The leading states for retail market development in the United States include Nevada, Texas, and New York. The Nevada legislature would be required to provide by law for an "open, competitive retail electric energy market with protections that entitle customers to safe, reliable, and competitively priced electricity" no later than July 1, 2023 [31]. In Texas, 85% of the power consumers can choose electricity service from different retail electric providers (REPs), which are approved by the Texas Public Utility Commission to sell electric services to residential and commercial customers. Each REP operates independently but competes with each other to entice potential customers. They offer a variety of rates and plan options, including green energy option, and are responsible for customer service, billing, and electricity sells. The New York State has proposed the Reforming the Energy Vision (REV) for the purpose of building a clean, more resilient, and affordable energy system for all customers within the state. The REV was driven by a transformation of the electric power industry moving to a new and sustainable energy future, with a proliferation of DERs and a focus on customer choice and participation. To achieve the goals

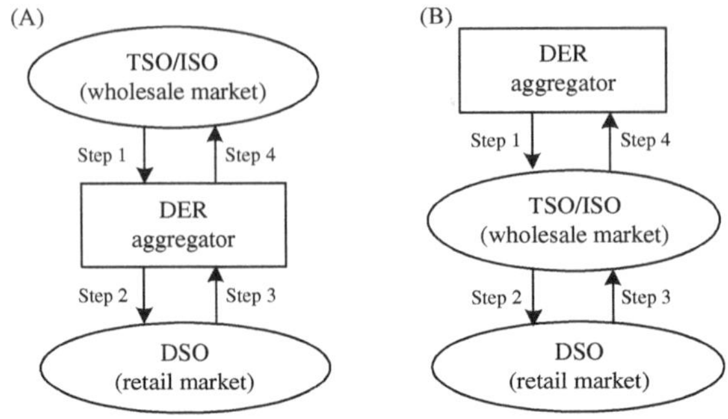

Figure 4.10 Two frameworks for the coordination of ISO, DSO, and DER. *DER*, Distributed energy resource; *DSO*, distribution system operator; *ISO*, independent system operator.

proposed by REV, many utilities plan to develop the DSP with the functionalities of forecasting, interconnecting, monitoring, controlling, and effectively managing the integration of DERs into the existing electric distribution system.

The coordination among the ISO, DER aggregator, and distribution system operator (DSO) is key for retail electricity market development. Fig. 4.10 describes two popular frameworks for the coordination process. One framework as shown in subplot (A) is that the DER aggregator can coordinate with both wholesale and retail markets, and the coordination process is as follows [32]: (1) the ISO/TSO requests for grid services from DER aggregators based on the DER status, availability, range of adjustability, and other information; (2) the aggregator informs the DSO about the services it is going to bid in the ISO/TSO; (3) the DSO determines the acceptable services and schedules by running economic dispatch and other security-check models; and (4) the aggregator updates the schedules and services and rebids into the ISO/TSO. The other framework as shown in subplot (B) is that the DER aggregator can only coordinate with the wholesale market. The corresponding coordination process is as follows: (1) the DER aggregator submits the commitment status and services to be provided in the wholesale market; (2) the ISO sends the DER aggregator's requests to the DSO; (3) the DSO conducts economic dispatch and security check to determine desired services and schedules for the DER aggregator; and (4) ISO updates the schedules and requests based on the decisions in step 3 and runs the system operation models and sends the final results to DER aggregator.

4.3.2 Design of retail electricity market operation framework

Increasing amounts of DERs located on the customer side of the electric grid represent both a challenge and an opportunity for grid operators. Modernizing and monetizing the distribution system requires the construction of a novel DSO, which operates and regulates the distributed grid independently, similar to an RTO/ISO on the transmission grid. The DSO operates a comprehensive decision-making platform with centralized and distributed resources, including generation, energy storage, distributed control devices, sensing devices, power flows on the transmission interfaces, as well as thermal energy storage. The DSO operates an auto-steered information–decision platform with four key functionalities: (1) fast and comprehensive data analytics and informatics, (2) monitoring real-time conditions and predicting near-future trends of the distributed grid with advanced demand management system (DMS)/SCADA systems, (3) forming the auto-steered decision-making framework to optimize distributed resources and power flow, and (4) increasing the utilization of renewable power and mitigating the load (both heat and electrical) variability by cooptimizing the electric and gas and heating/cooling (GHC) systems. These functionalities will form an intelligent network platform that will provide safe, reliable, and efficient electric services by integrating diverse resources to meet customers' and society's evolving needs. A proposed DSO framework and its connection with the transmission and GHC are shown in Fig. 4.11.

The construction of retail market under the DSO framework is a solution to address the challenges faced by today's power systems. Traditional power systems are designed and operated according to two fundamental assumptions: (1) the predominant flow of electric power is unidirectional, from central stations to customers and (2) electricity cannot be stored in significant quantities. However, the current state of affairs is beginning to rapidly change, especially in terms of electric distribution. With the emergence of large-scale distributed generation (DG) resources (e.g., rooftop PV, CHP, and microgrids) coupled to DR programs (e.g., load shedding and distributed storage), the bidirectional flows can originate from (and terminate at) nearly any point in the system. In New York State, utility companies are encouraged to shift their roles to "DSP providers". A "DSP" mainly plays two roles: (1) the aggregator of DG and other DERs, including energy efficiency, DR, energy storage, and electric vehicles and (2) the interface between the wholesale bulk power system and

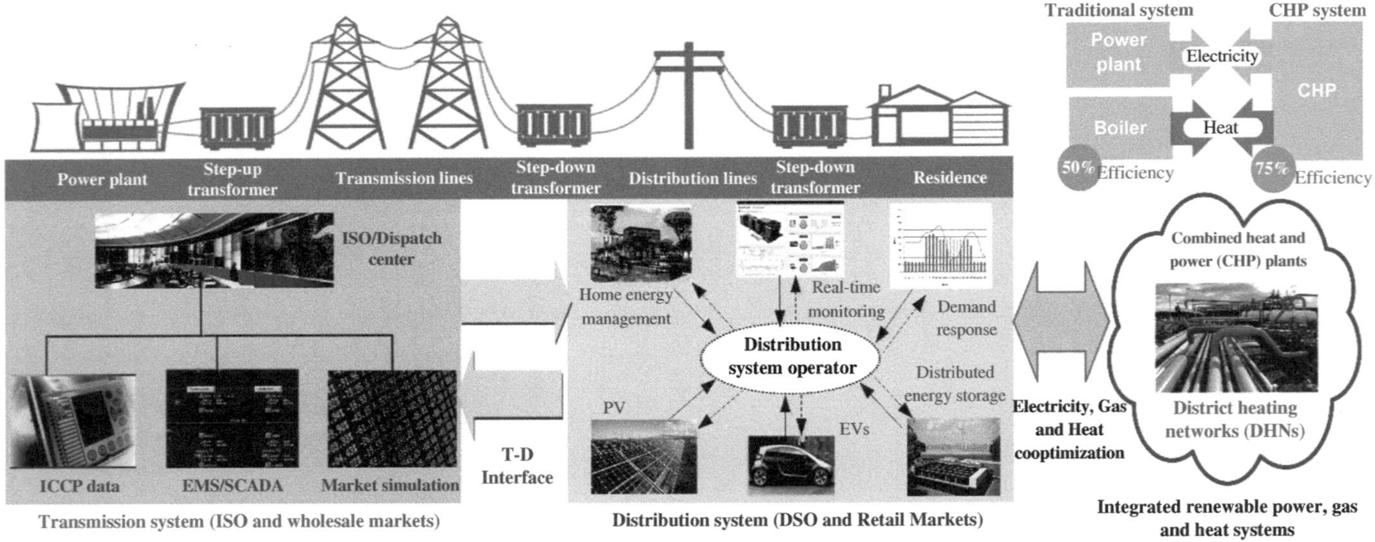

Figure 4.11 The proposed DSO framework and its connection with the transmission and gas systems. *DSO*, Distribution system operator.

increasingly diverse retail local markets, which are a mixture of burgeoning customer generation, modern net loads, and emerging energy services. In recent years the CAISO has proposed a similar idea called "integrated distributed electricity system (IDES)" to recognize that energy resources and operating decisions will be broadly decentralized and localized, and customers, microgrids, and larger DER continue to benefit from connections to the transmission grid [33].

Changes in the operational power system paradigm, with focus on the distribution grid, have brought about three fundamental needs for future, smart power grids:

- *Fast and comprehensive data analytics and informatics*: The evolution of smart meters and grid applications have provided utility companies with unprecedented capabilities for forecasting demand, scheduling outages, and shaping customer usage patterns. Technology innovations offer greater customer choice and functionality at decreasing costs that lead to changes in behavioral patterns on distribution systems. Utility companies need to develop a better decision-making platform capable of high-volume data management and advanced analytics designed to transform data into actionable insights.
- *Integrated renewable power, GHC coordination*: Currently, the electricity and natural gas sector are operated individually in most utilities. With the increasing utilization of DG technologies such as cogeneration, CHP, and reformer-based fuel cell generation, the coupling between electricity and natural gas becomes more common. Many utilities have built up the district heating system for distributing heat generated from one or more sources via a network of insulated pipes carrying steam or hot water to heat buildings. Large-campus district cooling systems are experiencing fast growth in recent years. On the other hand, many utilities in the United States (such as those in Texas and Colorado) are suffering from excessive generation of wind power and negative electricity price in the midnight, when thermal storage can be charged to consume the power. There are needs to construct an integrated platform to coordinate the renewable power and the GHC systems.
- *Novel DSO construct (from ISO to DSO):* During the last round of power industry disruption, the ISOs were instructed to manage the interface between IPPs and the wholesale grid. However, as grid operators, the ISOs have limited visibility and control over resources sited behind their meters. Today a new era has arrived for constructing operational models, that is, the DSO, which operate an IDES with a

mixture of centralized and distributed resources, for example, generation, energy storage, power flow and stability control devices, and control systems, including sensing devices and load management capabilities

The reliable and resilient operation of a DSO requires tight integration of the people, processes, and technology used to operate the distribution system. This is due to the highly dynamic nature of power flow and constantly changing local area topology on present distribution systems. In addition, the DSO would establish a secure clearing system to aggregate all participants' service offers and a transparent process for coordinating DER dispatch, including a pricing mechanism to reflect the dynamic constraints on the distribution system. The DSO framework is helpful for the accelerated implementation of advanced technology for the grid, along with an evolution of distribution system designs to create a node-friendly or "plug-n-play" grid that enables seamless integration of DERs and independently owned and operated microgrids. The benefits involve at least three aspects:

- better alignment of DER and GHC locational adoption to shape net load profiles and thereby improve system efficiency and load predictability;
- DERs providing services to the distribution grid and bulk power system; and
- localized controls of responsive devices and DERs to better balance the generation and dynamically evolving load.

One significant challenge for constructing the DSO is dynamically managing the large volume of two-way communication data. For example, fast distribution state estimation method is required to evaluate the system condition and provide starting points for the retail market optimization engine. The DSO's data management system has the ability to dynamically incorporate additional data into an executing application, and in reverse, ability of an application to dynamically steer the measurement process.

A typical DSO framework has four levels to realize different functionalities, as shown in Fig. 4.12. The techniques in each level are detailed as follows:

- Level 1: data management and analytics. The new paradigm of dynamic data-driven simulation asks for new approaches to incorporate real-time data in an effective and efficient manner. The operation condition in the distributed grid is highly stochastic, driven by the

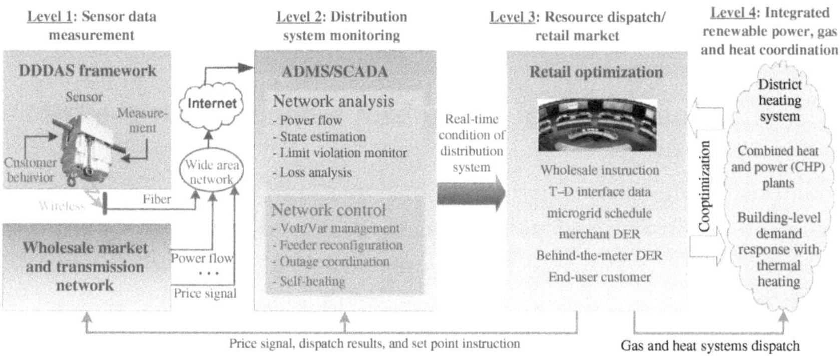

Figure 4.12 Functionalities in different levels of the DSO framework. *DSO*, Distribution system operator.

stochastic nature of renewable energy outputs, circuit faults, customer behaviors, and interchange flows. Thus it is difficult to apply inference techniques that rely on equation-based model representations. To overcome these challenges the Sequential Monte Carlo (SMC) method is selected as the data assimilation algorithm for supporting dynamic data-driven simulation.

- Level 2: advanced DMS/SCADA system. State estimation is a key function in DMSs. Traditionally, the least-squares estimation method has been widely used in the DMS engine. Advanced DMS techniques should not only measure the state of the current condition but also be able to predict the near-future states in a look-ahead manner. A promising approach is to use SMC-based method with Bayesian inference and stochastic sampling techniques to recursively estimate the states of dynamic systems from given observations and predicted near-future conditions.
- Level 3: resource dispatch system. As a dispatch coordinator, the proposed DSO will consolidate and coordinate the energy transactions and ancillary or reliability services offered by individual DER aggregators, services firms, and customers. A secure market optimization and clearing system is operated by the DSO to schedule operations, make real-time controls, and process settlements.
- Level 4: comprehensive energy system. DSO is a comprehensive platform to align the gas, heat, and electric schedules on the distribution system level. The cooptimization framework will increase the utilization of renewable power when the demand is low and enable the CHP to purchase power from the grid during high demand

periods/off-peak prices or sell surplus power to the grid during peak hours/high electricity prices.

4.3.3 Implementation of blockchain technology in electric power systems

Traditionally, the power system is operated in a centralized manner in which the transmission of power is carried over long distances before the generated power is available to consumers through distribution networks. In the past decade, with the rising share of DG resources such as PV, wind, storage and microgrid, the power system is becoming more decentralized. The complexity of the grid has increased significantly, which causes challenges for the grid operation. First, the scale of integrated devices such as DER, smart meters, inverters, and IoT equipment will expand exponentially. Ensuring efficient and reliable communication among them and managing the mass amount of data created by them is a great challenge to power companies. Second, the monitoring and control of broad distributed assets and the link of them with SCADA/DMS and other enterprise systems require new operating paradigm of utilities. New tools are needed to enable optimized control of the grid and DERs to realize efficient grid management with technologies such as Volt/VAR optimization, power quality management, and DER coordination. Finally, protecting the grid and customer privacy by enhancing the cyber security of grid-edge devices and networks is significant but difficult because many parties are involved and mutually impacted, including the utilities, customers, owners of home energy management devices, solar installers, and companies that use cloud-based software to monitor grid-edge devices [34]. The blockchain technology could be a promising way to address these challenges. A blockchain is a system in which the transaction records are made in cryptocurrency and maintained across several computers that are linked in a peer-to-peer network. Fig. 4.13 compares the centralized versus distributed transaction platforms. In the distributed transactional platform, every member in the network keeps its own coy of the ledger and can access in in the open cloud. As a result, all the members can access the log of the historical transactions and verify their validity.

The process of a transaction on the blockchain is shown in Fig. 4.14. When a transaction is requested in the system, it will be broadcasted to a peer-to-peer network that contains many distributed nodes. The transaction information needs to be validated on each node. The validated

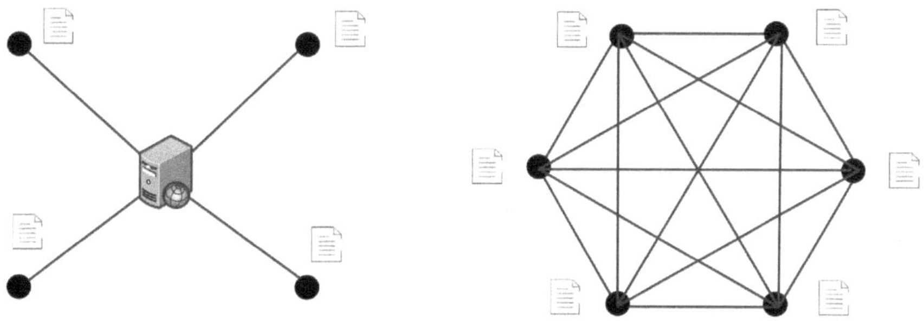

Figure 4.13 Centralized versus distributed transactional platform.

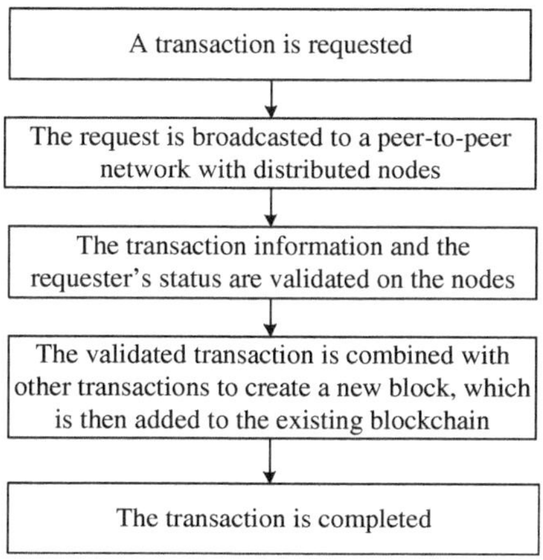

Figure 4.14 The process of a transaction on the blockchain.

transaction is combined with other transactions to create a new block, which is then added to the existing blockchain. Different algorithms can be used confirm transactions and produce new blocks to the chain, such as consensus algorithms, mining algorithms, and traceability chain algorithms [35]. The transaction completes after it is added to the existing blockchain.

In the retail electricity market, customers can buy or sell electricity at real-time prices based on their location. They may also be able to provide various types of grid service such as voltage and reactive power support in response to price signal. The blockchain technology can be used to develop smart contracts to facilitate the peer-to-peer transactions between retail customers. With smart contracts, transactions are made automatically in response to price signals and real-time renewable energy production information in the network.

4.4 Conclusion

The electric power industry is undergoing profound changes on all entities involved, including generation resources, transmission system, wholesale market, distribution system, retail merchant, and end use consumers/prosumers. This changing landscape leads to challenges for grid operations on both wholesale and retail level. This chapter discussed new market design and products on the wholesale level to deal with high renewables penetration. The bulk system and wholesale market operation under 100% renewables scenario is examined. Finally, the design of modern retail electricity market as well as how to implement blockchain technology on peer-to-peer transaction are proposed.

References

[1] Department of Energy, Public Utility Regulatory Policies Act of 1978 (PURPA), 1978. <https://www.energy.gov/oe/services/electricity-policy-coordination-and-implementation/other-regulatory-efforts/public>.
[2] Union of Concerned Scientists, Public Utility Regulatory Policy Act (PURPA). 2002. < https://www.ucsusa.org/clean_energy/smart-energy-solutions/strengthen-policy/public-utility-regulatory.html > .
[3] Federal Energy Regulatory Commission, 1996. Promoting Wholesale Competition Through Open Access Non-discriminatory Transmission Services by Public Utilities, and Recovery of Stranded Costs by Public Utilities and Transmitting Utilities (Order No. 888) and Open Access Same-Time Information System (formerly Real-Time Information Networks) and Standards of Conduct (Order No. 889). < https://www.ferc.gov/legal/maj-ord-reg/land-docs/rm95-8-0aj.txt > .
[4] Federal Energy Regulatory Commission, 1999. Regional Transmission Organizations (Order No. 2000). < https://www.ferc.gov/legal/maj-ord-reg/land-docs/RM99-2A.pdf > .
[5] Consolidated Edison, Distributed System Implementation Plan. July 31, 2018. < https://www.coned.com > .
[6] J. Zhu, K. Cheung, Flexible simultaneous feasibility test in energy market, in: IEEE Power and Energy Society General Meeting, Minneapolis, MN, July 2010.
[7] X. Luo, O. Obadina, Security assessment and enhancement in real-time operations of ERCOT nodal electricity market, in: 2010 IEEE Power and Energy Society General Meeting, July 25–July 29, 2010, Minneapolis, MN, 2010.
[8] J. Gubman, Draft proposed qualifying capacity and effective flexible capacity calculation methodologies, 2013. <https://www.cpuc.ca.gov/WorkArea/DownloadAsset.aspx?id = 6529>.
[9] N. Navid, G. Rosenwald, Ramp Capability Product Design for MISO Markets, 2013.
[10] California ISO, Business Requirements Specification: Flexible Ramping Product, Version 1.2, 2016.
[11] S. Lu, et al., Machine learning based multi-physical-model blending for enhancing renewable energy forecast—improvement via situation dependent error correction, in: Proc. 2015 Europ. Cont. Conf., Linz, Austria, July 15-17, 2015.

[12] Q. Wang, et al., The value of improved wind power forecasting: grid flexibility quantification, ramp capability analysis, and impacts of electricity market operation timescales, Appl. Energy 184 (2016) 696–713.
[13] Q. Wang, C. Brancucci, H. Wu, A.R. Florita, B.M. Hodge, Quantifying the economic and grid reliability impacts of improved wind power forecasting, IEEE Trans. Sustain. Energy 7 (4) (2016) 1525–1537.
[14] L.R. Chang-Chien, C.C. Sun, Y.J. Yeh, Modeling of wind farm participation in AGC, IEEE Trans. Power Syst. 29 (3) (2014) 1204–1211.
[15] C. Wang, P. Luh, N. Navid, Ramp requirement design for reliable and efficient integration of renewable energy, IEEE Trans. Power Syst. (2016). Available from: https://doi.org/10.1109/TPWRS.2016.2555855.
[16] V. Koritarov, et al., Modeling and Analysis of Value of Advanced Pumped Storage Hydropower in the United States, 2014.
[17] California ISO, Renewables Watch, May 16, 2017. <http://content.caiso.com/green/renewrpt/20170516_DailyRenewablesWatch.pdf>.
[18] Southwest Power Pool (March 16, 2018). SPP set a new wind-penetration record of 60.56 percent at 3:45 a.m. on March 16. <https://spp.org/>.
[19] EirGrid Group, All Time Record for Wind Generated Electricity Broken, Jan. 2017. <http://www.eirgridgroup.com/newsroom/wind-record/>.
[20] Colorado General Assembly (2018), Require 100% Renewable Energy By 2035 (SB18-064). <http://leg.colorado.gov/bills/sb18-064>.
[21] Hawaiian Electric, Maui Electric and Hawai'i Electric Light, Planning Hawaii's Grid for Future Generations, in: Integrated Grid Planning Report, Mar. 1, 2018.
[22] Federal Energy Regulatory Commission, Notice of Proposed Rulemaking, Electric Storage Participation in Markets Operated by Regional Transmission Organizations and Independent System Operators, 2016.
[23] Federal Energy Regulatory Commission, Electric Storage Participation in Markets Operated by Regional Transmission Organizations and Independent System Operators, 2018.
[24] Hawaiian Electric Companies, Planning Hawaii's Grid for Future Generations: Integrated Grid Planning Report, 2018.
[25] C. Loutan, et al., Demonstration of essential reliability services by a 300-MW solar photovoltaic power plant, in: Technical Report, NREL/TP-5D00-67799, 2017.
[26] V. Gevorgian, B. O'Neill, Advanced grid-friendly controls demonstration project for utility-scale PV power plants, in: Technical Report/TP-5D00-65368, National Renewable Energy Laboratory, 2016.
[27] NERC, Balancing and Frequency Control, NERC, Princeton, NJ, 2011.
[28] The 21st Century Power Partnership, An introduction to retail electricity choice in the United States, in: NREL/BR-6A50-68993, 2017.
[29] Environmental Protection Agency (2017). What is a Community Choice Aggregation (CCA)? <https://www.epa.gov/greenpower/community-choice-aggregation>
[30] D.R. Bohi, K.L. Palmer, Relative Efficiency Benefits of Wholesale and Retail Competition in Electricity: An Analysis and a Research Agenda. Prepared by Resources for the Future, National Renewable Energy Laboratory, Washington, DC, Golden, CO, 1996.
[31] State of Nevada, Statewide ballot questions, 2016. <https://www.nvsos.gov/sos/home/showdocument?id = 4434>.
[32] Brian Seal, Understanding DERMS, in: EPRI Tech. Report, Palo Alto, CA, 2018.
[33] L. Kristov, P. De Martini, 21st Century Electric Distribution System Operations, 2014.

[34] Scott Sowers, EPRI to Develop Secure 'Architectures' for Distributed Energy Resources, EPRI Journal, Jul./Aug. 2017. <http://eprijournal.com/securing-the-grids-edge/>.
[35] Zee Ali, A. Simple Introduction to Blockchain Algorithms, Mar 9, 2019.

CHAPTER FIVE

Wide-area monitoring and anomaly analysis based on synchrophasor measurement

Shutang You[1], Yu Su[1], Yong Liu[2] and Yilu Liu[1,3]
[1]University of Tennessee, Knoxville, TN, United States
[2]Pacific Gas and Electric Company, San Francisco, CA, United States
[3]Oak Ridge National Laboratory, Oak Ridge, TN, United States

Contents

5.1 Synchrophasor measurement technology introduction 144
 5.1.1 Situational awareness 146
 5.1.2 Advanced control 146
5.2 Wide-area measurement system example—FNET/GridEye 147
5.3 FNET/GridEye wide-area measurement system applications overview 151
 5.3.1 Visualization 152
 5.3.2 Disturbance detection and location 152
 5.3.3 Interarea oscillation detection and event-data-based oscillation modal analysis 153
 5.3.4 Online ambient-data-based oscillation modal analysis 154
 5.3.5 Islanding detection 154
 5.3.6 Event replay and postevent analysis 156
 5.3.7 Statistical analysis of historical data 156
 5.3.8 Model validation and parameter verification 157
 5.3.9 Machine learning–based inertia estimation 157
References 160
Further reading 161

Power system is one of the most complex, large-scale dynamic systems in the world. Modern power grids usually ramify across a county and even a whole continent to convey electric power from generators to consumers. Consequently, a wide-area measurement network is the prerequisite of modern power grid monitoring and operation. In addition, due to the increasing deployment of intermittent renewable generators and other new power grid components, power grid dynamic characteristics become more and more complicated. Conventional monitoring technologies, such

New Technologies for Power System Operation and Analysis.
DOI: https://doi.org/10.1016/B978-0-12-820168-8.00005-5

as those used in the existing supervisory control and data acquisition (SCADA) system that measures every 2–4 seconds without precise time synchronization, can only help operators understand the steady state situation of power grid. Comparatively, the synchrophasor technology-based wide-area monitoring measurement can monitor the power system dynamics in an unprecedented way. In this chapter, wide-area monitoring measurement and its various applications in modern power grids will be introduced.

5.1 Synchrophasor measurement technology introduction

The concept of phasor was first introduced by Charles Proteus Steinmetz in 1893. In one of his publications, he presented "phasor" to describe the waveforms of alternating current (AC) electricity. Mathematically, a phasor can be considered as a vector rotating about the origin in a complex plane, as shown in Fig. 5.1. In electrical engineering, phasor is usually utilized to represent the measurement of three-phase AC voltage or current using a common time source for time synchronization. Therefore phasor measurement sensors are also commonly referred to as synchrophasors.

The early prototype synchrophasor device to measure phasor, referred to as a phasor measurement unit (PMU) later, was invented by Dr. Arun G. Phadke and Dr. James S. Thorp in 1988 at Virginia Tech for transmission line protection purposes [1]. A PMU measures both AC voltage and current waveforms with accurate time synchronization at different locations of power grids. Fig. 5.2 shows the major components of a PMU

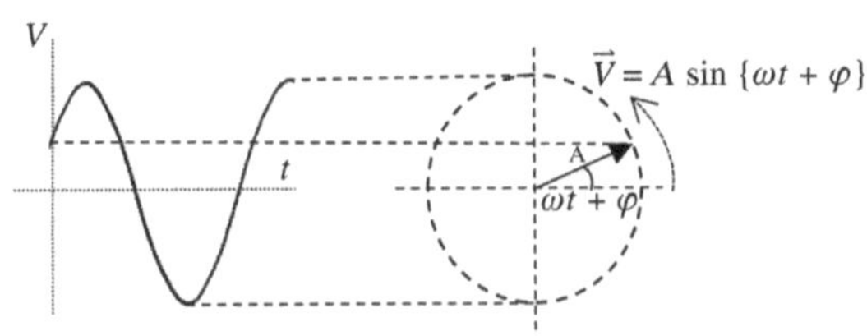

Figure 5.1 Phasor and sinusoidal/cosine function.

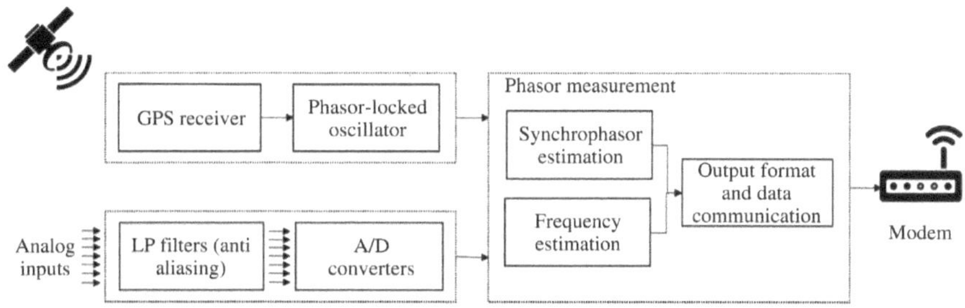

Figure 5.2 Major components of a PMU. *PMU*, Phasor measurement unit.

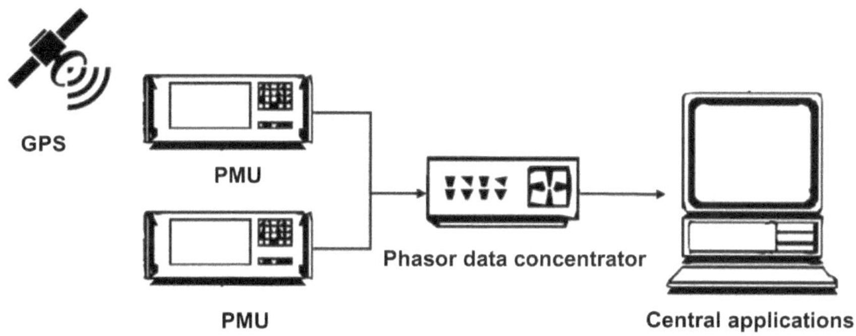

Figure 5.3 WAMS structure. *WAMS*, Wide-area measurement system.

device [2]. The three-phase analog inputs (voltages and currents) are digitized by analog-to-digital converters. A global positioning system (GPS) receiver connected with an outdoor antenna provides pulse-per-second signals with 1 μs or even better accuracy for both waveform sampling and time synchronization. A PMU calculates the voltage and current phasors and sends them to data centers via communication network, at the rate of 10–60 per second depending on the application.

Currently, a PMU-based wide-area measurement system (WAMS) usually consists of PMUs deployed across the power grid, phasor data concentrators collecting and aligning synchronized phasor data from PMUs, and data centers performing data storage and applications. A typical WAMS structure is shown in Fig. 5.3.

Since its birth 30 years ago, a wide variety of synchrophasor applications have been developed not only for transmission line protection purposes but also for enhancing power grid operators' situational awareness and control capabilities. All of these applications are further improving the security, reliability, and elasticity of modern power grids.

5.1.1 Situational awareness

With the invention of PMU in the late 1980s the PMU-based WAMS is able to reveal important insights into power grid dynamics by real time, synchronized, high-resolution measurement data. The difference between power grid monitoring using conventional SCADA systems and innovative PMU-based WAMS can be compared to a myopic people before and after vision correction, as shown in Fig. 5.4.

5.1.2 Advanced control

The increasing deployment and application of synchrophasor technology enables wide-area fast-response controls in power systems as well. Traditionally, controls are developed using mathematical models obtained off-line. In general, locally derived signals are used as the feedback, and controls based on SCADA are relatively slow, which are unable to track fast-system dynamics. For instance, frequency control (automatic generation control) is a periodic control at the rate of every several seconds. On the contrary, synchrophasors provide a great opportunity to bring the remote, synchronized, high-resolution, and real-time measurements of system variables to the controllers, and, thus, the advanced control becomes possible, which will significantly enhance system stability and security.

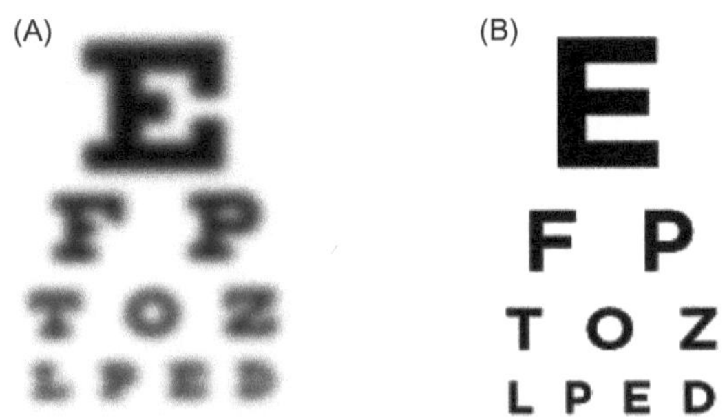

Figure 5.4 Metaphorical comparison of power system monitoring using SCADA and PMU/WAMS. (A) Before vision correction and (B) after vision correction. *PMU*, Phasor measurement unit; *SCADA*, supervisory control and data acquisition; *WAMS*, wide-area measurement system.

5.2 Wide-area measurement system example—FNET/GridEye

FNET/GridEye is a pioneering wide-area monitoring network operated by the authors of this chapter. It aims to help prevent power system blackouts and facilitate renewable energy integration. A typical FNET/GridEye system consists of the hardware component, which is a set of highly accurate monitoring devices, called frequency disturbance recorders (FDRs), and the software component, which is a high-volume data collection and processing center that supports both real time applications and off-line data analytics [3,4]. In this section, FNET/GridEye will be introduced as an example of WAMS in order to showcase how a WAMS will benefit the electric power grids in multiple aspects.

Fig. 5.5 demonstrates the key FNET/GridEye components. Measurements from the power grids (including frequency, voltage phase angle, and magnitude) are collected by the FDR sensors (which are equipped with GPS receivers in order to be GPS-synchronized) and are Internet-transmitted to the data center for processing, storage, and utilization.

In the data center the measurement data first come to the data concentrator. It conditions measurement data, including cleaning and aligning data. The conditioned data go to the data storage server and real time

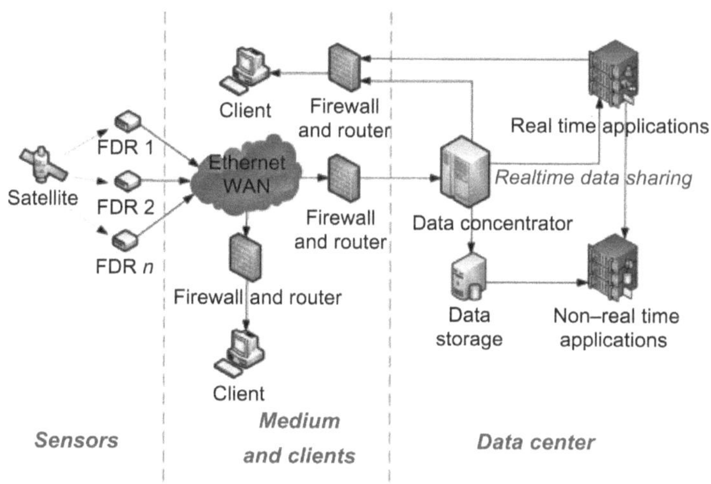

Figure 5.5 Architecture of FNET/GridEye.

application server simultaneously. The data storage server stores data for future retrieval and non–real time applications. The real time application server will execute multiple application modules, such as event location, oscillation monitoring, and event report. The real time application server can also provide visualized results to clients using the data from concentrator and generate event logs and event replay to the non–real time applications and data storage.

It is noted that FNET/GridEye's FDR sensors can be deployed at the distribution level. Since frequency and voltage phase measurement can be used in monitoring a variety of grid dynamics disturbances, including generation trip, load shedding, islanding, line tripping, and oscillations. The frequency and voltage phase measurement at the distribution level can enable many situational awareness applications of bulk power systems while significantly reducing WAMS deployment costs.

FNET/GridEye is now monitoring North American grids and over 20 major power grids worldwide, including United States, Canada, Mexico, Chile, Bahamas, Brazil, Finland, Sweden, The Netherlands, Italy, Spain, Russia, Denmark, Germany, Egypt, Turkey, United Kingdom, India, China, Taiwan, Japan, South Korea, Australia, and South Africa, as shown in Figs. 5.6 and 5.7.

Similar to its transmission level counterparts, phasor measurement at distribution level can be obtained from household outlets at variable sampling rates depending on specific applications. Then the phasor measurement will be GPS-time-stamped and transmitted to the data center through the Internet. FDR is now in its second generation (as shown in Fig. 5.8) and is based on an embedded system with GPS timing and Ethernet functions. FDR components are shown in Table 5.1 [5].

Fig. 5.9 is the hardware architecture of the first-generation FDR. Although advances are made in hardware and layout, both generations of FDR share the same principal design. First, the AC voltage signal from common wall outlets goes through a step-down transformer in the FDR and becomes 10 V. A low-pass filter subsequently blocks the high frequency noises. Then, an ADC module samples the 10 V signal based on a timing signal provided by GPS-synchronized oscillator pulses. The sampled values are fed to the central processing units to obtain phasor measurements including voltage magnitude, phase angle, and voltage frequency. Finally, the data are transmitted to the device server via Internet. The second-generation FDR separated the hardware for phasor calculation and communication. A digital signal processor is used for

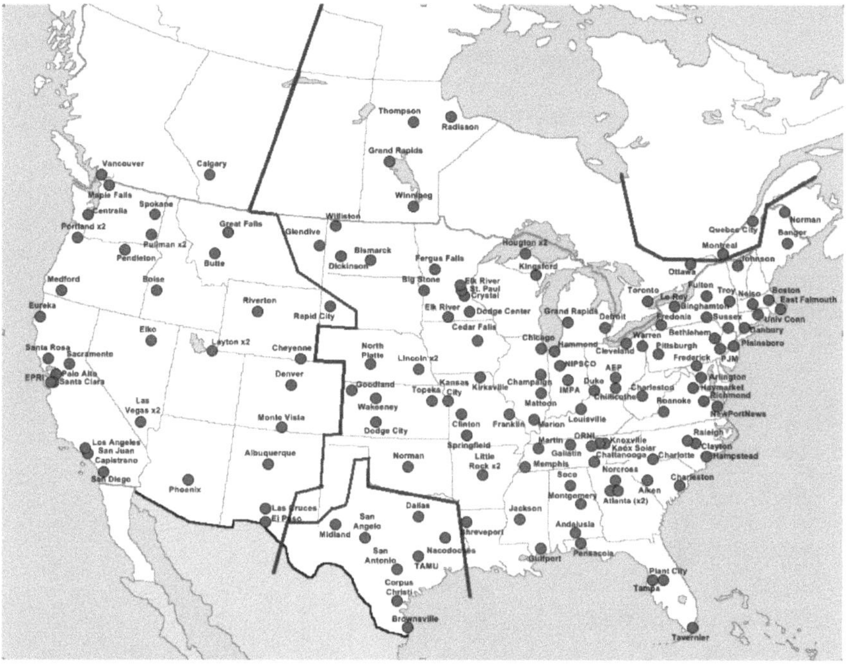

Figure 5.6 Map of FDR deployment in North America.

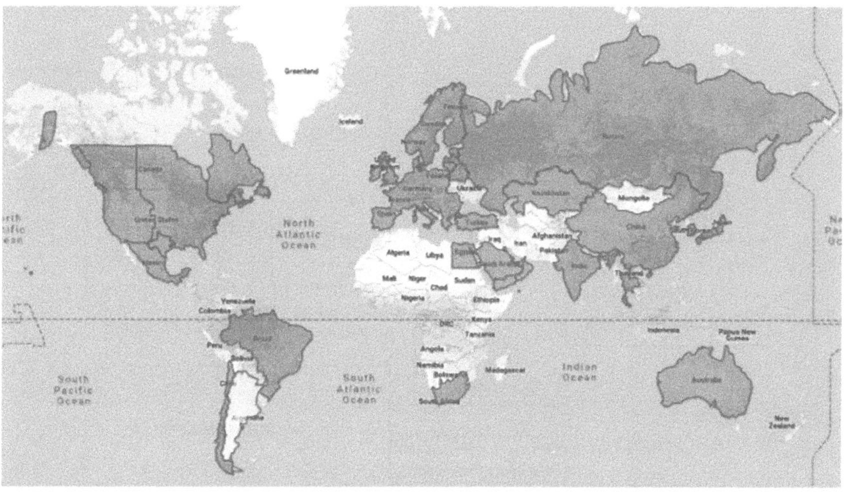

Figure 5.7 Worldwide power grid monitored by FNET/GridEye.

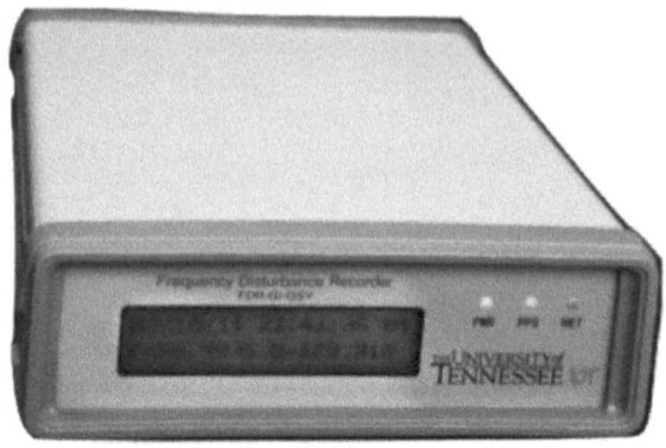

Figure 5.8 The second generation of FDR.

Table 5.1 FDR components.

No.	Component	Function
1	Instrument transformer	Power supply, voltage transform for measurement
2	Low-pass filter	Harmonics filtering
3	ADC	Signal transform for processing
4	GPS signal receiving module	Timing stamp
5	Microprocessor module	Synchrophasor data generation, sensor control and display
6	Ethernet communications module	Data transmission and communication

ADC, Analog-to-digital converter.

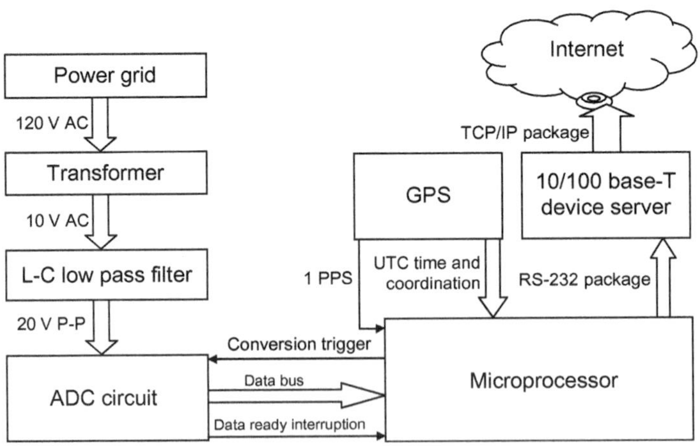

Figure 5.9 Hardware block diagram of first generation of FDR.

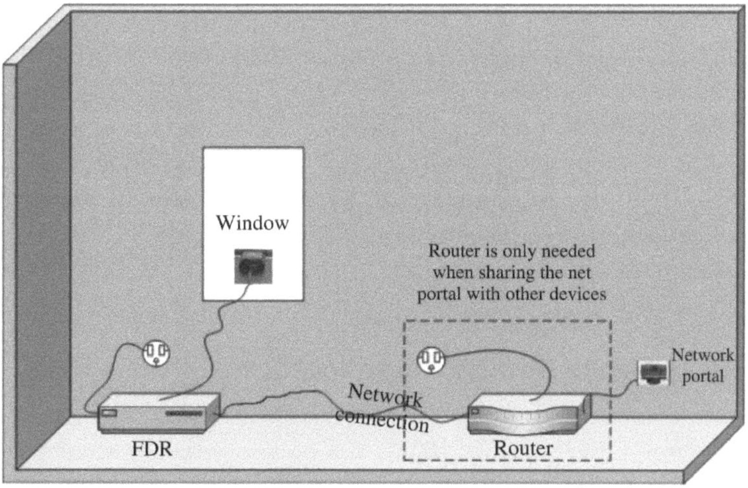

Figure 5.10 Illustration of FDR installation.

phasor calculation and a microcontroller to integrate GPS and Ethernet. The second-generation FDR also has better accuracy performance due to the active filter design replacing the LC filter in the first-generation FDR.

Low installation cost and high measurement accuracy are two important characteristics of FDR. Each FDR interfaces with the electric grid at low, easily accessible, voltage levels. The installation costs for an FDR are nearly nonexistent. This feature gives FDR an extreme cost advantage over PMU technology, which is designed to instrument high-voltage substations. In addition, taking the Internet as data collection solution makes the wide-area information communication and exchange possible and cost-effective. A typical FDR installation is illustrated in Fig. 5.10.

FDR can achieve high accuracy in measurement. For frequency measurements the error is less than 0.0005 Hz. A comparison study between an FDR and four commercial PMUs from 2003 was conducted by Wang et al. [5], showing that FDR is more accurate in measuring frequency than commercial PMUs.

5.3 FNET/GridEye wide-area measurement system applications overview

After more than 10 years of development and field testing, multiple function modules have been developed in FNET/GridEye to improve operators' situational awareness, as discussed in the following subsections.

5.3.1 Visualization

Frequency and phase angle are important indicators of power system health. By overlaying real time measurements from FNET/GridEye on a geographic information system, an intuitive visualization tool is created to help operators interpret system behavior in real time (Fig. 5.11). The FNET/GridEye system is not bound by electric power utility territories and provides system frequency observability of the entire North American power system.

5.3.2 Disturbance detection and location

Many disturbances, such as loss of generation or transmission lines, occur daily in a large power system. While the power system is resilient against most disturbances, it is still of great importance to have the ability to quickly detect and locate disturbances as they happen, to prevent an escalation into large-area blackouts under extreme conditions [6–8].

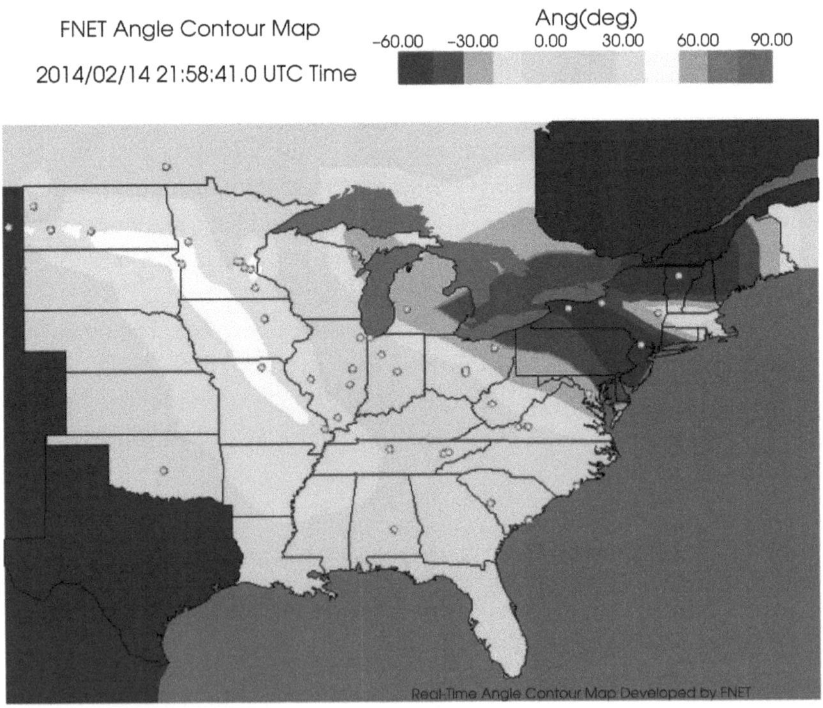

Figure 5.11 Real time visualization of measurement data.

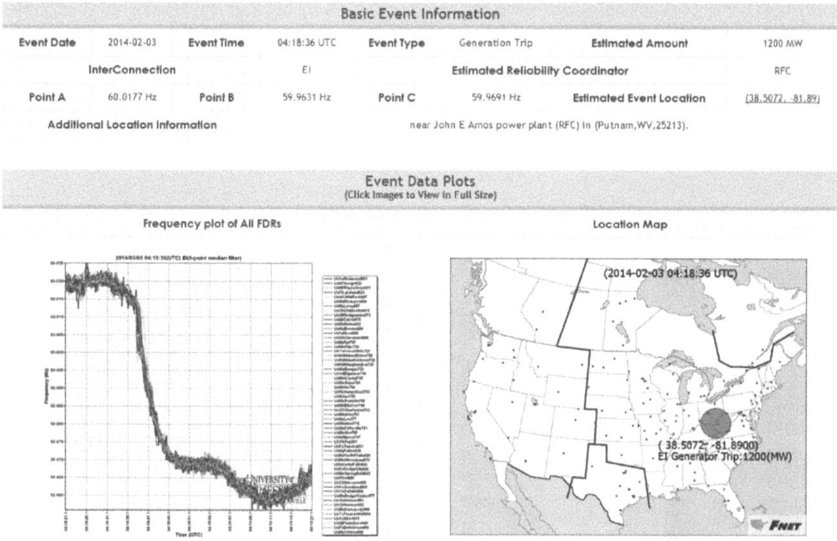

Figure 5.12 Disturbance detection and location.

By continuously screening real time measurement data the FNET/GridEye disturbance detection and location module (Fig. 5.12) sends alerts to operators and other reliability entities once any abrupt frequency or angle change is detected. Through the automated analysis of disturbances detected by the FNET/GridEye system, the geographic location of disturbances can be estimated. This location estimation technology is based on the premise that FDRs closest to the disturbance witness the frequency or angle change earlier than the others. Furthermore, after losing a generator or transmission line, operators are eager to know the magnitude of power loss so that they can take corrective actions. To that end, this function module also gives an estimate of such a generation-load mismatch in the disturbance alert.

5.3.3 Interarea oscillation detection and event-data-based oscillation modal analysis

Interarea oscillation is one of the potential threats to grid stability. Through a frequency/angle-based oscillation detection algorithm, FNET/GridEye monitors event-triggered interarea oscillations and provides grid operators with valuable oscillation modal information once an oscillation is detected (Fig. 5.13) [9].

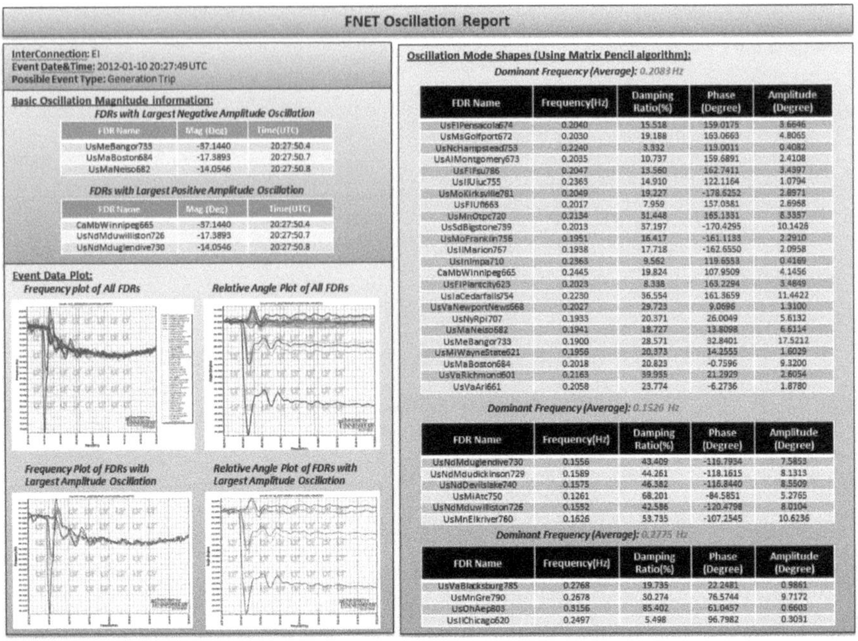

Figure 5.13 Interarea oscillation detection and event-data-based oscillation modal analysis.

5.3.4 Online ambient-data-based oscillation modal analysis

Considering the constant changing of operational conditions, continuous oscillation modal information is highly desirable so that appropriate measures can be taken in real time to prevent any serious interarea oscillations. The online oscillation modal analysis module uses real time ambient data (when there is no significant disturbance in the power system) to compute system oscillation modes and can assess power grid stability continuously in real time [10,11] (Fig. 5.14).

5.3.5 Islanding detection

An islanding event occurs when one or more generators lose synchronism with the rest of the power system. Critical intervention actions must be taken, or the islanding event can develop into blackouts. The FNET/GridEye system uses frequency signals to detect islanding events and send out islanding warnings to grid operators if an islanding event happens (Fig. 5.15).

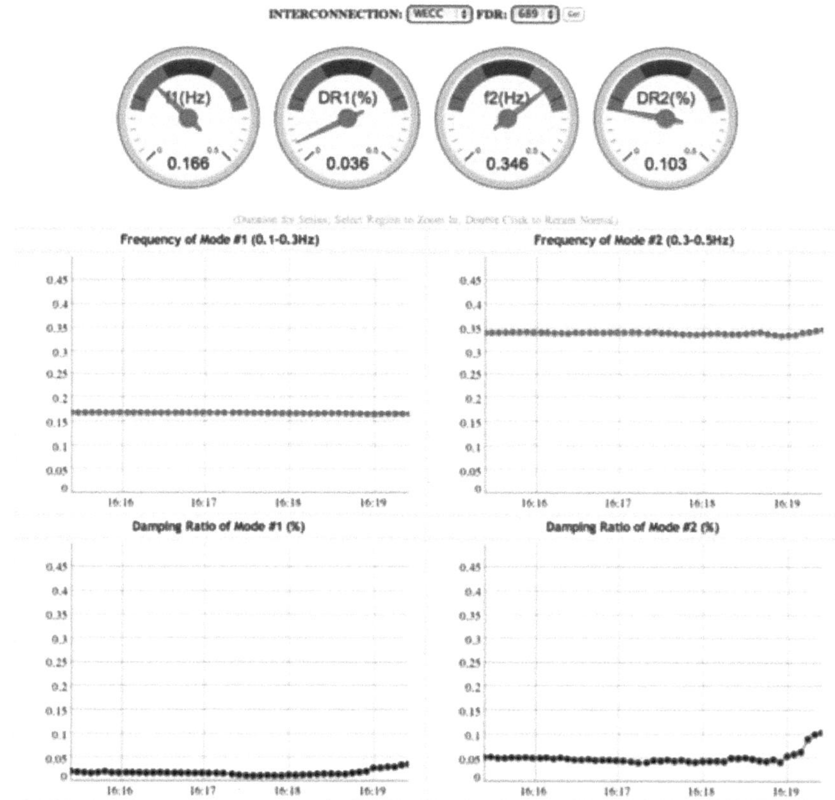

Figure 5.14 Online ambient-data-based oscillation modal analysis.

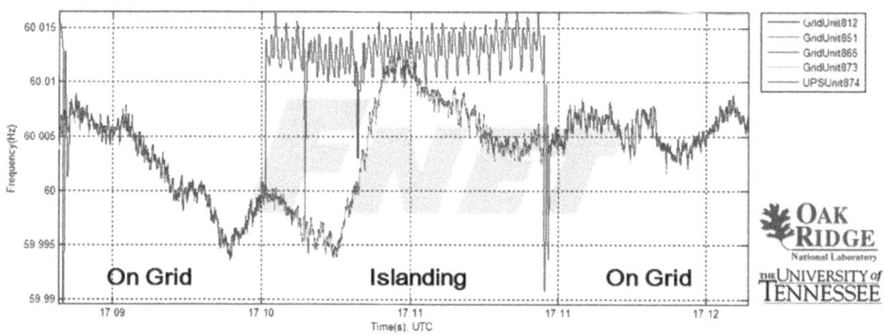

Figure 5.15 Islanding detection.

5.3.6 Event replay and postevent analysis

Blackouts over large areas are often disastrous. FNET/GridEye measurements of a blackout can be used to reconstruct the event scenario for postevent analysis. A blackout occurred in Florida in 2008 where 22 transmission lines, 4300 MW of generation, and 3650 MW of load were lost [12–16]. FNET/GridEye measurements captured the event and were used to construct a replay of the blackout event (The video of this event is available at http://www.youtube.com/watch?v=H7y-oJYpDkM&feature=c4-overview&list=UU0uqG3I8AxZ8ZMgB5NlFpFg). Fig. 5.16 shows several screenshots of the event replay. Event-replays like this facilitates postevent analyses and aids the development of tools for avoiding similar blackouts in the future.

5.3.7 Statistical analysis of historical data

Measurement data are collected from FNET/GridEye sensors on a 24/7 basis. With advanced data mining techniques, historical data can be extremely informative. For instance, disturbance statistical analysis performed with FNET/GridEye can reveal the distribution of disturbances

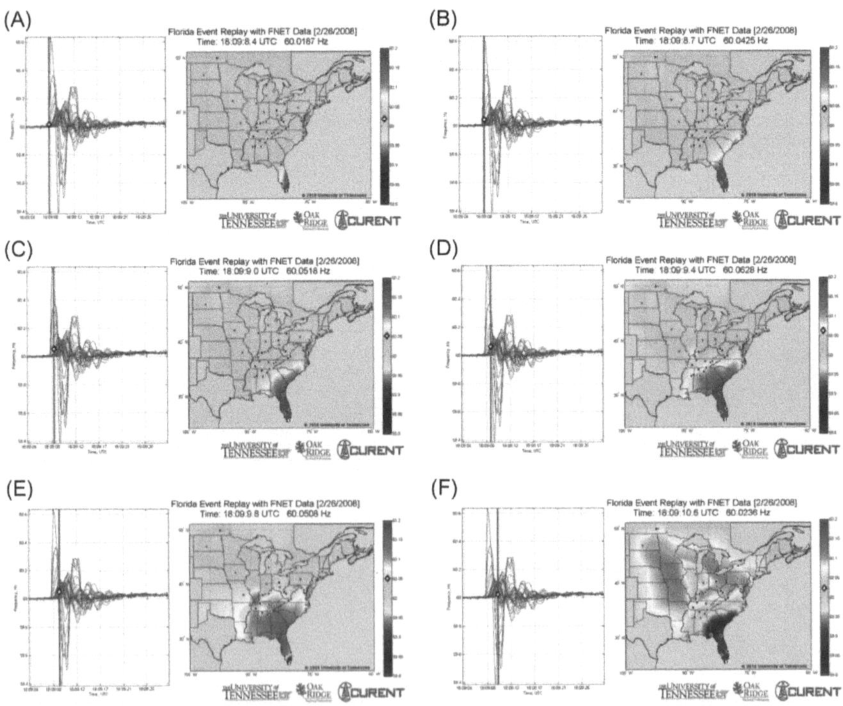

Figure 5.16 Event replay and postevent analysis. (A), 18:09:8.4 at UTC Time; (B), 18:09:8.7; (C), 18:09:9.0; (D), 18:09:9.4; (E), 18:09:9.8; (F), 18:09:10.6.

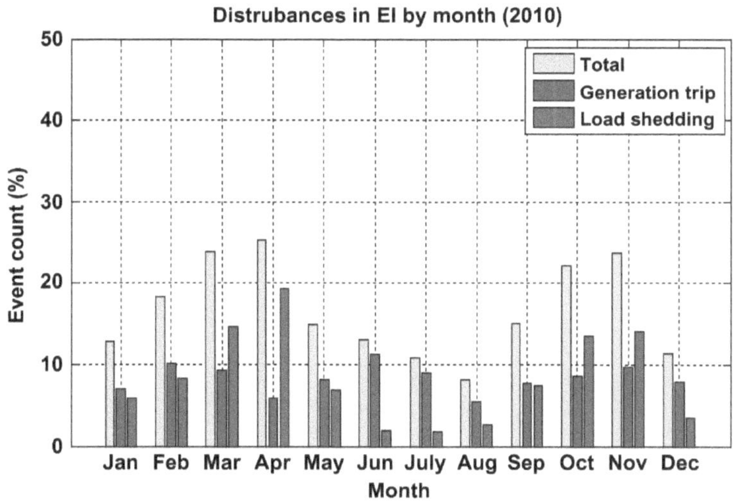

Figure 5.17 Statistical analysis of historical data.

over a certain time period. Shown in Fig. 5.17 is the 2010 monthly distribution of disturbances in the US Eastern Interconnection.

5.3.8 Model validation and parameter verification

Many of the advanced operation and planning functions in the power system's control center depend on the availability of accurate power grid models. Unfortunately, accurate models (and parameters) for many of the devices connected to the power grid, particularly with respect to dynamic behaviors, are unavailable.

Because FNET/GridEye sensors capture power grid behaviors accurately over the entire grid, FNET/GridEye measurement data can be used as the benchmark with which to validate power grid models. For instance, using the frequency measurement as the benchmark, FNET/GridEye has revealed that the current US Eastern Interconnection model is far from accurate for creditable simulation results and that a series of device parameters need to be tuned to match the actual system (Fig. 5.18). Moreover, FNET/GridEye can be used to tune the model to match the measurement.

5.3.9 Machine learning–based inertia estimation

With the development of machine learning technologies, many new applications using FNET/GridEye data become possible by taking full advantages of machine learning. An example of machine learning–based applications is

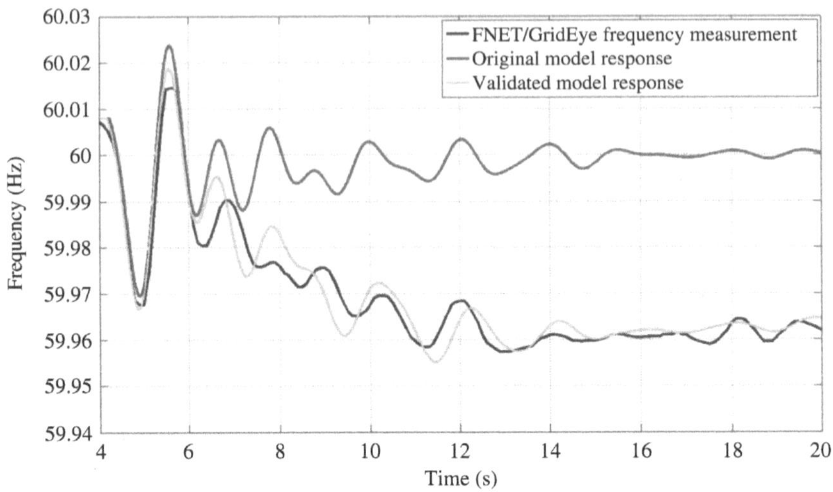

Figure 5.18 Model validation and parameter verification.

using FNET/GridEye data to estimate system inertia using ambient frequency measurement. Since the inertia level will influence the ambient frequency signatures, as shown in Fig. 5.19, system inertia can be estimated using FNET/GridEye data by extracting features from ambient frequency measurement. Fig. 5.20 shows the actual and estimated inertia profiles in the western electricity coordinating council (WECC) system in heavy and light load seasons. It can be seen that combining machine learning and FNET/GridEye measurement can achieve accurate estimation of system inertia. Estimated inertia information will facilitate advanced frequency control functions and benefit the estimation of event magnitudes as well.

In conclusion, as a WAMS, FNET/GridEye can increase the system situational awareness capability to a new level: (1) the FNET/GridEye sensor is a single-phase PMU with higher precision than commercial PMUs, but with reduced manufacturing and installation costs because it is installed in a common electrical wall outlet at the 120-V distribution level and (2) FNET/GridEye offers a full package of measurement data analytics functions for the purpose of improving operators' awareness of system status. FNET/GridEye is a cost-effective, multifunctional power grid monitoring network based on quickly deployable sensors and advanced data-utilization techniques. By overcoming the installation and maintenance difficulty of traditional PMUs, FNET/GridEye sensors are suitable for wide deployment at much higher density.

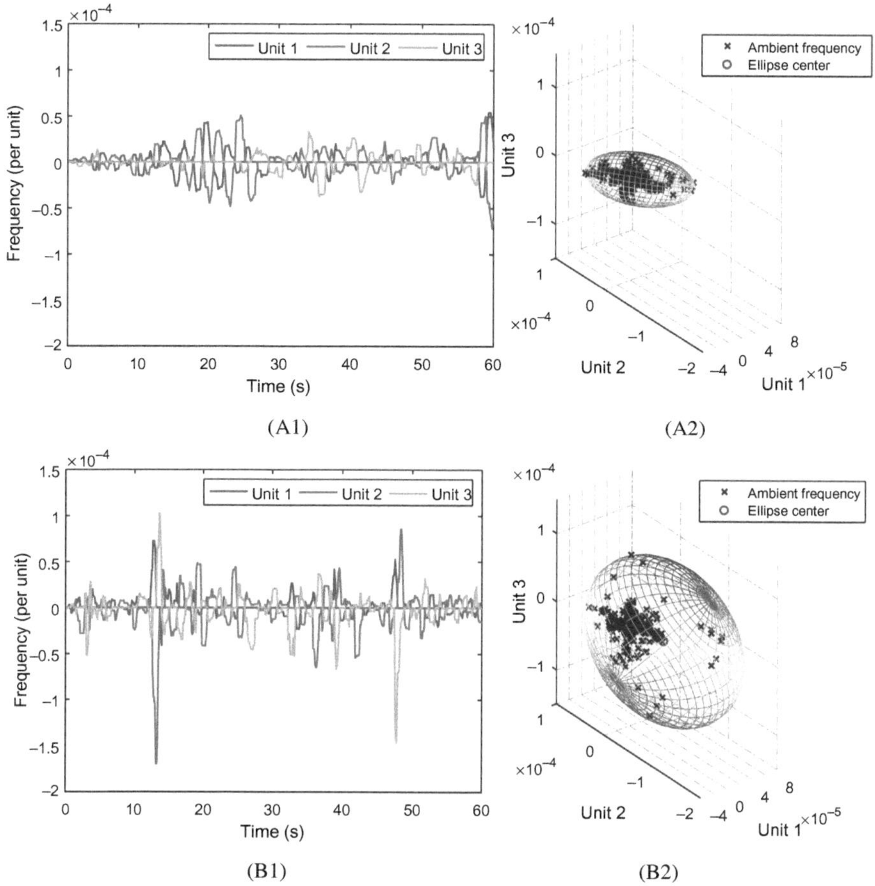

Figure 5.19 Ambient frequency variation (A1 and B1) and characteristic ellipsoids (A2 and B2) at (A1 and A2) 100% and (B1 and B2) 50% inertia level.

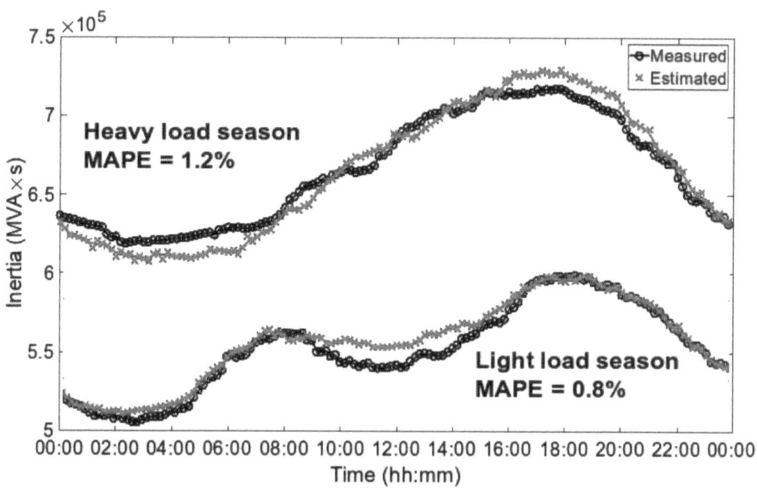

Figure 5.20 Comparison of measured and estimated inertia in WECC during heavy and light load seasons.

References

[1] A.G. Phadke, J.S. Thorp, History and applications of phasor measurements, in: IEEE PES Power Systems Conference and Exposition, Atlanta, GA, Oct. 29–Nov. 1, 2006.

[2] A.G. Phadke, J.S. Thorp, Synchronized Phasor Measurements and Their Applications, Springer, New York, 2008, p. 4.

[3] Z. Zhong, C. Xu, B.J. Billian, L. Zhang, et al., Power system frequency monitoring network (FNET) implementation, IEEE Trans. Power Syst. 20 (2005) 1914–1921. Available from: https://doi.org/10.1109/TPWRS.2005.857386.

[4] Y. Zhang, P.N. Markham, T. Xia, et al., Wide-area frequency monitoring network (FNET) architecture and applications, IEEE Trans. Smart Grid 1 (2) (2010) 159–167.

[5] L. Wang, J. Burgett, J. Zuo, C. Xu, B.J. Billian, R.W. Conners, Y. Liu, Frequency disturbance recorder design and developments, in: Proc. 2007 IEEE Power Engineering Society General Meeting, 2007, pp. 1–7.

[6] J.S. Thorp, C.E. Seyler, A.G. Phadke, Electromechanical wave propagation in large electric power systems, IEEE Trans. Circ. Syst. I: Fundam. Theory Appl. 45 (6) (1998) 614–622.

[7] S.S. Tsai, L. Zhang, A.G. Phadke, Y. Liu, M.R. Ingram, S.C. Bell, et al, "Study of global frequency dynamic behavior of large power systems," IEEE PES Power Systems Conference and Exposition, 2004, vol. 1, 10–13 Oct. 2004, 328–335.

[8] T. Xia, H. Zhang, R. Gardner, J. Bank, J. Dong, J. Zuo, Y. Liu, et al., Wide-area frequency based event location estimation, in: IEEE Power Engineering Society General Meeting, 2007, 24–28 June 2007, pp. 1–7.

[9] J.K. Wang, R.M. Gardner, Y. Liu, Analysis of system oscillations using wide-area measurements, in: Proc. 2006 IEEE Power Engineering Society General Meeting, 2006, pp. 6.

[10] R.M. Gardner, W. Li, J. West, J. Dong, Y, Liu, G. Zhang, Power system frequency oscillation characteristics, in: Proc. 2008 IEEE Power Engineering Society General Meeting, 2008, pp. 1–7.

[11] W. Li, R.M. Gardner, J. Dong, L. Wang, T. Xia, Y. Zhang, Y. Liu, G. Zhang, Y. Xue, Wide area synchronized measurements and inter-area oscillation study, in: Proc. 2009 Power Systems Conference and Exposition, 2009, pp. 1–8.

[12] T.J. Overbye, J.D. Weber, Visualizing the electric grid, Spectrum, IEEE 38 (2) (2001) 52–58.

[13] G. Pires de Azevedo, C. Sieckenius de Souza, B. Feijo, Enhancing the human-computer interface of power system applications, IEEE Trans. Power Syst. 11 (2) (1996) 646–653.

[14] J.D. Weber, T.J. Overbye, Voltage contours for power system visualization, IEEE Trans. Power Syst. 15 (1) (2000) 404–409.

[15] G. Zhang, P. Hirsch, S. Lee, Wide area frequency visualization using smart client technology, in: Power Engineering Society General Meeting, 2007. IEEE, 24–28 June 2007, pp. 1–8.

[16] G. Zhang, S. Lee, R. Carroll, J. Zuo; L. Beard, Y. Liu, Wide area power system visualization using real-time synchrophasor measurements, in: Power and Energy Society General Meeting, 2010 IEEE, 25–29 July 2010, pp. 1–7.

Further reading

J.N. Bank, R.M. Gardner, S.-J.S. Tsai, K.S. Kook, Y. Liu, Visualization of wide-area frequency measurement information, in: Power Engineering Society General Meeting, 2007, IEEE, 24–28 June 2007, pp. 1–8.

M. Paolone, Synchrophasor fundamentals: from computation to implementation, in: Tutorial – Synchrophasor Fundamentals and Applications: Leveraging the Investment, IEEE PES General Meeting, July 21–25, 2013.

S.-J.S. Tsai, J. Zuo, Y. Zhang, Y. Liu, Frequency visualization in large electric power systems, in: Power Engineering Society General Meeting, 2005, IEEE, Vol. 2, 12–16 June 2005, pp. 1467–1473.

CHAPTER SIX

Advanced grid operational tools based on state estimation

Yu Liu[1], Yuzhang Lin[2] and Junbo Zhao[3]

[1]School of Information Science and Technology, ShanghaiTech University, Shanghai, P.R. China
[2]Department of Electrical and Computer Engineering, University of Massachusetts, Lowell, MA, United States
[3]Department of Electrical and Computer Engineering, Mississippi State University, Starkville, MS, United States

Contents

6.1 Introduction 163
6.2 Model validation 165
  6.2.1 Largest Normalized Lagrange Multiplier test 166
  6.2.2 Computationally efficient implementation of the Largest Normalized Lagrange Multiplier test 173
  6.2.3 Detectability and identifiability of parameter and measurement errors 177
6.3 System monitoring 178
  6.3.1 Motivations for dynamic state estimation 178
  6.3.2 Problem formulation of dynamic state estimation 180
  6.3.3 Unified framework for Bayesian dynamic state estimation through nonlinear regression 184
6.4 Protective relaying 192
  6.4.1 Theoretical basis of dynamic state estimation–based protection 193
  6.4.2 Numerical experiments 197
6.5 Conclusion remarks 204
References 205
Further reading 208

6.1 Introduction

Accurate knowledge of operating states and models is essential for reliable and economic operation of electric power systems [1–5]. As a result, traditional data-acquisition systems are widely installed in present power systems, to obtain measurements, including voltage magnitudes, current magnitudes, real power, reactive power, and circuit breaker status

New Technologies for Power System Operation and Analysis.
DOI: https://doi.org/10.1016/B978-0-12-820168-8.00006-7

[6–11]. In addition, modern power systems are equipped with state-of-the-art measuring devices to provide high-quality and high-density measurements, including GPS-synchronized phasor measurements from phasor measurement units (PMUs) and GPS-synchronized instantaneous "points on wave" measurements from merging units (MUs) [12–14].

Nevertheless, even with these measurements, additional steps are still required for effective operation of electric power systems because of the following reasons. First, the models of the power system may not be accurate enough due to changes of operating conditions or even human errors. Second, the measurements are usually with errors due to imperfect instrumentation channels. Some measurements are with very large error (bad data) due to corrupted communication channels or incorrect configurations, cyberattacks, etc. Third, some important states and health conditions of the system could not be directly measured using today's measuring devices.

State estimation is a powerful method in solving the above issues. It systematically considers the measurement error and model uncertainty by taking full advantage of the measurement redundancy. State estimation can be further classified into static state estimation and dynamic state estimation (DSE) algorithms. For different applications the system of interest can be modeled using either static models with the assumption of pure sinusoidal waveforms and constant phasors while ignoring the dynamics governed by the differential equations, or the transient models using both differential and algebraic equations (DAEs). For the static models the static state estimator (SSE), tracking state estimator, and forecasting-aided state estimator have been widely investigated. By contrast, in the presence of large disturbances, the full consideration of DAEs is required, and this leads to the development of DSE methods. With increasing penetration of renewables and more complex power electronics–based components in modern power systems, different modeling procedures as well as state estimation algorithms should be properly selected for different applications.

This chapter focuses on advanced operational tools based on state estimation for modern power systems. Section 6.2 introduces the model validation tool, where the parameters of large-scale power system models can be effectively validated and calibrated in case they are identified as incorrect. Section 6.3 provides the system monitoring tool based on DSE, where the model uncertainties, measurement errors and bad data are systematically considered and filtered out, resulting in better situational awareness of the system operating states. Section 6.4 demonstrates the DSE-enabled protective relaying tool, where the health condition of the system is carefully examined in real time, and the component of interest can be reliably and quickly protected from faults.

6.2 Model validation

Most advanced applications in power system operation are heavily model-based today. Just to name a few, the performances of static security assessment, optimal power flow, and various control and protection schemes can be severely degraded when the quality of network parameters cannot be guaranteed. Therefore maintaining an accurate network parameter database is always of paramount importance. In practice, reliable network parameter values do not come naturally. A number of reasons can lead to parameter errors: flawed manufacturing data, human entry errors, status update failures, ambient condition changes, and even malicious cyberattacks. Furthermore, cleaning up errors from a practical power network parameter dataset is not a trivial task. There are two major challenges to be addressed:

1. Identification of erroneous parameters from a large model dataset. Large-scale power systems have thousands or tens of thousands of parameters, and the original data from generating these parameters may not be owned by the system operators. The effort can be prohibitively large to conduct an exhaustive search for the erroneous parameters, or even a reliable suspect set.
2. Differentiation between model parameter errors and measurement errors. The most effective avenue to identify parameter errors is to observe the inconsistencies between measurements and the network model. However, when such inconsistencies are observed, it is often difficult to determine whether the errors lie in the measurements or the network model. Misidentification of the error sources can lead to mistaken correction of good quantities, leading to worse situations.

In this section an effective tool for the detection, identification, and correction of model parameter errors based on state estimation will be presented. That is, the Largest Normalized Lagrange Multiplier (LNLM) test. Obviously, this is not the only avenue to tackle this problem. Since state estimation is the natural bridge between power system models and measurements, there has been a long history of exploiting its capability in improving the quality of power system models [12–20]. The LNLM test is one of the few approaches which systematically address the aforementioned two major challenges, thus having great potential for implementation in large-scale real-world power systems [15–20]. In this section the basic principles and procedures of the LNLM test will be presented. This will be followed by a computationally efficient implementation of the LNLM test, and the detectability and identifiability theory of model parameter errors.

6.2.1 Largest Normalized Lagrange Multiplier test

6.2.1.1 Extraction of Lagrange multipliers from state estimation problem

Suppose the parameter vector in the model is $\mathbf{p}$ but the true parameter vector is $\mathbf{p}_t$. Write the parameter error vector $\mathbf{p}_e$ as

$$\mathbf{p}_e = \mathbf{p} - \mathbf{p}_t \tag{6.1}$$

The measurement vector will then be written as a function of states and parameter errors as follows:

$$\mathbf{z} = \mathbf{h}(\mathbf{x}, \mathbf{p}_e) + \mathbf{e} \tag{6.2}$$

where $\mathbf{z}$ is the measurement vector, $\mathbf{x}$ is the state vector, and $\mathbf{e}$ is the measurement error vector, and $\mathbf{h}$ is the nonlinear function linking $\mathbf{x}$ and $\mathbf{p}_e$ to $\mathbf{z}$. The widely applied weighted least squares (WLS) state estimation problem can be formulated as the following form:

$$\min_{\mathbf{x},\ \mathbf{p}_e} \quad J(\mathbf{x}, \mathbf{p}_e) = \frac{1}{2}(\mathbf{z}-\mathbf{h}(\mathbf{x}, \mathbf{p}_e))^T \mathbf{R}^{-1}(\mathbf{z}-\mathbf{h}(\mathbf{x}, \mathbf{p}_e)) \quad \text{s.t.} \quad \mathbf{p}_e = \mathbf{0} \tag{6.3}$$

where $\mathbf{R}$ is the covariance matrix of the measurement vector. Initially, there is no reason to suspect any parameters, thus $\mathbf{p}_e$ is set equal to a zero vector. Note that Eq. (6.3) is completely equivalent to the traditional WLS state estimation formulation where parameter errors are not considered [21]

$$\min_{\mathbf{x}} J(\mathbf{x}) = \frac{1}{2}(\mathbf{z}-\mathbf{h}(\mathbf{x}))^T \mathbf{R}^{-1}(\mathbf{z}-\mathbf{h}(\mathbf{x})) \tag{6.4}$$

In Eq. (6.4) the constraints can be eliminated by forming the Lagrangian:

$$L(\mathbf{x}, \mathbf{p}_e, \boldsymbol{\lambda}) = \frac{1}{2}(\mathbf{z}-\mathbf{h}(\mathbf{x}, \mathbf{p}_e))^T \mathbf{R}^{-1}(\mathbf{z}-\mathbf{h}(\mathbf{x}, \mathbf{p}_e)) - \boldsymbol{\lambda}^T \mathbf{p}_e \tag{6.5}$$

where $\boldsymbol{\lambda}$ is the Lagrange multiplier vector associated with $\mathbf{p}_e$. The first-order necessary condition for optimality will yield:

$$\frac{\partial L}{\partial \mathbf{p}} = \mathbf{H}_p^T \mathbf{R}^{-1}(\mathbf{z}-\mathbf{h}(\mathbf{x}, \mathbf{p}_e)) + \boldsymbol{\lambda} = \mathbf{0} \tag{6.6}$$

where $\mathbf{H}_p$ is the Jacobian matrix of measurement function $\mathbf{h}(\mathbf{x}, \mathbf{p}_e)$ with respect to network parameter vector $\mathbf{p}$. At the solution point $\hat{\mathbf{x}}$, define the residual vector $\mathbf{r}$ as

$$\mathbf{r} = \mathbf{z} - \mathbf{h}(\hat{\mathbf{x}}, \mathbf{p}_e) \tag{6.7}$$

Then the Lagrange multiplier vector can be recovered by

$$\boldsymbol{\lambda} = -\mathbf{H}_p^T \mathbf{R}^{-1} \mathbf{r} \tag{6.8}$$

Mathematically, Lagrange multipliers show the increment of the objective function in response to disturbances of the constraints. Hence, if the Lagrange multiplier corresponding to a model parameter error is large, it infers that changing the estimate of this parameter error will lead to significant reduction of the objective function, that is, the initial assumption that this error is zero may be invalid. Therefore the Lagrange multipliers convey useful information for detection of model parameter errors [15,17].

6.2.1.2 Normalized Lagrange multipliers and hypothesis testing

In order to develop a rigorous procedure for parameter error detection using the information conveyed by Lagrange multipliers, the quantitative relationship between Lagrange multipliers and parameter errors should be derived. A linearized measurement model can be used for this purpose:

$$\Delta \mathbf{z} = \mathbf{H}\Delta \mathbf{x}' + \mathbf{H}_p \mathbf{p}_e + \mathbf{e} \tag{6.9}$$

where $\Delta \mathbf{x}'$ is the increment of the erroneous state vector $\mathbf{x}'$ when the parameter errors are not properly processed (initial assumption of zero parameter errors), $\mathbf{H}$ is the Jacobian matrix of measurement function

$\mathbf{h}(\mathbf{x}, \mathbf{p}_e)$ with respect to network parameter vector $\mathbf{p}$. The WLS estimate is given by

$$\Delta\hat{\mathbf{x}}' = \mathbf{G}^{-1}\mathbf{H}^T\mathbf{R}^{-1}(\Delta\mathbf{z} - \mathbf{H}_p\mathbf{p}_e) \tag{6.10}$$

where $\mathbf{G}$, the so-called gain matrix, is defined by

$$\mathbf{G} = \mathbf{H}^T\mathbf{R}^{-1}\mathbf{H} \tag{6.11}$$

Hence, the estimated measurement vector can be written as

$$\begin{aligned}\Delta\hat{\mathbf{z}}' &= \mathbf{H}\Delta\hat{\mathbf{x}}' + \mathbf{H}_p\mathbf{p}_e \\ &= \mathbf{H}\mathbf{G}^{-1}\mathbf{H}^T\mathbf{R}^{-1}\Delta\mathbf{z} - \mathbf{H}\mathbf{G}^{-1}\mathbf{H}^T\mathbf{R}^{-1}\mathbf{H}_p\mathbf{p}_e + \mathbf{H}_p\mathbf{p}_e \\ &= \mathbf{K}\Delta\mathbf{z} + \mathbf{S}\mathbf{H}_p\mathbf{p}_e\end{aligned} \tag{6.12}$$

where

$$\mathbf{K} = \mathbf{H}\mathbf{G}^{-1}\mathbf{H}^T\mathbf{R}^{-1} \tag{6.13}$$

$$\mathbf{S} = \mathbf{I} - \mathbf{K} \tag{6.14}$$

Meanwhile, the true linearized measurement model, which is free of parameter errors, can be written as

$$\Delta\mathbf{z} = \mathbf{H}\Delta\mathbf{x} + \mathbf{e} \tag{6.15}$$

Substituting Eq. (6.9) into Eq. (6.12) will yield

$$\begin{aligned}\mathbf{r} &= \Delta\mathbf{z} - \Delta\hat{\mathbf{z}}' \\ &= \Delta\mathbf{z} - \mathbf{K}\Delta\mathbf{z} - \mathbf{S}\mathbf{H}_p\mathbf{p}_e \\ &= \mathbf{S}(\Delta\mathbf{z} - \mathbf{H}_p\mathbf{p}_e) \\ &= \mathbf{S}(\mathbf{H}\Delta\mathbf{x} + \mathbf{e} - \mathbf{H}_p\mathbf{p}_e) \\ &= \mathbf{S}\mathbf{e} - \mathbf{S}\mathbf{H}_p\mathbf{p}_e \\ &= \mathbf{S}\mathbf{e} + \mathbf{B}\mathbf{p}_e\end{aligned} \tag{6.16}$$

where

$$\mathbf{B} = -\mathbf{S}\mathbf{H}_p \tag{6.17}$$

The second to last step of Eq. (6.15) is due to the fact that

$$\mathbf{S}\cdot\mathbf{H} = (\mathbf{I} - \mathbf{K})\mathbf{H} = \mathbf{H} - \mathbf{H} = \mathbf{0} \tag{6.18}$$

which can be easily verified.

Eq. (6.16) gives insight into the relationship between measurement residual and the two types of errors: measurement errors and parameter errors. It is noted that besides the familiar term $\mathbf{S} \cdot \mathbf{e}$ from conventional bad data analysis [21], there is an extra term $-\mathbf{S} \cdot \mathbf{H}_p \cdot \mathbf{p}_e$ that accounts for the impact of parameter errors on the measurement residual vector $\mathbf{r}$.

Combining Eq. (6.16) with Eq. (6.8) will yield

$$\begin{aligned} \boldsymbol{\lambda} &= -\mathbf{H}_p^T \mathbf{R}^{-1} \mathbf{r} \\ &= -\mathbf{H}_p^T \mathbf{R}^{-1} \left(\mathbf{Se} - \mathbf{SH}_p \mathbf{p}_e \right) \\ &= \mathbf{H}_p^T \mathbf{R}^{-1} \mathbf{SH}_p \mathbf{p}_e - \mathbf{H}_p^T \mathbf{R}^{-1} \mathbf{Se} \\ &= \mathbf{\Lambda} \mathbf{p}_e + \mathbf{Ae} \end{aligned} \tag{6.19}$$

where

$$\mathbf{A} = -\mathbf{H}_p^T \mathbf{R}^{-1} \mathbf{S} \tag{6.20}$$

$$\mathbf{\Lambda} = \mathbf{H}_p^T \mathbf{R}^{-1} \mathbf{SH}_p \tag{6.21}$$

Eq. (6.19) shows the relationship between Lagrange multipliers and the two types of errors. In particular, $\mathbf{\Lambda}$ is the sensitivity matrix linking $\boldsymbol{\lambda}$ to $\mathbf{p}_e$. Similar to measurement residuals, Lagrange multipliers are also linear combinations of the errors. The major difference lies in the fact that $\boldsymbol{\lambda}$ has one-to-one correspondence to $\mathbf{p}_e$, whereas $\mathbf{r}$ has one-to-one correspondence to $\mathbf{e}$. The sensitivity matrix between $\boldsymbol{\lambda}$ and $\mathbf{p}_e$, $\mathbf{\Lambda}$, and the sensitivity matrix between $\mathbf{r}$ and $\mathbf{e}$, $\mathbf{S}$, are both square matrices. This property implies that it is more convenient to use measurement residuals to analyze measurement errors and use Lagrange multipliers to analyze parameter errors.

After deriving the relationship between Lagrange multipliers and parameter errors, hypothesis testing can be formulated for the detection of parameter errors. Unlike analog measurements that commonly carry random noise, network parameters are relatively constant. Therefore parameter errors can be considered as deterministic variables, and $\mathbf{p}_e$ can be considered as a vector whose entries have either zero or nonzero deterministic values.

When $\mathbf{p}_e = \mathbf{0}$, Eq. (6.19) reduces to

$$\boldsymbol{\lambda} = -\mathbf{H}_p^T \mathbf{R}^{-1} \mathbf{Se} \tag{6.22}$$

Assuming Gaussian measurement errors, Lagrange multipliers will also obey Gaussian distribution with zero mean, whose covariance matrix is given by

$$\begin{aligned}\text{cov}(\boldsymbol{\lambda}) &= E\left(\boldsymbol{\lambda}\boldsymbol{\lambda}^T\right)\\ &= E\left(\mathbf{H}_p^T\mathbf{R}^{-1}\mathbf{r}\mathbf{r}^T\mathbf{R}^{-T}\mathbf{H}_p\right)\\ &= \mathbf{H}_p^T\mathbf{R}^{-1}\text{cov}(\mathbf{r})\mathbf{R}^{-T}\mathbf{H}_p\\ &= \mathbf{H}_p^T\mathbf{R}^{-1}(\mathbf{SR})\mathbf{R}^{-T}\mathbf{H}_p\\ &= \mathbf{H}_p^T\mathbf{R}^{-1}\mathbf{S}\mathbf{H}_p = \boldsymbol{\Lambda}\end{aligned} \tag{6.23}$$

The third to last step uses the result $\text{cov}(\mathbf{r}) = \mathbf{SR}$ [21]. It is interesting to note that the covariance matrix of $\boldsymbol{\lambda}$ happens to be the sensitivity matrix of $\boldsymbol{\lambda}$ with respect to $\mathbf{p_e}$.

For a Lagrange multiplier (taking the *i*th one as example), normalization can be performed as follows:

$$\lambda_i^N = \frac{\lambda_i}{\sqrt{\Lambda_{ii}}} \tag{6.24}$$

where Λ_{ii} is the *i*th diagonal entry of $\boldsymbol{\Lambda}$. The resulting variables are referred to as the normalized Lagrange multipliers (NLMs). They will obey standard normal distribution:

$$H_0: \quad \lambda_i^N \sim N(0,\ 1) \tag{6.25}$$

This is the null hypothesis of the LNLM test.

If $\mathbf{p}_e \neq 0$ but is deterministic, Lagrange multipliers will obey Gaussian distribution with nonzero mean values:

$$E(\boldsymbol{\lambda}) = E\left(\boldsymbol{\Lambda}\mathbf{p}_e\right) + E\left(-\mathbf{H}_p^T\mathbf{R}^{-1}\mathbf{Se}\right) = \boldsymbol{\Lambda}\mathbf{p}_e \tag{6.26}$$

and the covariance is given by

$$\begin{aligned}\text{cov}(\boldsymbol{\lambda}) &= \text{cov}\left(\mathbf{H}_p^T\mathbf{R}^{-1}\mathbf{S}\mathbf{H}_p\mathbf{p}_e + \mathbf{H}_p^T\mathbf{R}^{-1}\mathbf{Se}\right)\\ &= \text{cov}\left(\mathbf{H}_p^T\mathbf{R}^{-1}\mathbf{Se}\right)\\ &= \mathbf{H}_p^T\mathbf{R}^{-1}(\mathbf{SR})\mathbf{R}^{-T}\mathbf{H}_p\\ &= \mathbf{H}_p^T\mathbf{R}^{-1}\mathbf{S}\mathbf{H}_p = \boldsymbol{\Lambda}\end{aligned} \tag{6.27}$$

Comparing Eq. (6.27) and Eq. (6.23) it is clear that the covariance is independent of parameter errors. Suppose only the *i*th parameter carries error. Normalizing λ_i by Eq. (6.24) yields

$$H_1: \quad \lambda_i^N \sim N\left(\sqrt{\Lambda_{ii}}p_{e,i}, 1\right) \tag{6.28}$$

This is the alternative hypothesis of the test. NLMs should be checked against a critical value to determine whether the null hypothesis should be accepted. Since the distribution of H_0 is symmetric with respect the vertical axis, a positive threshold t can be chosen and compared against $|\lambda_i^N|$.

6.2.1.3 Detection, identification, and correction of model parameter errors

The detection of model parameter errors is accomplished by differentiating between the null hypothesis H_0 and the alternative hypothesis H_1. Choosing a desirably low probability of creating false alarms, α, the threshold for parameter error detection can be chosen as

$$t = \Phi^{-1}\left(\frac{\alpha}{2}\right) \tag{6.29}$$

where Φ is the cumulative distribution function of standard normal distribution.

When an error is detected, the next step is to identify the source of the error, that is, which parameter carries the error. From Eq. (6.19), it is known that the Lagrange multipliers are correlated, so a single parameter error not only leads to a large value of its corresponding NLM but also causes large values of the NLMs corresponding to other parameters. Thus observing a large NLM does not always imply that its corresponding parameter is erroneous. For the identification of parameter errors the "LNLM" criterion can be used: the parameter corresponding to the LNLM is identified as erroneous. The theoretical foundation of the LNLM criterion is the theorem shown below.

Theorem 6.1. *Suppose there is a single erroneous parameter, and all other parameters and measurements are correct. The normalized Lagrange multiplier corresponding to the error will have the largest absolute value among all the normalized Lagrange multipliers of parameters in the system.*

The proof of the theorem can be found in Ref. [17].

As mentioned at the beginning of the subsection, one of the major challenges in the parameter error identification problem is the differentiation between parameter errors and measurement error. This problem can be addressed by applying the LNLM test jointly with the largest normalized residual (LNR) test. The LNR test is a well-known approach for measurement error detection and identification. At the solution point of the WLS state estimation problem, the normalized residual (NR) of a measurement (taking the ith measurement as example) can be obtained by

$$r_i^N = \frac{r_i}{\sqrt{\Omega_{ii}}} \tag{6.30}$$

where Ω_{ii} is the ith diagonal entry of the covariance matrix of the measurement residual vector [21]:

$$\mathbf{\Omega} = \mathbf{SR} \tag{6.31}$$

The measurement errors are identified using the "LNR" criterion, namely, the measurement corresponding to the LNR is identified as erroneous. The theoretical foundation is provided by the following theorem.

Theorem 6.2. *Suppose there is a single erroneous measurement, and all other measurements and parameters are correct. The normalized residual corresponding to the error has the largest absolute value among all the normalized residuals in the system.*

The proof of the theorem can be found in Ref. [21].

Combining the LNLM and LNR criteria, model parameter errors and analog measurement errors can be differentiated. In the combined approach the NLMs and NRs for all the model parameters and analog measurements are computed, respectively. Subsequently, the quantity corresponding to the largest normalized variable is identified as erroneous. In other words, if the largest normalized variable is a NLM, it is a parameter error; otherwise, it is a measurement error. This interesting property is given by the following two theorems.

Theorem 6.3. *Suppose there is a single erroneous parameter, and all other parameters and measurements are correct. The normalized Lagrange multiplier corresponding to the error will have a no smaller absolute value than any of the normalized residuals in the system.*

Theorem 6.4. *Suppose there is a single erroneous measurement, and all other parameters and measurements are correct. The normalized residual corresponding to the error will have a no smaller absolute value than any of the normalized Lagrange multipliers in the system.*

The proof of the above two theorems can be found in Ref. [17].

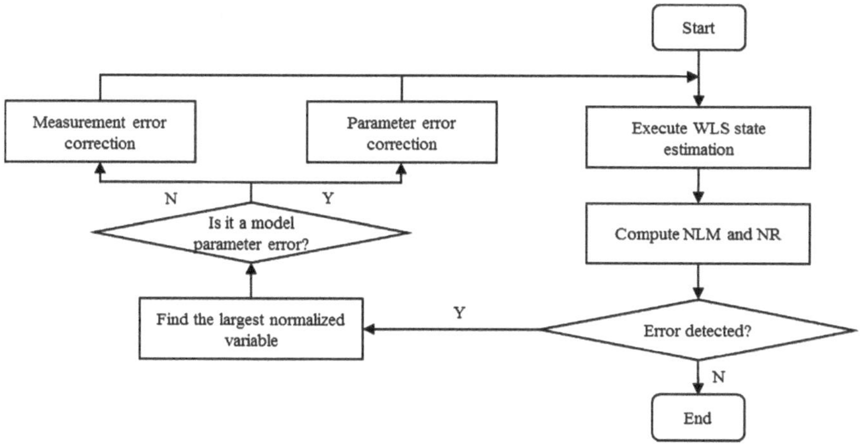

Figure 6.1 Flowchart for model parameter error processing.

After the erroneous parameter or measurement is identified, it needs to be corrected. It is known that an erroneous measurement can be corrected as follows [21]:

$$z_{i,corr} = z_{i,bad} - \frac{R_{ii}}{\Omega_{ii}} r_i \tag{6.32}$$

where $z_{i,corr}$ is the corrected value of the ith measurement, $z_{i,bad}$ is the erroneous value of the ith measurement.

For an erroneous parameter, following Eq. (6.19), the equation below can be used for correction [18]:

$$p_{i,corr} = p_{i,bad} - \frac{1}{\Lambda_{ii}} \lambda_i \tag{6.33}$$

Obviously, using the LNLM/NR criteria, only one erroneous quantity will be identified and corrected. After the correction of the identified error the WLS state estimation needs to be executed again to obtain the unbiased results. In the presence of multiple errors, this process should be repeated until no error is detected. The high-level flowchart of the entire procedure is illustrated in Fig. 6.1.

6.2.2 Computationally efficient implementation of the Largest Normalized Lagrange Multiplier test

In the LNLM test–based approach described in the last subsection, the main computational effort is spent on the computation of the covariance matrix entries. Defining the weight matrix in the WLS state estimation problem as $\mathbf{W} = \mathbf{R}^{-1}$, It can be known from Eqs. (6.13), (6.14), (6.19), and (6.21) that

$$\mathbf{\Lambda} = \mathbf{H}_p^T \ \mathbf{W}\mathbf{H}_p - \mathbf{H}_p^T \ \mathbf{W}\mathbf{H}^T\mathbf{G}^{-1}\mathbf{H}^T\mathbf{W}^T\mathbf{H}_p^T \tag{6.34}$$

Note that the second term on the right-hand side of Eq. (6.34) involves inversion of the gain matrix $\mathbf{G}$, and the product of the inverse and two Jacobian matrices. Both the matrix inversion and multiplication operations consume a significant amount of CPU time and memory. However, since only the diagonal entries of $\mathbf{\Lambda}$ are actually needed, and $\mathbf{H}_p$, $\mathbf{H}$, and $\mathbf{W}$ are all highly sparse, only a small subset of the entries in $\mathbf{G}^{-1}$ are actually needed and used. In this subsection the structure of this subset will be determined first, and then a method for computing this subset efficiently will be presented.

Denote the second term of Eq. (6.34) as

$$\mathbf{\Lambda}' = \mathbf{H}_p^T \ \mathbf{W}\mathbf{H}^T\mathbf{G}^{-1}\mathbf{H}^T\mathbf{W}^T\mathbf{H}_p \tag{6.35}$$

and

$$\mathbf{\Omega}' = \mathbf{H}\mathbf{G}^{-1}\mathbf{H}^T \tag{6.36}$$

such that

$$\mathbf{\Lambda}' = \mathbf{H}_p^T \ \mathbf{W}\mathbf{\Omega}'\mathbf{W}^T\mathbf{H}_p \tag{6.37}$$

To obtain the necessary subset of $\mathbf{G}^{-1}$, first the necessary subset of $\mathbf{\Omega}'$ should be derived. Suppose that the system has n states, m measurements, and u parameters. Using Eq. (6.37), the diagonal entries of $\mathbf{\Lambda}'$ can be expressed as

$$\Lambda'_{ii} = \sum_{k=1}^{u}\sum_{l=1}^{u} H_{p,ki} W_{kk} \Omega'_{kl} W_{ll} H_{p,li} \tag{6.38}$$

Since $\mathbf{H}_p$ is super sparse, only a few terms of the right-hand side are nonzero. Thus in $\mathbf{\Omega}'$, only the entries in the nonzero terms need to be computed. Hence, the necessary subset of $\mathbf{\Omega}'$ can be written as

$$\Omega'_{nec} = \left\{\Omega'_{kl} | H_{p,il} \neq 0; H_{p,ik} \neq 0; i = 1, 2, \ldots, u\right\} \tag{6.39}$$

The required entries in $\mathbf{\Omega}'$ for computing the variance of λ_i are those corresponding to the covariance of measurements related to λ_i, that is, the measurements in whose equations p_i is present. For example, if there are

m_i measurement equations associated with p_i, for computing the variance of λ_i, only $m_i \times m_i = m_i^2$ entries in $\Omega'^{(q)}$ need to be computed. Since only the local measurements are associated with a specific parameter, the number m_i is typically small and the number of necessary entries in $\boldsymbol{\Omega}'$ is also small.

Following similar logic, the structure required subset of $\mathbf{G}^{-1}$ entries can be found based on the structure of the necessary subset of $\boldsymbol{\Omega}'$. Entries of $\boldsymbol{\Omega}'$ can be expressed as

$$\Omega'_{ij} = \sum_{k=1}^{m} \sum_{l=1}^{m} H_{ik} \left[\mathbf{G}^{-1}\right]_{kl} H_{jl} \tag{6.40}$$

Because $\mathbf{H}$ is highly sparse, only a few terms of the right-hand side is nonzero. Thus in $\mathbf{G}^{-1}$, only the entries in the nonzero terms need to be computed. This being given the necessary subset of $\mathbf{G}^{-1}$ can be written as

$$\left(\mathbf{G}^{-1}\right)_{nec} = \left\{ \left[\mathbf{G}^{-1}\right]_{kl} | H_{il} \neq 0; \;\; H_{jk} \neq 0; \;\; \Omega'_{ij} \in \Omega'_{nec} \right\} \tag{6.41}$$

Note that $\mathbf{G}^{-1}$ is the covariance matrix of the state variables $\mathbf{x}$. The required entries in $\mathbf{G}^{-1}$ for computing Ω'_{ij}, that is, the covariance of r_i and r_j, are those corresponding to the covariance of states related to r_i and r_j, that is, the states that appear in the equations of r_i and r_j. For example, if there are n_i states associated with r_i, and n_j states associated with r_j, then for computing Ω'_{ij}, only $n_i \times n_j$ entries in $\boldsymbol{\Omega}'$ need to be computed. Since only the local state variables are associated with a specific measurement, the number n_i and n_j are typically small, so the number of necessary entries in $\mathbf{G}^{-1}$ is also small.

From the procedure described above, it can be seen that the variances of Lagrange multipliers rely only on local information, that is, the covariance of neighboring state variables. Therefore the number of necessary entries in $\mathbf{G}^{-1}$ per parameter is independent of the system size.

Given $(\mathbf{G}^{-1})_{nec}$, Λ'_{ii} can be obtained by Eqs. (6.40) and (6.38) in the reverse direction. Next, the procedure for efficiently computing $(\mathbf{G}^{-1})_{nec}$ will be presented [20].

An efficient algorithm for computing a specific subset of entries in the inverse of a sparse matrix is presented in Refs. [22,23]. After minor modifications, this method can be used to compute any desired subset conveniently.

In a fully observable system the gain matrix, $\mathbf{G}$, is symmetric, nonsingular, and positive definite. Applying Cholesky factorization, it can be decomposed as

$$\mathbf{G} = \mathbf{L}\mathbf{D}\mathbf{L}^T \tag{6.42}$$

where $\mathbf{L}$ is a unit lower triangular matrix, and $\mathbf{D}$ is a diagonal matrix. It is easy to verify that

$$\mathbf{G}^{-1} = \mathbf{D}^{-1}\mathbf{L}^{-1} + \left(\mathbf{I} - \mathbf{L}^T\right)\mathbf{G}^{-1} \tag{6.43}$$

Note that on the right-hand side, $\mathbf{D}^{-1}\mathbf{L}^{-1}$ is lower triangular, and $\mathbf{I} - \mathbf{L}^T$ is strictly upper triangular. Therefore $\mathbf{D}^{-1}\mathbf{L}^{-1}$ is irrelevant to the computation of the upper triangular entries of $\mathbf{G}^{-1}$.

Let us define the sparse inverse subset of $\mathbf{G}^{-1}$ as

$$\left(\mathbf{G}^{-1}\right)_{sp} = \left\{\left[\mathbf{G}^{-1}\right]_{kl} \mid Z_{kl} \neq 0; k, l = 1, 2, \ldots, n\right\} \tag{6.44}$$

where $\mathbf{Z} = \mathbf{L} + \mathbf{D} + \mathbf{L}^T$. It is proven in Ref. [22] that the elements of $(\mathbf{G}^{-1})_{sp}$ can be computed in terms of the nonzero entries of $\mathbf{L}$ and other entries in $(\mathbf{G}^{-1})_{sp}$:

$$\left[\mathbf{G}^{-1}\right]_{ij} = -\sum_{k=i+1}^{n} L_{ki}\left[\mathbf{G}^{-1}\right]_{kj} \quad (i<j) \tag{6.45}$$

$$\left[\mathbf{G}^{-1}\right]_{ii} = D_{ii}^{-1} - \sum_{k=i+1}^{n} L_{ki}\left[\mathbf{G}^{-1}\right]_{ki} \tag{6.46}$$

$$\left[\mathbf{G}^{-1}\right]_{ij} = \left[\mathbf{G}^{-1}\right]_{ji} \quad (i>j) \tag{6.47}$$

In other words the computation of the entries in $(\mathbf{G}^{-1})_{sp}$ is independent of the rest of $\mathbf{G}^{-1}$. Starting from $[\mathbf{G}^{-1}]_{nn}$, all the entries in $(\mathbf{G}^{-1})_{sp}$ can be computed using Eqs. (6.45)–(6.47) iteratively.

In general, the subsets $(\mathbf{G}^{-1})_{nec}$ and $(\mathbf{G}^{-1})_{sp}$ are not equal, and there is no guarantee that entries of $(\mathbf{G}^{-1})_{nec}$ necessarily belong to $(\mathbf{G}^{-1})_{sp}$. However, it is fortunate that all the nonzero entries in $\mathbf{G}$ have locations corresponding to entries belonging to $(\mathbf{G}^{-1})_{sp}$. In other words, if $G_{ij} \neq 0$, we have $[\mathbf{G}^{-1}]_{ij} \in (\mathbf{G}^{-1})_{sp}$. Hence, to let $(\mathbf{G}^{-1})_{nec} \in (\mathbf{G}^{-1})_{sp}$, simply let the entries in $\mathbf{G}$ with locations corresponding to elements in $(\mathbf{G}^{-1})_{nec}$ be recorded as nonzeros (zeros in sparse matrices are generally not stored), while they may remain zero numerically. In this way, all the entries of $(\mathbf{G}^{-1})_{nec}$ will belong to the expanded $(\mathbf{G}^{-1})_{sp}$ and computed by Eqs. (6.45)–(6.47).

The algorithm of computing NLMs is summarized as below.

1. Perform state estimation Eq. (6.3).

2. Compute $\boldsymbol{\lambda}$ by Eq. (6.8).
3. Determine the structure of Ω'_{nec} by Eq. (6.39).
4. Determine the structure of $(\mathbf{G}^{-1})_{nec}$ by Eq. (6.41).
5. Add the entries in $\mathbf{G}$ corresponding to the elements of $(\mathbf{G}^{-1})_{nec}$ as nonzero.
6. Factorize $\mathbf{G}$ as Eq. (6.42).
7. Compute the elements of $(\mathbf{G}^{-1})_{nec}$ by Eqs. (6.45)–(6.47).
8. Compute the elements of Ω'_{rec} by Eq. (6.40).
9. Compute the diagonals of $\boldsymbol{\Lambda}'$ by Eq. (6.38).
10. Compute the diagonals of $\boldsymbol{\Lambda}$ by Eq. (6.34).
11. Compute every entry of $\boldsymbol{\lambda}^N$ by Eq. (6.24).

6.2.3 Detectability and identifiability of parameter and measurement errors

The general conditions for detectability and identifiability of parameter errors are derived in Ref. [17]. These can be viewed as natural extensions of the corresponding conditions derived for analog measurement errors in Ref. [21].

Definition 6.1. *If* $\boldsymbol{\Lambda}$ *matrix in* Eq. (6.21) *has a "null" column, the corresponding parameter will be defined as "critical."*

Theorem 6.5. *Errors in critical parameters are not detectable.*

Definition 6.2. *If two columns of* $\boldsymbol{\Lambda}$ *in* Eq. (6.21) *are linearly dependent, the corresponding parameters will be defined as a "critical parameter pair." In a similar vein, if a column of* $\boldsymbol{\Lambda}$ in Eq. (6.21) and a column of $\mathbf{A}$ in Eq. (6.20) are linearly dependent, the corresponding parameter and measurement will be defined as a "critical parameter-measurement pair."

Theorem 6.6. *Errors in a critical pair are not identifiable.*

Definition 6.3. *Assume that* $k = p + q$*, then* p parameters and q measurements will be defined as a critical k-tuple, if the corresponding p columns of $\boldsymbol{\Lambda}$ defined by Eq. (6.21) and q columns of $\mathbf{A}$ defined by Eq. (6.20) are linearly dependent.

Theorem 6.7. *Errors in a critical parameter-measurement k-tuple are not identifiable.*

The detailed proof of the theorems given earlier can be found in Ref. [17].

When the error in a parameter is not detectable, it is impossible to evaluate the NLM, since the denominator of Eq. (6.24) will be zero. When the errors in a group of parameters and/or measurements are not identifiable, their NLMs and/or NRs will always have the same value; thus it is impossible to find the truly erroneous quantities using the LNLM/NR criteria.

6.3 System monitoring

The widespread deployment of PMUs on power transmission grids has made possible the real-time monitoring and control of power system dynamics. However, these functions cannot be reliably achieved without the development of a fast and robust DSE method. In this subsection the motivations, the concepts, as well as the methods used for DSE are discussed.

6.3.1 Motivations for dynamic state estimation

With the increasing penetration of distributed energy resources (DERs), flexible loads, and microgrids, the power system has been subject to emerging dynamics. For example, the stochastic and intermittent characteristics of DERs increase the probability of sudden changes in the bus voltage phasors in a short time-frame [24]. These changes are mainly driven by the abrupt changes in active and/or reactive power injections, which further cause changes in the generator's state variables, that is, rotor speeds and rotor angles. The majority of today's monitoring and control tools at the control center EMS are based on steady-state power system models, such as SSE, which cannot capture the system dynamics [25]. This limitation is primarily due to the fact that the EMS functions rely on the SCADA systems that have slow scan rates and no timestamps. Therefore the state estimates are updated every few seconds to minutes, and most of the control schemes associated with the generators or the FACTS devices are based on locally available information and measurements. To address these challenges the SSE needs to be revisited and new monitoring tool, that is, DSE is imperative. The benefits of using DSE include but are not limited to the following:

- Improved oscillations monitoring: The estimated dynamic state variables can be used to carry out modal analysis [26], and the identified modes can then be utilized to adaptively tune power system stabilizers (PSS), thereby achieving better damping of interarea modes of oscillation and improving system stability. Recall that the effectiveness of conventional PSS using local measurements may be limited by the observability of the modes in the signal. Using the estimated states of entire regions as opposed to local measurements will increase the stabilizer's response to interarea modes if the generator has significant influence on such modes. Note that there are several ways for monitoring the entire regions/systems via DSE, namely, the hierarchical and distributed DSE and the centralized DSE using high-performance computing technique [27,28] or reduced order model of the interested areas of the power system [29]. Hierarchical and distributed DSE methods are first implemented locally to monitor small areas, and their results are submitted to the control center for further processing. This is the widely used strategy in the current literature. The high-performance computing technique-based DSE for large-scale systems is in its infancy and merits further research.

- Enhanced hierarchical decentralized control [30–32]: The availability of local and wide-area dynamic states obtained from DSE enables the design of effective local and wide-area controls; for instance, the estimated rotor speed and other states can be used as input signals to control excitation systems of synchronous machines [31,32] or of FACTS devices [30] so as to damp out oscillations. The implementation can be in either fully decentralized or hierarchically decentralized manner; on the other hand, the raw measurements are typically used as direct control inputs, and if the noise level is high, the controller performance could be significantly influenced; if the measurements are corrupted with gross errors, the controllers may even fail and provide misleading results. With the design of robust DSE the measurement noise can be effectively filtered out and the bad data are detected and processed, enhancing the reliability of the developed controllers.
- Improved dependability and reliability of protection systems [33–35]: DSE combines the model outputs as well as the real-time measurements to obtain an optimal estimate of the most likelihood of the interested system states. The systems may be a synchronous machine, a transformer, and a DER unit. By checking the consistency between the model outputs and the measured response, the innovation vector is able to tell us if there is any abnormality. This criterion has been leveraged by researchers to design novel setting-less protection scheme. In particular, by testing the consistency between the PMU or MUs

measurements and the dynamical model of the protection zone, both internal and external faults can be effectively detected without any a priori protection relay settings, yielding more reliable protection systems compared with the traditional coordinated settings-based schemes; the estimated online dynamic states can be utilized to initiate effective generator out-of-step protections [35] and transient stability monitoring based on the extended equal-area criterion or the energy function approach [35]; furthermore, fast state estimation is a prerequisite for the implementation of system integrity protection schemes that can prevent blackouts.

- Enhanced reliability of the system models and initial states utilized for dynamic security assessment (DSA) [36]: DSA requires the availability of accurate models of the generators and their associated controllers, of the composite loads and of the special protection schemes, to name a few. By developing DSE, both the static and dynamic models can be validated [37], and if incorrect parameter values are identified, they can be included as additional state variables in DSE for parameter estimation and calibration, yielding improved system models [38] and more reliable DSA; in addition, with the increasing penetration of DERs and flexible loads, the steady-state assumption may not yield reasonable results of the power flow—based state initiation for transient dynamic simulations. As a result, the DSA results can be misleading and lead to unsafe planning and control coordination. With DSE, the system dynamic states are continuously monitored and can be directly leveraged for DSA initialization.
- Other applications include but are not limited to improved synchrophasor data quality and cyber security leveraging the model information, such as filtering out measurement error that is modeled as Gaussian or non-Gaussian distribution [39,40], detecting bad and delayed measurements or recovering missing data [41]; enhanced synchronous generators coherency identification and dynamic model reduction [42] using the estimated dynamic states and parameters; enhanced bus frequency and center of inertia estimation [43,44]; and the detection of failures in controllers, such as exciter, PSS, and governor.

6.3.2 Problem formulation of dynamic state estimation

For a power system, its characteristics can be described by the following DAEs:

$$\begin{cases} \dot{\mathbf{x}} = \mathbf{f}(\mathbf{x}, \mathbf{y}, \mathbf{u}, \mathbf{p}) \\ \mathbf{0} = \mathbf{c}(\mathbf{x}, \mathbf{y}, \mathbf{u}, \mathbf{p}) \end{cases} \tag{6.49}$$

where $\mathbf{x}$ and $\mathbf{y}$ represent system dynamic and algebraic (or static) state vectors, respectively; note that the algebraic equations include those that are associated with *power flow and generator stators* [29]; $\mathbf{u}$ is the system input vector that drives the state transition; $\mathbf{p}$ represents system parameters; and $\mathbf{f}$ and $\mathbf{c}$ are nonlinear functions. An equivalent but more convenient form of (6.49) for state estimation is expressed as

$$\begin{cases} \dot{\mathbf{x}} = \mathbf{f}(\mathbf{x}, \mathbf{y}, \mathbf{u}, \mathbf{p}) \\ \mathbf{y} = \mathbf{g}(\mathbf{x}, \mathbf{u}, \mathbf{p}) \end{cases} \tag{6.50}$$

where $\mathbf{g}$ is the inverse function of $\mathbf{y}$ that is derived from Eq. (6.49). The use of Eq. (6.50) is because some of the algebraic variables can be measured by PMUs or intelligent electronic devices, and as a result, it is easy to build the measurement functions of dynamic state variables. For example, if the detailed two-axis model with IEEE-DC1A exciter and TGOV1 turbine-governor is considered, we will obtain a ninth-order model expressed by the following DAEs:

Two-axis generator differential equations:

$$\begin{aligned} \frac{2H_i}{\omega_s}\frac{d\omega_i}{dt} &= T_{Mi} - P_{ei} - D(\omega_i - \omega_s)/\omega_s \\ \frac{d\delta_i}{dt} &= \omega_i - \omega_s \\ T'_{doi}\frac{dE'_{qi}}{dt} &= -E'_{qi} - \left(X_{di} - X'_{di}\right)I_{di} + E_{fdi} \\ T'_{qoi}\frac{dE'_{di}}{dt} &= -E'_{di} - \left(X_{qi} - X'_{qi}\right)I_{qi} \end{aligned} \tag{6.51}$$

IEEE-DC1A exciter differential equations:

$$\begin{aligned} T_{Ei}\frac{dE_{fdi}}{dt} &= -\left(K_{Ei} + S_{Ei}\left(E_{fdi}\right)\right)E_{fdi} + V_{Ri} \\ T_{Fi}\frac{dV_{Fi}}{dt} &= -V_{Fi} + \frac{K_{Fi}}{T_{Ei}}V_{Ri} - \frac{K_{Fi}}{T_{Ei}}\left(K_{Ei} + S_{Ei}\left(E_{fdi}\right)\right)E_{fdi} \\ T_{Ai}\frac{dV_{Ri}}{dt} &= -V_{Ri} + K_{Ai}\left(V_{refi} - V_{Fi} - V_i\right) \end{aligned} \tag{6.52}$$

TGOV1 turbine-governor differential equations:

$$\begin{aligned} T_{CHi}\frac{dT_{Mi}}{dt} &= -T_{Mi} + P_{SVi} \\ T_{SVi}\frac{dP_{SVi}}{dt} &= -P_{SVi} + P_{Ci} - \frac{1}{R_{Di}}\left(\frac{\omega_i}{\omega_s} - 1\right) \end{aligned} \tag{6.53}$$

Algebraic equations:

$$\begin{aligned} &V_{di} = V_i\cos\theta_i\sin\delta_i - V_i\sin\theta_i\cos\delta_i \\ &V_{qi} = V_i\cos\theta_i\cos\delta_i + V_i\sin\theta_i\sin\delta_i \\ &I_{di} = \frac{E'_{qi} - V_{qi}}{X'_{di}} \qquad I_{qi} = \frac{V_{di} - E'_{di}}{X'_{qi}} \\ &P_{ei} = V_{di}I_{di} + V_{qi}I_{qi} \qquad Q_{ei} = -V_{di}I_{qi} + V_{qi}I_{di} \end{aligned} \tag{6.54}$$

where ω_i and δ_i are generator rotor speed and angle; E'_{di} and E'_{qi} generator d-axis and q-axis transient voltages; E_{fdi}, V_{Fi}, V_{Ri}, T_{Mi} and P_{SVi} are field voltage, scaled output of the stabilizing transformer and scaled output of the amplifier, synchronous machine mechanical torque, and steam valve position, respectively; X_{di}, X_{qi}, E'_{di}, and X'_{qi} are generator parameters; T'_{doi}, T'_{qoi}, T_{Ei}, T_{Fi}, T_{Ai}, T_{CHi}, and T_{SVi} are time constants; K_{Ei}, K_{Fi}, and K_{Ai} are exciter gains; R_{Di} is governor droop constant; V and θ are the terminal bus voltage magnitude and phase angle, respectively; P_e and Q_e are the active and reactive electrical power outputs; I_d and I_q are the d- and q-axis currents, respectively. V_{refi} and P_{Ci} are the reference voltage and the governor input power. It is clear from Eq. (6.54) that instead of expressing the zero-equality constraints such as Eq. (6.49), the formulation Eq. (6.50) for the real and reactive power of generators can be directly measured by PMUs. This makes the construction of measurement equations for the state-space model much more straightforward.

When the system is mainly driven by small changes of loads and DERs, the generators and other controllers are able to absorb almost "instantly" these slow changes, yielding negligible changes of the dynamic states $\mathbf{x}$, that is, $\dot{\mathbf{x}} \cong \mathbf{0}$. As a result, the system is mathematically characterized by

$$\begin{cases} \mathbf{0} \cong \mathbf{f}(\mathbf{x}, \mathbf{y}, \mathbf{u}, \mathbf{p}) \\ \mathbf{y} = \mathbf{g}(\mathbf{x}, \mathbf{u}, \mathbf{p}) \end{cases} \tag{6.55}$$

where slowly varying algebraic variables $\mathbf{y}$ are of interest. In real-time operation, since system inputs are usually not perfectly known and the parameters are always inaccurate to a certain extent, state estimators capable of processing measurement snapshots are developed, including static, forecasting-aided, and tracking state estimators. It is worth pointing out that at the distribution system levels, individual loads or DERs may change abruptly enough for the local controllers to be subject to disturbances, requiring the dynamic model Eq. (6.50) to be resorted to.

For practical applications the continuous-time models are transformed into their discrete-time state-space forms through numerical integration methods, such as Euler, Trapezoidal, or Runge–Kutta-based ones [30]. For example, after time-discretization, Eq. (6.50) can be approximated as

$$\begin{cases} \mathbf{x}_k = \mathbf{f}\left(\mathbf{x}_{k-1}, \mathbf{y}_{k-1}, \mathbf{u}_k, \mathbf{p}\right) + \mathbf{w}_k \\ \mathbf{y}_k = \mathbf{g}(\mathbf{x}_k, \mathbf{u}_k, \mathbf{p}) + \mathbf{v}_k \end{cases} \tag{6.56}$$

where $\mathbf{w}_k$ and $\mathbf{v}_k$ are error terms. Specifically, $\mathbf{w}_k$ includes model time-discretization errors, model uncertainties, as well as uncertain/unknown inputs. $\mathbf{v}_k$ may include model time-discretization errors, model uncertainties, and measurement errors. Except for the direct measurements of algebraic variables, there are other ones that show either linear or nonlinear relationship with the dynamic state variables, such frequency and current phasors, yielding a complete measurement vector $\mathbf{z}_k$. Then, a more general dynamic state-space model for state estimation is

$$\begin{cases} \mathbf{x}_k = \mathbf{f}\left(\mathbf{x}_{k-1}, \mathbf{y}_{k-1}, \mathbf{u}_k, \mathbf{p}\right) + \mathbf{w}_k \\ \mathbf{z}_k = \mathbf{h}(\mathbf{x}_k, \mathbf{u}_k, \mathbf{p}) + \mathbf{e}_k \end{cases} \tag{6.57}$$

where $\mathbf{h}$ is the nonlinear measurement function; $\mathbf{e}_k$ accounts for measurement error. The noise/errors $\mathbf{w}_k$ and $\mathbf{e}_k$ are usually assumed to be normally distributed with zero means and covariance matrices, $\mathbf{Q}_k$ and $\mathbf{R}_k$. Due to the complex combination of various error terms, they typically do not follow a Gaussian distribution. Note that the conventional RTU measurements can only be used for quasisteady state estimation. For wide-area estimation under transient conditions, synchrophasor measurements may be the only useful source of information. At the local level (e.g., a generator station, a power transformer, or a FACTS controller), digital recorders or dedicated protection devices can also provide the required synchronized information to carry out a DSE.

6.3.3 Unified framework for Bayesian dynamic state estimation through nonlinear regression

In this section the Bayesian state estimators for a general dynamic system DSE is formulated. Then, a unified DSE framework is developed that includes the well-known extended Kalman filter (EKF), unscented Kalman filter (UKF), ensemble Kalman filter (EnKF), particle filter (PF), and their variants. To further enhance their robustness to non-Gaussian noise, bad data, etc., a unified robust DSE framework is proposed.

6.3.3.1 Bayesian state estimators for nonlinear dynamic system models

To estimate the system dynamic states a two-step procedure is applied, namely, prediction step using the first equation in Eq. (6.57), which is a Markov model, and a filtering/update step using the second equation in Eq. (6.57). Given a state estimate at time step $k-1$, $\hat{\mathbf{x}}_{k-1|k-1}$ with its covariance matrix $\mathbf{P}_{k-1|k-1}$, the predicted state is calculated from the first equation in Eq. (6.57) directly or through a set of points drawn from the distribution that are following the probability distribution of $\hat{\mathbf{x}}_{k-1|k-1}$, which is dependent on the assumed probability distributions of $\mathbf{w}_k$. As for the filtering step, the predictions are used together with the observations at time sample k to estimate the state vector and its covariance matrix. This can be casted as a Bayesian estimation approach as explained next. Note that $\mathbf{u}$ and $\mathbf{p}$ have been dropped for simplicity in the sequel.

From a Bayesian perspective the DSE problem consists of recursively calculating some degree of belief in the state vector $\mathbf{x}_k$ at time k, given the data $\mathbf{z}_{1:k}$ up to time k. Thus it is required to calculate the conditional probability density function (pdf), $p(\mathbf{x}_k|\mathbf{z}_{1:k})$. In principle, $p(\mathbf{x}_k|\mathbf{z}_{1:k})$ may be obtained recursively in two steps, which are the prediction and the update step. Suppose that $p(\mathbf{x}_{k-1}|\mathbf{z}_{1:k-1})$ at time $k-1$ is available. Thus the prediction step involves applying the first equation in Eq. (6.57) to obtain the prior *pdf* of the state at time k using the Chapman–Kolmogorov equation

$$p(\mathbf{x}_k|\mathbf{z}_{1:k-1}) = \int p(\mathbf{x}_k|\mathbf{x}_{k-1})p(\mathbf{x}_{k-1}|\mathbf{z}_{1:k-1})d\mathbf{x}_{k-1}. \quad (6.58)$$

The probabilistic model of the state evolution $p(\mathbf{x}_k|\mathbf{x}_{k-1})$ is defined by the system equation and the assumed probability distribution of $\mathbf{w}_k$. At time k, when the measurements/observations are available, the update step can be performed via Bayes' rule expressed as

$$p(\mathbf{x}_k|\mathbf{z}_{1:k}) = \frac{p(\mathbf{z}_k|\mathbf{x}_k)p(\mathbf{x}_k|\mathbf{z}_{1:k-1})}{p(\mathbf{z}_k|\mathbf{z}_{1:k-1})}, \tag{6.59}$$

where the normalizing constant is given by

$$p(\mathbf{z}_k|\mathbf{z}_{1:k-1}) = \int p(\mathbf{z}_k|\mathbf{x}_k)p(\mathbf{x}_k|\mathbf{z}_{1:k-1})d\mathbf{x}_k, \tag{6.60}$$

depends on the likelihood function $p(\mathbf{z}_k|\mathbf{x}_k)$, which is defined by the observation model in the second equation of Eq. (6.57) and the assumed probability distribution of $\mathbf{e}_k$.

The recurrence relations Eqs. (6.58) and (6.59) form the basis for the optimal Bayesian solution. However, this recursive propagation of the posterior density is only a conceptual solution in general; it cannot be determined analytically for the nonlinear dynamic system model. It is in this sense that the EKF [45,46], UKF [47–49], EnKF [50,51], PF [52–54], and their variants only approximate the optimal Bayesian solution under the Gaussian noise assumption. Furthermore, when the dynamic system model is correct, the better the statistical efficiency (accuracy) and the robustness of the previous state vector estimates and its error covariance matrix, the better the efficiency and robustness of the state prediction. On the other hand, the quality of the filtered/updated state vector relies heavily on the quality of the received observations. Therefore to obtain reliable state estimates for dynamic systems, robust state prediction and state filtering with good efficiency are required.

6.3.3.2 Proposed unified framework for nonlinear Bayesian dynamic state estimation

Let us define $\hat{\mathbf{x}}_{k|k-1} = \mathbf{x}_k - \zeta_k$, where $\hat{\mathbf{x}}_{k|k-1}$ is the predicted state from iterated EKF (IEKF), UKF, EnKF, PF or their variants; $\mathbf{x}_k$ is the true state at time sample k; ζ_k is the error between the true state and its prediction and $\mathrm{E}\left[\zeta_k\zeta_k^T\right] = \mathbf{P}_{k|k-1}$ is the prediction error covariance matrix. The objective of the *maximum a posterior* estimation is to find the state vector $\boldsymbol{x}_k$ that maximizes the probability $p(\mathbf{z}_k|\mathbf{x}_k)p(\mathbf{x}_k|\mathbf{z}_{1:k-1})$, yielding the following maximum likelihood objective function under Gaussian assumption:

$$J(\mathbf{x}_k) = (\mathbf{z}_k - \mathbf{h}(\boldsymbol{x}_k))^T \mathbf{R}_k^{-1} (\mathbf{z}_k - \mathbf{h}(\boldsymbol{x}_k)) + \left(\mathbf{x}_k - \hat{\mathbf{x}}_{k|k-1}\right)^T \mathbf{P}_{k|k-1}^{-1} \left(\mathbf{x}_k - \hat{\mathbf{x}}_{k|k-1}\right) \tag{6.61}$$

where $\hat{\mathbf{x}}_{k|k-1}$ is the predicted state that is calculated using Eq. (6.58) and $\mathbf{P}_{k|k-1}$ is the prediction error covariance matrix. Instead of solving for Eq. (6.61) as it is usually performed, we first put the regression model as

$$\begin{bmatrix} \mathbf{z}_k \\ \hat{\mathbf{x}}_{k|k-1} \end{bmatrix} = \begin{bmatrix} \mathbf{h}(\boldsymbol{x}_k) \\ \boldsymbol{x}_k \end{bmatrix} + \begin{bmatrix} \mathbf{e}_k \\ -\zeta_k \end{bmatrix}, \tag{6.62}$$

which can be put in matrix form as

$$\tilde{\mathbf{z}}_k = \tilde{\mathbf{h}}(\mathbf{x}_k) + \tilde{\mathbf{w}}_k. \tag{6.63}$$

Let $\mathbf{E}\left[\tilde{\mathbf{w}}_k \tilde{\mathbf{w}}_k^T\right] = \begin{bmatrix} \mathbf{R}_k & \mathbf{0} \\ \mathbf{0} & \mathbf{P}_{k|k-1} \end{bmatrix} = \mathbf{S}_k \mathbf{S}_k^T$, where $\mathbf{S}_k$ is calculated using the Cholesky decomposition. We premultiply Eq. (6.63) by $\mathbf{S}_k{}^{-1}$ to get a new regression model $\mathbf{y}_k = \mathbf{m}(\boldsymbol{x}_k) + \boldsymbol{\eta}_k$ with an uncorrelated additive error vector $\boldsymbol{\eta}_k$. Finally we estimate the state vector $\mathbf{x}_k$ by minimizing a quadratic objective function defined as

$$J(\mathbf{x}_k) = \left(\mathbf{y}_k - \mathbf{m}(\boldsymbol{x}_k)\right)^T \left(\mathbf{y}_k - \mathbf{m}(\boldsymbol{x}_k)\right) = \left\|\mathbf{y}_k - \mathbf{m}(\boldsymbol{x}_k)\right\|^2. \tag{6.64}$$

Applying a Gauss–Newton's iterative method to minimize (6.64), we get

$$\hat{\mathbf{x}}_{k|k}^{i+1} = \hat{\mathbf{x}}_{k|k-1} + \left(\mathbf{M}_k^T \mathbf{M}_k\right)^{-1} \mathbf{M}_k^T \left(\mathbf{y}_k - \mathbf{m}\left(\hat{\mathbf{x}}_{k|k}^i\right)\right), \tag{6.65}$$

where $\mathbf{M}_k = \partial \mathbf{m}/\partial \boldsymbol{x}|_{\boldsymbol{x}=\hat{\mathbf{x}}_{k|k}^i} = \mathbf{S}_k^{-1} \begin{bmatrix} \mathbf{H}_k \\ \mathbf{I} \end{bmatrix}$, $\mathbf{H}_k = \partial \mathbf{h}/\partial \boldsymbol{x}|_{\mathbf{x}=\hat{\mathbf{x}}_{k|k}^i}$, $\mathbf{I}$ is the identity matrix and i is the iteration counter. Upon convergence the state error covariance matrix is calculated as $\mathbf{P}_{k|k} = \left(\mathbf{M}_k^T \mathbf{M}_k\right)^{-1}$.

By using the inverse matrix lemma, we can show that Eq. (6.65) is the so-called IEKF [55]. All what we need to do is to apply, in the state filtering step, a first-order Taylor series approximation to the nonlinear measurement function $\mathbf{h}(\boldsymbol{x}_k)$. If the statistical linearization [39,40] is applied using the sigma point, it can be shown this would yield the UKF method. While, if the statistical linearization is applied using the ensembles, it can be shown this would result in the EnKF method.

A general nonlinear regression framework for IEKF, UKF, and EnKF: For these filters, we have the following general expression of the state vector estimate at sample time k:

$$\begin{aligned}\hat{\mathbf{x}}_{k|k}^{i+1} &= \hat{\mathbf{x}}_{k|k-1} + \mathbf{P}_{k|k-1}^{\mathbf{xz}}\left(\mathbf{P}_{k|k-1}^{\mathbf{zz}} + \mathbf{R}_k\right)^{-1} \\ &\left(\mathbf{z}_k - \mathbf{h}\left(\hat{\mathbf{x}}_{k|k}^{i}\right) - \left(\mathbf{P}_{k|k-1}^{\mathbf{xz}}\right)^T \mathbf{P}_{k|k-1}^{-1}\left(\left(\hat{\mathbf{x}}_{k|k-1} - \hat{\mathbf{x}}_{k|k}^{i}\right)\right)\right),\end{aligned} \tag{6.66}$$

with its error covariance matrix given by

$$\mathbf{P}_{k|k} = \mathbf{P}_{k|k-1} - \mathbf{P}_{k|k-1}^{\mathbf{xz}}\left(\mathbf{P}_{k|k-1}^{\mathbf{zz}} + \mathbf{R}_k\right)^{-1}\left(\mathbf{P}_{k|k-1}^{\mathbf{xz}}\right)^T. \tag{6.67}$$

For IEKF, we have

$$\begin{cases}\mathbf{P}_{k|k-1}^{\mathbf{xz}} = \mathbf{P}_{k|k-1}\mathbf{H}_k^T \\ \mathbf{P}_{k|k-1}^{\mathbf{zz}} = \mathbf{H}_k\mathbf{P}_{k|k-1}\mathbf{H}_k^T\end{cases}. \tag{6.68}$$

For UKF, we have

$$\begin{cases}\mathbf{P}_{k|k-1}^{\mathbf{xz}} = \sum_{i=0}^{2n} w_i\left(\chi_{k|k-1}^{i} - \hat{\mathbf{x}}_{k|k-1}\right)\left(\mathbf{Z}_{k|k-1}^{i} - \hat{\mathbf{Z}}_{k|k-1}\right)^T \\ \mathbf{P}_{k|k-1}^{\mathbf{zz}} = \sum_{i=0}^{2n} w_i\left(\mathbf{Z}_{k|k-1}^{i} - \hat{\mathbf{Z}}_{k|k-1}\right)\left(\mathbf{Z}_{k|k-1}^{i} - \hat{\mathbf{Z}}_{k|k-1}\right)^T\end{cases}, \tag{6.69}$$

where n is the number of state variables, $\chi_{k|k-1}^{i}$ are the propagated sigma points through the nonlinear function with their weights w_i, $\mathbf{Z}_{k|k-1}^{i}$ are the predicted observations and $\mathbf{Z}_{k|k-1}^{i} = \mathbf{h}\left(\chi_{k|k-1}^{i}\right)$, and $\hat{\mathbf{Z}}_{k|k-1} = \sum_{i=0}^{2n} w_i\mathbf{Z}_{k|k-1}^{i}$.

For EnKF, we have

$$\begin{cases}\mathbf{P}_{k|k-1}^{\mathbf{xz}} = \mathbf{P}_{k|k-1}\mathbf{H}_k^T = \frac{1}{N}\sum_{i=1}^{N}\left(\xi_{k|k-1}^{i} - \hat{\mathbf{x}}_{k|k-1}\right)\left(\mathbf{Z}_{k|k-1}^{i} - \hat{\mathbf{Z}}_{k|k-1}\right)^T \\ \mathbf{P}_{k|k-1}^{\mathbf{zz}} = \mathbf{H}_k\mathbf{P}_{k|k-1}\mathbf{H}_k^T = \frac{1}{N}\sum_{i=1}^{N}\left(\mathbf{Z}_{k|k-1}^{i} - \hat{\mathbf{Z}}_{k|k-1}\right)\left(\mathbf{Z}_{k|k-1}^{i} - \hat{\mathbf{Z}}_{k|k-1}\right)^T\end{cases}, \tag{6.70}$$

where N is the number of ensembles; $\boldsymbol{\xi}^{i}_{k|k-1}$ are the propagated ensembles through the nonlinear function expressed as

$$\hat{\mathbf{x}}_{k|k-1} = \frac{1}{N}\sum_{i=1}^{N}\boldsymbol{\xi}^{i}_{k|k-1};\ \mathbf{Z}^{i}_{k|k-1} = \mathbf{h}\left(\boldsymbol{\xi}^{i}_{k|k-1}\right);\ \hat{\mathbf{Z}}_{k|k-1} = \frac{1}{N}\sum_{i=1}^{N}\mathbf{Z}^{i}_{k|k-1}. \tag{6.71}$$

As for the PF, it is based on Monte Carlo simulations and the resampling technique, which seems to be a different approach from that of EnKF. However, it has been shown in Ref. [56] that the adaptive Gaussian mixture filter can provide a bridge between the EnKF and the PF. Therefore the general framework can be resorted to as well. In the future work, we will investigate the specific form for PF.

Remark:. The least squares estimator on which the foregoing filters are based on assumes Gaussian noise and is vulnerable to outliers. We will build next a framework that allows us to develop robust DSEs that exhibit good performance when the process and observation noises follow thick-tailed probability distributions.

6.3.3.3 Proposed framework for robustifying the Bayesian dynamic state estimation

If we directly premultiply Eq. (6.63) by the matrix $\mathbf{S}_k$ as proposed to derive the classical DSE, the outliers present in the data will strongly bias the state estimates [55]. Therefore it is necessary to detect and suppress the outliers before that step. This is carried out by calculating the projection statistics [55], PS_i, for each of the two-dimensional row vectors $\mathbf{l}_i^T$, $i = 1, \ldots, m + n$ of the innovation matrix $\tilde{\mathbf{Z}}$ given by

$$\tilde{\mathbf{Z}} = \left[\mathbf{z}_{k-1} - \mathbf{h}(\hat{\mathbf{x}}_{k-1|k-1})\mathbf{z}_k - \mathbf{h}\left(\hat{\mathbf{x}}_{k|k-1}\right)\hat{\mathbf{x}}_{k-1|k-1}\hat{\mathbf{x}}_{k|k-1}\right]. \tag{6.72}$$

Recall that the projection statistics are defined as

$$\mathrm{PS}_i = \max_{\|\mathbf{v}_j\|=1} \frac{|\mathbf{l}_i^T\boldsymbol{\nu}_j - \mathrm{med}_k\left(\mathbf{l}_k^T\mathbf{v}_j\right)|}{1.4826\mathrm{med}_\gamma|\mathbf{l}_\gamma^T\mathbf{v}_j - \mathrm{med}_k\left(\mathbf{l}_k^T\mathbf{v}_j\right)|}, \tag{6.73}$$

where the maximum is calculated over $j = 1, \ldots, m + n$. Here $\mathbf{v}_j$ is the vector of unit length that originates from the coordinate medians, M, and points in the direction of the jth data point, $\mathbf{l}_j$. Once the PS_i are calculated, the outliers that are flagged by the decision rule, $\mathrm{PS}_i > \chi^2_{2,0.975}$, are downweighted using the following weight function:

$$\varpi_i = \min\left(1, \frac{d^2}{\mathrm{PS}_i^2}\right), \tag{6.74}$$

where the value of d is chosen to achieve a good statistical efficiency at the Gaussian distribution (i.e., $d = 1.5$) while ensuring high robustness under contamination [39,40]. Now, we can execute the prewhitening step by premultiplying the regression model given by Eq. (6.63) by $\mathbf{S}_k^{-1}$, yielding

$$\mathbf{S}_k^{-1}\tilde{\mathbf{z}}_k = \mathbf{S}_k^{-1}\tilde{\mathbf{h}}(\mathbf{x}_k) + \mathbf{S}_k^{-1}\tilde{\mathbf{w}}_k, \tag{6.75}$$

which can be rewritten in a compact form as

$$\mathbf{y}_k = \mathbf{m}(\boldsymbol{x}_k) + \boldsymbol{\eta}_k \tag{6.76}$$

where $\mathrm{E}\left[\boldsymbol{\eta}_k\boldsymbol{\eta}_k^T\right] = \mathbf{I}$ and $\mathbf{I}$ is the identity matrix. It should be noted that in the prewhitening step, it is the robust estimation of the covariance matrix $\mathbf{P}_{k-1|k-1}$ based on the total influence function and PS that makes the prediction error covariance matrix $\mathbf{P}_{k|k-1}$ robust, which in turn results in a robust prewhitening matrix $\mathbf{S}_k^{-1}$ against outliers.

At this stage a robust version of the IEKF, UKF, EnKF, PF, and their variants can be derived using the regression model Eq. (6.76). To estimate the state vector $\mathbf{x}_k$, we propose to use a generalized maximum (GM)-likelihood-estimator defined as the minimum of an objective function $J(\mathbf{x})$

$$J(\mathbf{x}) = \sum_{i=1}^{m} \varpi_i^2 \rho\left(r_{S_i}\right), \tag{6.77}$$

where $\rho(.)$ is a nonlinear convex function; ϖ_i is calculated by Eq. (6.74); $r_{S_i} = r_i/s\varpi_i$ is the standardized residual; and $r_i = y_i - m(\boldsymbol{x})$ is the residual. The scale estimator s is defined as $s = b\cdot\mathrm{median}_i|r_i|$ for a symmetric probability distribution and as $s = c\cdot\ \mathrm{lomed}_{i=1,\ldots,m}\,\mathrm{lomed}_{j\neq i}\,|r_i - r_j|$ for an asymmetric probability distribution, where the constant b and c are correction factors to achieve Fisher consistency and unbiasedness at a given probability distribution. Here, lomed denotes the low median defined as the $[(m+1)/2]$-th order statistic out of m numbers and [.] is the integer operator. Regarding the $\rho(\cdot)$ function, a good choice would be the Huber-ρ function [55] expressed as

$$\rho\left(r_{S_i}\right) = \begin{cases} \frac{1}{2} r_{S_i}^2, & \text{for } |r_{S_i}| < \lambda \\ \lambda |r_{S_i}| - \lambda^2/2, & \text{elesewhere} \end{cases}, \tag{6.78}$$

where the breakpoint λ between the quadratic and the linear segment of $\rho(.)$ will be chosen such that a good statistical efficiency is achieved at both the Gaussian and the Laplacian probability distribution. Note that the quadratic segment enables the estimator to discriminate between the good and the bad leverage point since in the case of a good leverage point, $\varpi_i^2 \rho(r_{si}) = (r_i/s)^2/2$, which is the conventional least squares criterion. Recall that a leverage point is a data point distant from the bulk of the point cloud in the space spanned by the row vectors of the Jacobian matrix $\mathbf{M}_k$. The reader is referred to Ref. [55] for further details. In power systems, leverage points are power injection measurements on buses with many incident branches and power flow measurements on lines with relatively small reactance compared to the others.

To minimize Eq. (6.77), one takes its partial derivative and set it equal to zero, yielding

$$\frac{\partial J(\mathbf{x})}{\partial \mathbf{x}} = \sum_{i=1}^{m} -\frac{\varpi_i \mathbf{a}_i}{s} \psi\left(r_{S_i}\right) = \mathbf{0}, \tag{6.79}$$

where $\psi\left(r_{S_i}\right) = \partial \rho\left(r_{S_i}\right)/\partial r_{S_i}$ is the so-called ψ function and $\mathbf{a}_i$ is the ith row of the Jacobian matrix $\mathbf{M}_k$. By dividing and multiplying standardized residual, r_{S_i}, each term of the summation in Eq. (6.79) and putting the expression in a matrix form, we get

$$\mathbf{M}_k{}^T \tilde{\mathbf{Q}}(\mathbf{y} - \mathbf{m}(\boldsymbol{x})) = \mathbf{0}, \tag{6.80}$$

where $\tilde{\mathbf{Q}} = \operatorname{diag}\left(q\left(r_{S_i}\right)\right)$ and $q\left(r_{S_i}\right) = \psi\left(r_{S_i}\right)/r_{S_i}$. Applying a first-order Taylor series expansion to $\mathbf{m}(\boldsymbol{x})$, we get the iteratively reweighted least squares algorithm [55]. As an example, the correction of the state vector at the jth iteration for the GM-IEKF is given by

$$\Delta \hat{\mathbf{x}}_{k|k}^{j+1} = \left(\mathbf{M}^T \tilde{\mathbf{Q}}^{(j)} \mathbf{M}\right)^{-1} \mathbf{M}^T \tilde{\mathbf{Q}}^{(j)} \left(\mathbf{y} - \mathbf{m}\left(\boldsymbol{x}^j\right)\right). \tag{6.81}$$

A good stopping rule is

$$\left\| \Delta \hat{\mathbf{x}}_{k|k}^{j+1} \right\| = \left\| \hat{\mathbf{x}}_{k|k}^{j+1} - \hat{\mathbf{x}}_{k|k}^{j} \right\|_{\infty} \leq 10^{-2}. \tag{6.82}$$

Upon convergence of the iterative algorithm Eq. (6.82) the error covariance matrix $\mathbf{P}_{k|k}$ needs to be updated so that the state forecasting for the next time sample can be performed. To this end the theorem proved in Ref. [55] can be used. It states that when the number of the observation is very large, $\mathbf{P}_{k|k}$ is approximately equal to the asymptotic covariance matrix of the total influence function $\mathbf{IF}(r_S; F)$ of the GM-estimator at the probability distribution F, that is $\mathbf{P}_{k|k} = E[\mathbf{IF}(r_S; F)\mathbf{IF}(r_S; F)^T]$. Recall that the total influence function is defined as the derivative of the estimator, $\hat{\mathbf{x}}_k$, with respect to the fraction of contamination, ε, at the assumed probability distribution F, of the good data points. In the derivations the ε-contamination model, $F^* = (1-\varepsilon)F + \varepsilon\ \Delta_{\mathbf{r}}$, is used. Note that F can be chosen to be the Gaussian distribution, Φ, or the Laplacian distribution or the Cauchy distribution, or any other distribution of the process and observation noise that may be assumed. For instance, for the GM-IEKF, we get

$$\mathbf{P}_{k|k} = \mathbf{E}\left[\mathbf{IF}\cdot\mathbf{IF}^T\right] = \frac{\mathbf{E}_{\Phi}\left[\psi^2\left(r_{S_i}\right)\right]}{\left\{\mathbf{E}_{\Phi}\left[\psi'\left(r_{S_i}\right)\right]\right\}^2}\left(\mathbf{M}_k^T\mathbf{M}_k\right)^{-1}\left(\mathbf{M}_k^T\mathbf{Q}_{\varpi}\mathbf{M}_k\right)\left(\mathbf{M}_k^T\mathbf{M}_k\right)^{-1}, \tag{6.83}$$

where $\mathbf{M}_k$ is the Jacobian matrix at time sample k and $\mathbf{Q}_{\varpi} = \mathrm{diag}\left(\varpi_i^2\right)$. In the proposed GM-estimator framework for robustifying DSEs, the matrix $\tilde{\mathbf{Q}}$ and the weights ϖ_i are used to bound the influence of the residual and of the structural outliers, respectively, guaranteeing a robust estimation of the state vector, while the matrix $\mathbf{Q}_{\varpi}$ contributes to the robust estimation of estimation error covariance matrix.

6.3.3.4 Decentralized versus centralized dynamic state estimation for power system

There are two ways of DSE implementations, namely, the centralized and decentralized ones. Centralized DSE assumes that the system is observable by PMUs and Kron reduction can be carried out to reduce the system to the generators' terminals. In addition, it requires accurate knowledge of each system component parameters as well as the real-time wide-area PMU measurements. By contrast the decentralized DSE is implemented using only local PMU measurements. It assumes that the terminal bus of the interested dynamic components is observed by PMU measurements and local observability of dynamic states is satisfied. If a generator terminal bus is not equipped with PMU, a PMU-based linear state estimator for that local system should be performed first. Then, the estimated measurements at the interested terminal buses can be obtained and the decentralized DSE is executed. Note that the local phasor data concentrator is in charge of communicating and processing the PMU measurements. Centralized DSE allows us to achieve global monitoring and control applications and has good robustness to data quality and security issues as the measurement redundancy is high. However, it has large computational burden and strong assumptions about the accuracy of the whole dynamic system models. By contrast, decentralized DSE only needs local measurements at the terminal buses and the dynamic model of interested components, which is fast to execute and not impacted by the model inaccuracy of other system components. However, the local measurement redundancy is low, and therefore decentralized DSE has difficulties in dealing with PMU data quality and security issues. On the other hand, with a decentralized DSE, only local controls are implemented. If coordinated control is deployed between different local DSEs, additional communication bandwidth is required, making the comparison with the communication cost of the centralized DSE difficult to assess. Choosing between the two implementation schemes depends on the applications and the communication infrastructures being used. It should be noted that for both centralized and decentralized DSE-based applications, time tags of the estimated dynamic state variables are required when the communication network is involved.

6.4 Protective relaying

Sometimes faults may occur inside power systems. When the faults occur, protective relays should quickly operate to isolate the faulted devices (or the protection zone), to maximize service continuity, minimize the extent and time of the power outage, and ensure safety of human beings as well as the overall power system. The basic objectives of a protective relays mainly includes the following aspects. First, the relay should operate reliability, which means that the relay should operate corresponding to all kinds of internal faults in its protection zone (also known as "dependability"), and should not operate during all other scenarios such as external faults or system transients (also known as "security"). Second, the relay should isolate the faulted device as soon as possible (in the order tens of milliseconds, especially in high voltage power systems).

Over the years, protective relays are evolving with higher reliability. Nevertheless, statistically speaking, the industry in the United States and abroad is still experiencing an average of 10% misoperations [57]. The challenges of the present protective relaying methods are summarized as follows: first, the protection functions are based on limited information, that is, the relays are lack of an overall picture of the health condition of the device under protection, which may result in compromised reliability of the relays. Second, complex coordination among different protection functions is often required, which increases the possibility of relay miscoordination and misoperation. Third, there are certain types of faults that are hard to be quickly detected using present protective relays, such as downed conductors in distribution systems, high impedance faults, and faults near the neutral of solid grounded systems.

With the development of substation automation the data-acquisition system could be achieved using the MUs that are separated from the protective relays. Instead of being physically connected to the analog channels, the protective relays are able to directly access the digital measurement data from the MUs through the process bus. This brings new opportunities for improving the state-of-the-art protective relays. First, the protective relays could easily access more measurement channels from the process bus. Second, since the MUs obtain "points on wave" measurements instead of phasor measurements, the information inside each measurement channel is also enhanced.

6.4.1 Theoretical basis of dynamic state estimation–based protection

Therefore with aforementioned challenges and opportunities, a new group of approaches has been proposed in literatures: DSE-based protection [57–66]. The method detects internal faults of the protection zone by checking the consistency between the "points on wave" (instantaneous) measurements and the dynamic model of the protection zone through DSE method. The "points on wave" measurements can include many physical quantities such as voltage, current, torque, rotating speed, and temperature. The dynamic model is simply a set of DAEs describing all the physical laws that the protection zone should obey. With the DSE procedure, a high consistency indicates a healthy condition of the protection zone, while a low consistency indicates an internal fault. In fact, the well-known current differential protection is one of the most reliable and straightforward protection approaches, and it does not require complex coordination among different protection functions. The DSE-based protection method can be treated as a generalization of the current differential protection: instead of only checking the Kirchhoff's Current Laws (KCLs) of the protection zone, the method systematically checks all the physical laws that the protection zone should obey through the DSE procedure. As a result, the method provides a more complete monitoring of the overall protection zone and therefore better protection performances. The DSE-based protection approaches have been successfully applied in many power system devices, including transmission lines [57,58,60,61], microgrid circuits [59,65], transformers [62,64,66], and capacitor banks [63].

6.4.1.1 Dynamic model of the protection zone

The dynamic model of the protection zone can be expressed using a set of DAEs describing all the physical laws that the protection zone should obey. If the equations exhibit nonlinearities higher than second order, they are "quadratized" by the introduction of additional variables. The quadratized model is then cast into a standard syntax, as follows:

$$\begin{aligned}
\mathbf{i}(t) &= \mathbf{Y}_{eqx1}\mathbf{x}(t) + \mathbf{D}_{eqxd1}\frac{d\mathbf{x}(t)}{dt} + \mathbf{C}_{eqc1} \\
\mathbf{0} &= \mathbf{Y}_{eqx2}\mathbf{x}(t) + \mathbf{D}_{eqxd2}\frac{d\mathbf{x}(t)}{dt} + \mathbf{C}_{eqc2} \\
\mathbf{0} &= \mathbf{Y}_{eqx3}\mathbf{x}(t) + \left\{ \begin{matrix} \vdots \\ \mathbf{x}(t)^T \left\langle \mathbf{F}^i_{eqxx3} \right\rangle \mathbf{x}(t) \\ \vdots \end{matrix} \right\} + \mathbf{C}_{eqc3}
\end{aligned} \tag{6.84}$$

where $\mathbf{x}(t)$ and $\mathbf{i}(t)$ represent the states and the terminal measurements of the model, respectively; other matrices ($\boldsymbol{Y}$, $\boldsymbol{D}$, $\boldsymbol{F}$, and $\boldsymbol{C}$) are coefficient matrices. Usually terminal voltages are included in $\mathbf{x}(t)$ as well as other internal states.

Next, the measurements are expressed as functions of the protection zone state:

$$\mathbf{z}(t) = \mathbf{Y}_{zx}\mathbf{x}(t) + \left\{ \begin{matrix} \vdots \\ \mathbf{x}(t)^T \left\langle \mathbf{F}^i_{zx} \right\rangle \mathbf{x}(t) \\ \vdots \end{matrix} \right\} + \mathbf{D}_{zx}\frac{d\mathbf{x}(t)}{dt} + \mathbf{C}_z \tag{6.85}$$

where $\mathbf{z}(t)$ is the measurements of the protection zone, where the zeros (equality constraints) are included as well.

6.4.1.2 Quantification of consistency through dynamic state estimation algorithm

To solve Eq. (6.85), one can use *EKF*, *constraint optimization method* or *unconstraint optimization method*. Numerical experiments show that all three methods provide similar results. Here the unconstraint optimization method is selected as an example.

The first step is to convert the set of DAEs in Eq. (6.85) to a set of algebraic equation without differential terms. Here the quadratic integration method [67] is adopted over a time step of $2\Delta t$. Specifically, the differential equations are integrated over the two intervals $[t-2\Delta t, t]$ and $[t-2\Delta t, t_m]$, where Δt is the sampling interval, and $t_m = t - \Delta t$. This process converts Eq. (6.85) into Eq. (6.86):

$$\begin{aligned}
&\mathbf{z}(t) = \mathbf{Y}_{1m,x}\mathbf{x}(t,t_m) + \left\{ \begin{array}{c} \vdots \\ \mathbf{x}(t,t_m)^T \mathbf{F}^i_{1m,x}\mathbf{x}(t,t_m) \\ \vdots \end{array} \right\} + \mathbf{C}_{1m}(t-2\Delta t) \\
&\mathbf{C}_{1m}(t-2\Delta t) = \mathbf{N}_{1m,x}\mathbf{x}(t-2\Delta t) + \mathbf{M}_{1m}i(t-2\Delta t) + \mathbf{K}_{1m} \\
&\mathbf{z}(t_m) = \mathbf{Y}_{2m,x}\mathbf{x}(t,t_m) + \left\{ \begin{array}{c} \vdots \\ \mathbf{x}(t,t_m)^T \mathbf{F}^i_{2m,x}\mathbf{x}(t,t_m) \\ \vdots \end{array} \right\} + \mathbf{C}_{2m}(t-2\Delta t) \\
&\mathbf{C}_{2m}(t-2\Delta t) = \mathbf{N}_{2m,x}\mathbf{x}(t-2\Delta t) + \mathbf{M}_{2m}i(t-2\Delta t) + \mathbf{K}_{2m}
\end{aligned} \tag{6.86}$$

The compact notation can be obtained as follows:

$$\mathbf{z}(t,t_m) = h(\mathbf{x}(t,t_m)) \tag{6.87}$$

where $\mathbf{z}(t,t_m) = \begin{bmatrix} \mathbf{z}(t) \\ \mathbf{z}(t_m) \end{bmatrix}$ and $\mathbf{x}(t,t_m) = \begin{bmatrix} \mathbf{x}(t) \\ \mathbf{x}(t_m) \end{bmatrix}$.

The second step is to solve the best estimates of state vector $\mathbf{x}(t,t_m)$. The estimation problem is

$$\text{Min}\, J = (h(\mathbf{x}(t,t_m)) - \mathbf{z}(t,t_m))^T \mathbf{W}(h(\mathbf{x}(t,t_m)) - \mathbf{z}(t,t_m)) \tag{6.88}$$

where $\mathbf{W} = \text{diag}\{1/\sigma_1^2, 1/\sigma_2^2, \ldots\}$, and σ_i is the standard deviation of measurement i. Note that the zeros (equality constraints) in $\mathbf{z}(t,t_m)$ are treated as measurements with a very small error (very small standard deviation).

The best state estimate $\hat{\mathbf{x}}(t,t_m)$ is computed by the following iterative algorithm until convergence:

$$\hat{\mathbf{x}}(t,t_m)^{\nu+1} = \hat{\mathbf{x}}(t,t_m)^{\nu} - (\mathbf{H}^T\mathbf{W}\mathbf{H})^{-1}\mathbf{H}^T\mathbf{W}(h(\hat{\mathbf{x}}(t,t_m)^{\nu}) - \mathbf{z}(t,t_m)) \tag{6.89}$$

where the Jacobian matrix $\mathbf{H} = \partial h(\mathbf{x})/\partial \mathbf{x}$.

After solving $\hat{\mathbf{x}}(t, t_m)$ the residuals $\hat{\mathbf{r}}(t, t_m)$ and NRs $\hat{\mathbf{s}}(t, t_m)$ are computed using the estimated state $\hat{\mathbf{x}}(t, t_m)$:

$$\hat{\mathbf{r}}(t, t_m) = h(\hat{\mathbf{x}}(t, t_m)) - \mathbf{z}(t, t_m) \tag{6.90}$$

$$\hat{\mathbf{s}}(t, t_m) = \sqrt{\mathbf{W}} \cdot \hat{\mathbf{r}}(t, t_m) \tag{6.91}$$

Subsequently, the chi-square value $\zeta(t)$ and the confidence level $P_{\text{conf}}(t)$ can be computed from the NRs $\hat{\mathbf{s}}(t, t_m)$:

$$\zeta(t) = \hat{\mathbf{s}}(t, t_m)^T \hat{\mathbf{s}}(t, t_m) \tag{6.92}$$

$$P_{conf}(t) = P\left(\chi^2 \geq \zeta(t)\right) = 1 - P(\zeta(t), m_\nu) \tag{6.93}$$

where $P(\zeta(t), m_\nu)$ is the probability of χ^2 distribution given $\chi^2 \leq \zeta(t)$ with $m_\nu = m_\mathbf{z} - m_\mathbf{x}$ degrees of freedom; $m_\mathbf{z}$ and $m_\mathbf{x}$ are total number of measurements and states in Eq. (6.87).

In fact, both the chi-square value and the confidence level can be utilized to quantify the consistency between the measurements and dynamic model. A low chi-square value (or a high confidence level) indicates that the measurements and the dynamic model are consistent, while a high chi-square value (or a low confidence level) implies that, assuming no instrumentation error, there exists an internal fault.

6.4.1.3 Trip decision

Here the confidence level is selected as an example to quantify the consistency between the measurements and the dynamic model, and for constructing the trip signal. For dependable and secure protection a user defined time delay for tripping is introduced. We use the following trip signal that includes a user defined delay τ_{delay} and a reset time τ_{reset}:

$$\text{TripValue}(t) = \int_{t-\tau_{reset}}^{t} \left(1 - P_{conf}(t)\right) dt \tag{6.94}$$

$$\text{Trip}(t) = \begin{cases} 1, & \text{if TripValue}(t) \geq \tau_{delay} \\ 0, & \text{if TripValue}(t) < \tau_{delay} \end{cases} \tag{6.95}$$

Above equations guarantee that the trip signal is issued only when the confidence level remains consistently low for a time period until the trip signal goes to 1.0.

6.4.2 Numerical experiments

This section demonstrates the performances of the DSE-based protection approach in an example test system [57].

6.4.2.1 Example test system: series compensated transmission line

Here we use the protection function on a series compensated transmission line as an example. We compare the performance of the DSE-based protection method with typical legacy protection schemes for a series compensated power line. The compensation is located at one terminal of the line. The line ratings are 60 Hz, 500 kV, 83-mi long, 4330 MVA. The series capacitors are 0.148 mF (42.4% compensation). Three-phase voltage and current instantaneous measurements are installed at both terminals of the series compensated line, as shown in Fig. 6.2. The rest of the network is not shown. The measurement sampling rate is 80 samples per cycle. The source impedance is shown in Table 6.1. The R, L, G, and C matrices of the transmission line are provided in Table 6.2. The sequence parameters of the line are shown in Table 6.3.

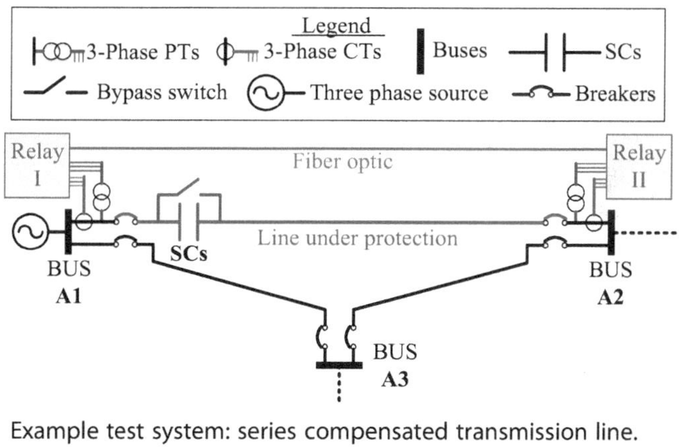

Figure 6.2 Example test system: series compensated transmission line.

Table 6.1 Source impedances.

Parameter	Value
Positive sequence impedance	$13.59\angle 86.99°\Omega$
Negative-sequence impedance	$12.88\angle 86.82°\Omega$
Zero-sequence impedance	$5.76\angle 82.88°\Omega$

Table 6.2 R, L, and C matrices of the line.

Parameter matrices	Value
Series resistance (per mile)	$\begin{bmatrix} 0.104 & 0.089 & 0.089 & 0.088 & 0.088 \\ 0.089 & 0.104 & 0.089 & 0.088 & 0.088 \\ 0.089 & 0.089 & 0.104 & 0.088 & 0.088 \\ 0.088 & 0.088 & 0.088 & 2.357 & 0.087 \\ 0.088 & 0.088 & 0.088 & 0.087 & 2.357 \end{bmatrix} \Omega/\text{mi}$
Series inductance (per mile)	$\begin{bmatrix} 2.79 & 1.48 & 1.26 & 1.47 & 1.26 \\ 1.48 & 2.89 & 1.48 & 1.41 & 1.41 \\ 1.26 & 1.48 & 2.79 & 1.26 & 1.47 \\ 1.47 & 1.41 & 1.26 & 5.55 & 1.36 \\ 1.26 & 1.41 & 1.47 & 1.36 & 5.55 \end{bmatrix} \text{mH/mi}$
Shunt capacitance (per mile)	$\begin{bmatrix} 9.01 & -2.11 & -0.75 & -1.33 & -0.50 \\ -2.11 & 9.55 & -2.11 & -0.94 & -0.94 \\ -0.75 & -2.11 & 9.01 & -0.50 & -1.33 \\ -1.33 & -0.94 & -0.50 & 5.52 & -0.58 \\ -0.50 & -0.94 & -1.33 & -0.58 & 5.52 \end{bmatrix} \text{nF/mi}$
Shunt conductance (per mile)	$\mathbf{0}_{5\times5}$ mho/mi

Table 6.3 Sequence parameters of the line.

Parameter	Value
Positive (negative) sequence series impedance (Z_1)	$42.53\angle 88.34°\Omega$
Zero-sequence series impedance (Z_0)	$142.71\angle 74.56°\Omega$
Positive (negative) sequence shunt susceptance (Y_1)	$0.341\angle 90.0°$ mΩ
Zero-sequence shunt susceptance (Y_0)	$0.186\angle 90.0°$ mΩ

6.4.2.2 Dynamic model of the series compensated transmission line

The dynamic model of the series compensated transmission line consists of the following two parts, as shown in Fig. 6.3. The first part is the multisection transmission line model where each section is a π-equivalent transmission line model. The second part is the model of the three-phase series capacitors.

6.4.2.2.1 Dynamic model of section *k* in the multisection line

The dynamic model of section k in the multisection transmission line is shown in Fig. 6.4. Definitions of variables are shown as follows: $\mathbf{v}_k(t)$ and $\mathbf{v}_{k+1}(t)$ are the three-phase and neutral voltage vectors at both terminals of section k, $\mathbf{i}_{ak}(t)$ and $\mathbf{i}_{bk}(t)$ are the three-phase and neutral current

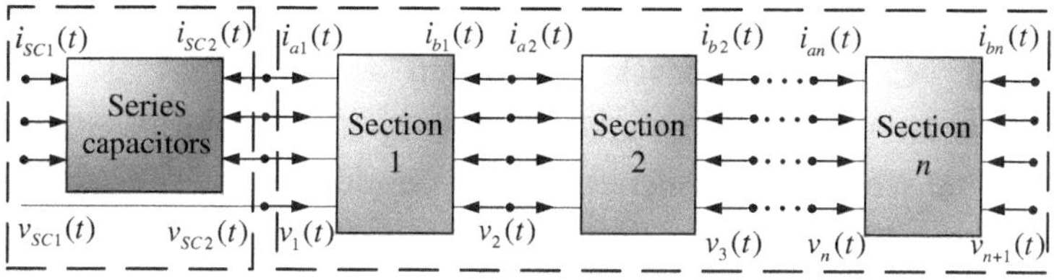

Figure 6.3 Dynamic model of the series compensated transmission line.

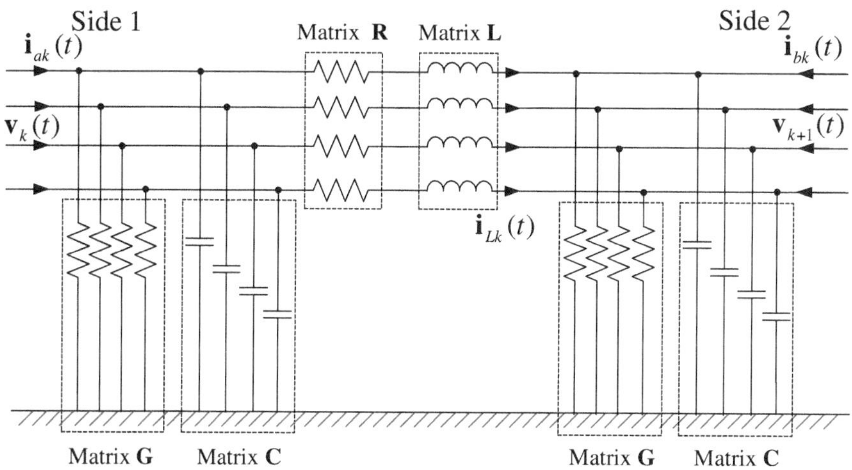

Figure 6.4 Dynamic model of the section *k* in the multisection transmission line.

vectors at both terminals of section k, $\mathbf{i}_{Lk}(t)$ is the three-phase and neutral current vector through the inductors; $\boldsymbol{R}$, $\boldsymbol{L}$, $\mathbf{G}$, and $\boldsymbol{C}$ are series resistance, series inductance, shunt conductance, and shunt capacitance matrices of section k.

The dynamic model can be formulated in compact form as follows:

$$
\begin{aligned}
\mathbf{i}_{ak}(t) &= \mathbf{G}\cdot\mathbf{v}_k(t) + \mathbf{C}\cdot\frac{d\mathbf{v}_k(t)}{dt} + \mathbf{i}_{Lk}(t) \\
\mathbf{i}_{bk}(t) &= \mathbf{G}\cdot\mathbf{v}_{k+1}(t) + \mathbf{C}\cdot\frac{d\mathbf{v}_{k+1}(t)}{dt} - \mathbf{i}_{Lk}(t) \\
\mathbf{0} &= -\mathbf{v}_k(t) + \mathbf{v}_{k+1}(t) + \mathbf{R}\cdot\mathbf{i}_{Lk}(t) + \mathbf{L}\cdot\frac{d\mathbf{i}_{Lk}(t)}{dt}
\end{aligned}
\tag{6.96}
$$

In the syntax shown in Eq. (6.84) the corresponding matrices are $\mathbf{i}(t) = \begin{bmatrix}\mathbf{i}_{ak}(t)\\ \mathbf{i}_{bk}(t)\end{bmatrix}$, $\mathbf{x}(t) = \begin{bmatrix}\mathbf{v}_k(t)\\ \mathbf{v}_{k+1}(t)\\ \mathbf{i}_{Lk}(t)\end{bmatrix}$, $\mathbf{Y}_{eqx1} = \begin{bmatrix}\mathbf{G} & \mathbf{0} & \mathbf{I}_n\\ \mathbf{0} & \mathbf{G} & -\mathbf{I}_n\end{bmatrix}$, are exceeded. Prior to the fault, the ratio is in the restraint region, and none of the thresholds are exceeded (character "o"). During the fault the ratio exits the restraint region at 0.504 second and the threshold asserts (character "x") at 0.520 second. The line is tripped at 0.520 + 0.015 second = 0.535 second.

For the DSE-based protective relay the results are depicted in Fig. 6.9. The first two sets of traces show the residuals and NRs of three-phase currents of side A2. The confidence level and the trip signal are given in the next two traces. The confidence level drops to near zero 0.2 ms after the fault initiation. The trip command is issued at 0.5154 second.

Therefore for this high impedance internal fault, the distance relay fails to detect the fault, the differential relay and the DSE-based protective relay issue a trip command at 0.535 and 0.5154 second, respectively. We can observe that the DSE-based protective relay reliably detects this internal fault and trip this internal fault faster than legacy protection schemes.

6.5 Conclusion remarks

In power system operation, state estimation serves as the core function for extracting reliable information from abundant but imperfect sensor measurement data. By fully exploiting the capabilities of estimation theories, operational tools for various objectives can be developed under the framework of state estimation. This chapter describes in detail the theories and implementations of such three state estimation–based applications for different aspects of system operation: modeling, monitoring, and protection.

First, a systematic framework for the detection, identification, and correction of power system model parameters is presented. The core of this framework is the LNLM test. It provides a practical avenue for the cleaning of large-scale power system model dataset, and effective differentiation between modeling errors and measurement errors. A highly computationally efficient implementation of the approach is described, and the theoretical conditions for detecting and identifying model parameter errors are provided.

Second, the real-time monitoring tool for a power system is developed based on DSE. Its motivations and concepts are discussed in detail. Then, the Bayesian framework motivated general DSE is formulated. Then, a unified DSE framework is developed that includes the well-known EKF, UKF, EnKF, PF, and their variants. To further enhance their robustness to non-Gaussian noise, bad data, etc., a unified robust DSE framework is proposed.

Third, the DSE-based protective relay is demonstrated to ensure the safety and service continuity of the power system. It systematically examines the health condition of components of interest in real time. The relay can reliably and quickly detect internal faults in the protection zone by identifying any inconsistency between the "points on wave" measurements and the dynamic model of the protection zone. The dynamic model of the protection zone is a set of DAEs in time domain describing all physical laws that the zone should obey. Numerical experiments on an example series compensated transmission line prove its advantages toward legacy protection schemes.

References

[1] D. Fletcher, W. Stadlin, Transformer tap position estimation, IEEE Trans. Power Appar. Syst. PAS-102 (11) (1983) 3680–3686.

[2] W. Liu, F. Wu, S. Lun, Estimation of parameter errors from measurement residuals in state estimation, IEEE Trans. Power Syst. 7 (1) (1992) 81–89.

[3] B. Mukherjee, G. Fuerst, S. Hanson, C. Monroe, Transformer tap estimation – field experience, IEEE Trans. Power Appar. Syst. PAS-103 (6) (1984) 1454–1458.

[4] R. Smith, Transformer tap estimation at Florida power corporation, IEEE Trans. Power Appar. Syst. PAS-104 (12) (1985) 3442–3445.

[5] T. Van Cutsem, V. Quintana, Network parameter estimation using online data with application to transformer tap position estimation, Proc. Inst. Elect. Eng. 135 (1988) 31–40. Jan.

[6] O. Alsac, N. Vempati, B. Stott, A. Monticelli, Generalized state estimation, IEEE Trans. Power Syst. 13 (3) (1998) 1069–1075.

[7] W. Liu, S. Lim, Parameter error identification and estimation in power system state estimation, IEEE Trans. Power Syst. 10 (1) (1995) 200–209.

[8] K. Clements, R. Ringlee, Treatment of parameter uncertainly in power system state estimation, IEEE Trans. Power Appar. Syst. (1974).

[9] E. Handschin, E. Kliokys, Transformer tap position estimation and bad data detection using dynamic signal modeling, IEEE Trans. Power Syst. 10 (2) (1995) 810–817.

[10] I. Slutsker, K. Clements, Real time recursive parameter estimation in energy management systems, IEEE Trans. Power Syst. 11 (3) (1996) 1393–1399.

[11] M.R.M. Castillo, J.B.A. London, N.G. Bretas, S. Lefebvre, J. Prevost, B. Lambert, Offline detection, identification, and correction of branch parameter errors based on several measurement snapshots, IEEE Trans. Power Syst. 26 (2) (2011) 870–877.

[12] X. Bian, X.R. Li, H. Chen, D. Gan, J. Qiu, Joint estimation of state and parameter with synchrophasors – Part I: State tracking, IEEE Trans. Power Syst. 26 (3) (2011) 1196–1208.

[13] X. Bian, X.R. Li, H. Chen, D. Gan, J. Qiu, Joint estimation of state and parameter with synchrophasors – Part II: Parameter tracking, IEEE Trans. Power Syst. 26 (3) (2011) 1209–1219.

and is assisting in autonomous driving, robotic surgery, and legal judgment. In the energy sector, AI is also making profound impacts.

With the deployment of smart grid technologies, power systems have benefited from AI techniques, as shown in Fig. 7.1. The power system data has characteristics of high-volume, high-velocity, and high-variety. For example, sensors such as phasor measurement units (PMUs) take measurements at a millisecond resolution. The advanced metering infrastructure (AMI) in the New York state collects more than 127 TB of consumption data per day [1]. With the big data in power systems, AI provides new solutions to system planning, operation, maintenance, market monitoring, and risk management [2]. For example, reinforcement learning (RL) has been used for power system stability control [3], automatic generation control (AGC) [4], and optimal power flow control [5]. Deep learning has been applied to energy resource assessment [6], injection attack detection [7], and fault diagnosis [8]. A more comprehensive review on AI applications to power systems can be found in Ref. [9].

Among various AI applications in power systems, forecasting is one of the most popular use cases. Forecasting in power systems is to predict the future load, renewable generation output, or electricity and energy price, which are used to assist power system operations at different timescales [10,11]. Therefore forecasting has been widely studied and adopted in power systems. For example, most of the independent system operators (ISO) in the United States have adopted load, wind, and solar power forecasting to assist their system operations [12]. ISO New England utilizes day-ahead forecasts for the dispatch scheduling of generating capacity, reliability analysis, and maintenance planning for the generators [13]. A number of forecasting projects have been or is being conducted to promote renewable energy forecasting, such as Wind Forecast Improvement Project [14], WindView [15], Watt-Sun [16], and SUMMER-GO [17].

Another emerging topic in power systems that relies on AI techniques is system state estimation. Power system state estimation is to retrieve system dynamics, for example, voltage magnitude and phase angles, from available measurements. Traditional methods have challenges in solving state estimation problems with large scale and nonconvexity [18]. Therefore AI techniques have been introduced to help solve state estimation problems through a learning-based optimization manner [19]. For example, a feedforward neural network was used to initialize the Gauss–Newton algorithm to solve the distribution system state estimation in Ref. [20]. An autoencoder was adopted to estimate the voltage magnitude and angle in Ref. [21].

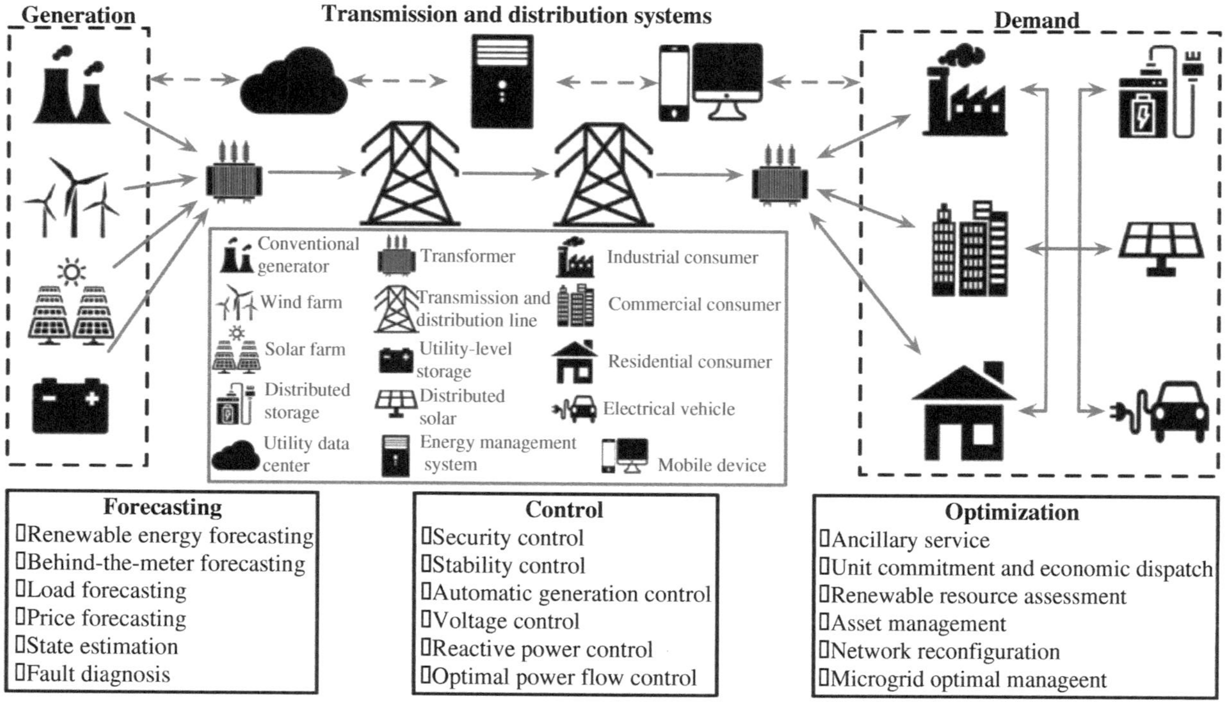

Figure 7.1 Artificial intelligence applications to smart grid.

In addition, a recurrent neural network (RNN) was trained to estimate the states of the IEEE 57- and 118-bus systems [18]. Bad data detection and cyberattack detection are also recognized to impact the system states and, thus, have also been widely researched. For instance, Hink et al. [22] successfully detected cyberattacks with several different machine learning methods. A support vector machine (SVM) model was used to detect the stealthy false data injection with nearly 95% accuracy [23].

AI has also been applied to building occupancy detection, which is critical to building energy management and building-to-grid integration. The building sector accounts for over 70% of the total electricity consumption in the United States, making the building integration a critical part of the smart grid. The building occupancy information helps building management in several ways, such as occupancy-driven demand response and building energy management. For example, an occupancy-based feedback control algorithm was applied to a heating, ventilation, and air-conditioning system and achieved 29%–80% energy savings [24]. Korkas et al. [25] developed a control optimization method for demand response management in microgrids considering occupancy information and reduced energy costs by 20%. Due to the remarkable benefits of the occupancy information, accurate occupancy detection is recognized as a crucial factor and has received growing attention [26,27].

To demonstrate the AI applications to power systems, this chapter reviews the state-of-the-art machine learning methods, including ensemble learning and deep learning, in renewable energy forecasting, power system network reconfiguration, and smart building occupancy detection.

7.2 Modern forecasting technology

This section reviews the state-of-the-art forecasting methods and discusses the details of two types of ensemble learning-based forecasting techniques.

7.2.1 Prior research work

Time series forecasting can be categorized based on different criteria, such as forecasting task (e.g., load, wind, solar, and price), time horizon (e.g., short-term, midterm, and long-term), or methods. Fig. 7.2 summarizes a rough classification of forecasting methods, which are mainly divided into deterministic forecasting and probabilistic forecasting. The former can obtain accurate forecasting results after a specific time horizon, while the latter can provide probabilistic and confidence levels for the uncertainty of desired forecasts.

7.2.1.1 Deterministic forecasting methods

Generally, deterministic forecasting models can be further divided into three categories: statistical [28], intelligent [29], and ensemble models [30]. Statistical methods refer to the utilization of mathematical theory knowledge such as mathematical statistics, probability theory, and stochastic processes. Intelligent models refer to AI models using machine learning and deep learning. Ensemble models refer to the combination of two different algorithms or methods. Ensemble models can combine the merits and characteristics of different methods, which normally perform better than single models [31]. Conventional statistical models include the autoregressive model, the autoregressive moving average model, and the autoregressive integrated moving average model. The most popular machine learning algorithms in renewable energy forecasting are artificial neural networks (ANNs), support vector regression (SVR) model, random forest (RF), and gradient boosting machine (GBM). For example, Feng et al. [32] proposed a short-term solar forecasting method based on sky imaging and pattern recognition. A least squares SVR model was proposed in Ref. [33] to model the nonlinearity of electric load. A GBM model was developed to quantify the dataset diversity for short-term wind forecasting [34], which helps system operational scheduling such as economic dispatch and

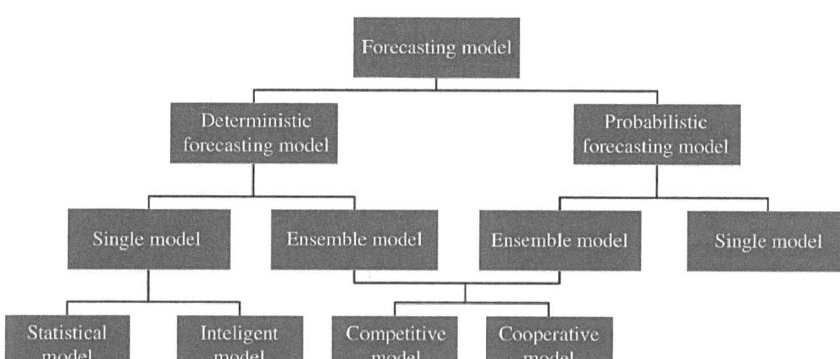

Figure 7.2 Machine learning–based forecasting methods categorization.

parameters of the multimodel ensemble were solved by an expectation maximizing algorithm and continuous ranked probability score optimization. Kim and Hur [54] developed an enhanced ensemble method for probabilistic wind power forecasting. The wind speed spatial ensemble was built by using correlation-based weight and kriging models, and the temporal ensemble was built through an average ensemble of three models (i.e., an exogenous variable model, a polynominal regression model, and an analog ensemble model).

7.2.2 Ensemble learning forecasting methodologies

Both competitive and cooperative ensemble learning rely on single models. Therefore single-machine learning algorithms, including ANN, SVR, GBM, and RF, are first introduced in this section. Then, two wind forecasting methodologies based on competitive or cooperative ensemble learning are described.

7.2.2.1 Single-machine learning algorithm models

ANN is a popular machine learning algorithm in speech recognition, target tracking, signal analysis, and nonlinear regression problems (such as time series forecasting). ANN mimics the action of the human neurons, and each neuron is a weighted sum of its inputs and is connected to the neurons in the next layer. A weight decides how much influence the input will have on the output. The ANN architecture contains one input layer, one or more hidden layer(s), and one output layer. The output signal of ANN can be either 0, 1, or any real value between 0 and 1 depending on whether we are dealing with "binary" or "real valued" artificial neurons. The configuration of the ANN model needs to be well designed to avoid overfitting issues. ANN can be classified into different types with different activation functions and learning algorithms. Sigmoid and hyperbolic tangent are two commonly used activation functions. Deep learning is also a configuration of ANN, where multiple hidden layers are built. The mathematical description of ANN is expressed as:

$$y_i^{(n)} = f\left(\sum_{j=1}^{N} w_{ij}^{(n,n-1)} y_j^{(n-1)} + \theta_i^n\right) \tag{7.1}$$

where i is a neuron of the nth layer, w_{ij} is the weight from the neuron j in the layer $(n-1)$ to the neuron i in layer n, and θ_i^n is the threshold of the neuron i in layer n.

SVM is originally a supervised linear classifier proposed by Cortes and Vapnik [55]. As one of the most popular classification methods, SVM has been applied in text categorization, image classification, and other recognition tasks. When dealing with linearly inseparable data, nonlinear mapping-based kernel methods, $\mathrm{K}(x){:}R^n \longrightarrow R^{n_h}$, are used to map the nonlinear data into the high-dimensional feature space. Then, a linear hyper plane can be found by maximizing the distance between support vectors and the hyper plane. The SVM algorithm can also be applied in regression problems, which is called SVR. However, the compute and storage requirements increase significantly with the data dimension. The hyper plane function, also called the SVR function, is described as [56]:

$$f(x) = \omega^T \mathrm{K}(x) + b \tag{7.2}$$

where ω and b are variables solved by minimizing the empirical risk, which is given by:

$$R(f) = \frac{1}{n}\sum_{i=1}^{n} \Theta(y_i, f(x)) \tag{7.3}$$

where $\Theta_\varepsilon(y_i, f)$ is the ε-insensitive loss function, expressed as:

$$\Theta_\varepsilon(y_i, f) = \begin{cases} \|f - y\| - \varepsilon, \text{if} \|f - y\| \geq \varepsilon \\ 0, \text{otherwise} \end{cases} \tag{7.4}$$

Then the optimal hyper plane can be found by solving the inequality-constrained quadratic optimization problem.

GBM is a highly customizable learning algorithm, which is widely used in the regression and classification fields. GBM model relies on the combination of "weak learners" to create an accurate learner and, therefore, is able to generate both deterministic and probabilistic results in time series forecasting. The combination is achieved by adding the weighted base learner to the previous model iteratively. The principle of GBM is illustrated by the pseudo-code in Algorithm 1. In each iteration the negative gradient of the chosen loss function is calculated and used to estimate the split variables a by Eqs. (7.5) and (7.6). Then the multiplier β is optimized by Eq. (7.7). The weak learner $\beta h(x;\ a)$ is added to the previous model, where $h(x;a)$ is a learning function.

RF is another supervised ensemble learning method that consists of many single classification and regression trees (CARTs):

Algorithm 1 Gradient boosting machine

Initialize $f_0(x)$ to be a constant, $f_0(x) = \text{argmin}_\rho \sum_{i=1}^{n} \Psi(y_i, \rho)$
for $i = 1$ to M do
Compute the negative gradient of the loss function:

$$\overline{y_i} = -\left[\frac{\partial\Psi(y_i, F(x_i))}{\partial F(x_i)}\right]_{f(x)} = f_{i-1}(x), \quad i = \{1, 2, \ldots, n\} \tag{7.5}$$

Fit a model to $\overline{y}$ by least squares to get a_t:

$$a_t = \arg\min_{\alpha,\beta} \sum_{i=1}^{n} [\overline{y_i} - \beta h(x_i, a)]^2 \tag{7.6}$$

Calculate β_t by:

$$\beta_t = \arg\min_{\beta} \sum_{i=1}^{n} \Psi\left(y_i, f_{t-1}(x_i) + \beta h(x_i, a_t)\right)\Big) \tag{7.7}$$

Update the model by:

$$f_t(x) = f_{t-1}(x) + \beta_t h(x; a_t)\Big) \tag{7.8}$$

end for
Output $\hat{f}(x) = f_T(x\Big)$

$$T = \left\{t\left(X, s_{\Lambda_1}\right), t\left(X, s_{\Lambda_2}\right), \ldots, t\left(X, s_{\Lambda_n}\right)\right\} \tag{7.9}$$

where T is the set of CARTs, t is a single CART, X is the input to the RF model, and s_{Λ_i} is the random vector to extract bootstrap samples that are determined by the bagging algorithm. The robustness of the RF model is enhanced by the randomness of the bagging algorithm and the best splits search process. Since RF is a combination of various regressions, the model is generally free from overfitting [57].

7.2.2.2 Competitive ensemble learning

A competitive ensemble methodology is developed for short-term wind forecasting with both deterministic and probabilistic forecasts, which is called the machine learning–based multi-model forecasting framework (M3), as shown in Fig. 7.3. The M3 deterministic model has two layers (note that this is different from ANN layers). The first layer consists of several machine learning models, that is, ANN, SVM, GBM, and RF, which are built based on the historical data. These models forecast wind

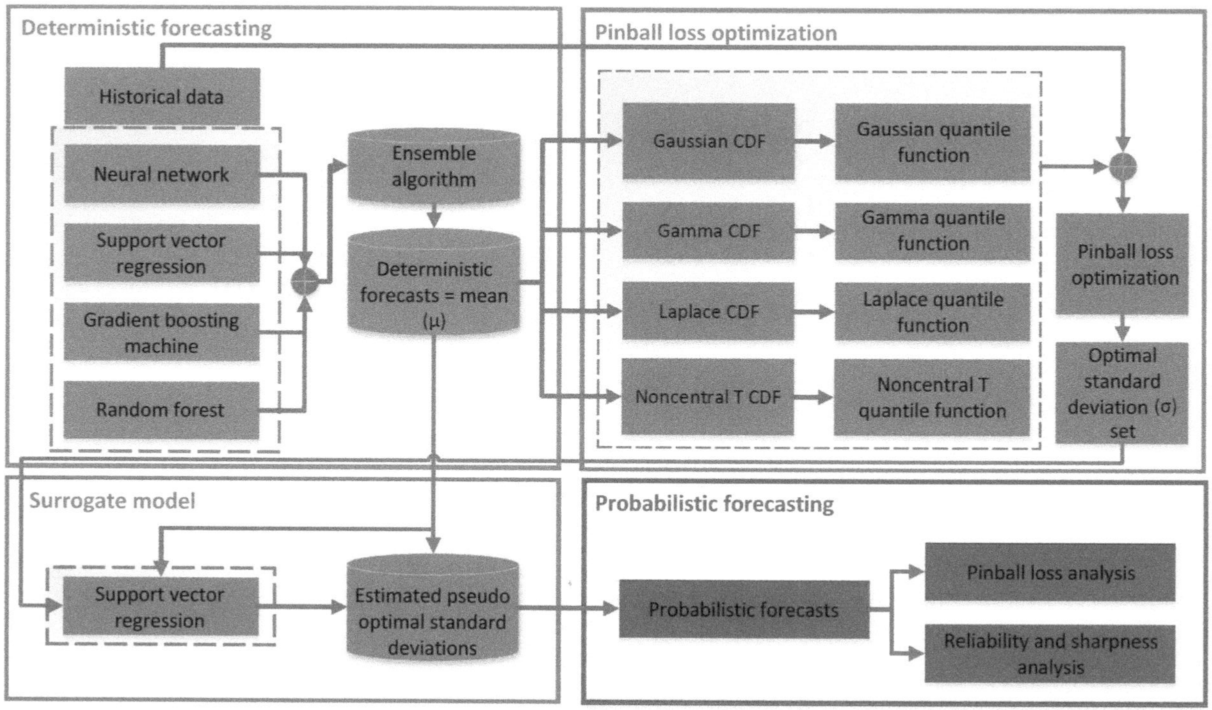

Figure 7.3 Competitive ensemble learning framework based on the M3 model.

speed or wind power as the output. A blending model is developed in the second layer to combine the forecasts produced by different algorithms from the first layer, and to generate both deterministic and probabilistic forecasts. This blending model is expected to take advantage of different algorithms by canceling or smoothing the local forecasting errors. The mathematical description is shown as:

$$y_i = f_i(x_1, x_2, \ldots, x_p) \tag{7.10}$$

$$\hat{y} = \Phi(y_1, y_2, \ldots, y_m) \tag{7.11}$$

where $f_i(^*)$ is the ith algorithm and y_i is the wind speed forecasted by $f_i(^*)$. $\hat{y}$ is the second-layer blending algorithm.

The M3 deterministic forecasts can also be transferred to probabilistic forecasts. Specifically, after obtaining deterministic forecasts, a set of unknown parameters in the predictive distribution are determined by minimizing the pinball loss that is an evaluation metric of probabilistic forecasts. Note that the optimal distribution parameters are adaptively and dynamically updated based on the deterministic forecast value at each time stamp. The optimal adaptive predictive distribution parameters are first determined offline with the historical training data. Then a surrogate model is developed to represent the optimized distribution parameter as a function of the deterministic forecast. At the real-time forecasting stage the surrogate model is used together with deterministic forecasts to adaptively predict the unknown distribution parameters and thereby generate probabilistic forecasts.

7.2.2.3 Cooperative ensemble learning

In addition to deterministic forecasts, the ensemble methods could also be applied to probabilistic forecasts directly. A cooperative ensemble methodology of probabilistic wind power forecasts from different type of predictive distributions is introduced in this section, and the overall framework is illustrated in Fig. 7.4.

In the deterministic forecasting stage, instead of integrating multiple single-machine learning algorithms together to get a refined forecast, a Q-learning-enhanced deterministic forecasting method [58] is adopted to select the best forecasting model from a pool of state-of-the-art machine learning–based forecasting models (i.e., ANN, SVR, GBM, and RF) at each time step. To be more specific, the developed method trains Q-learning agents based on the rewards of transferring from the current model to the next model. For example, a Q-learning agent will receive a

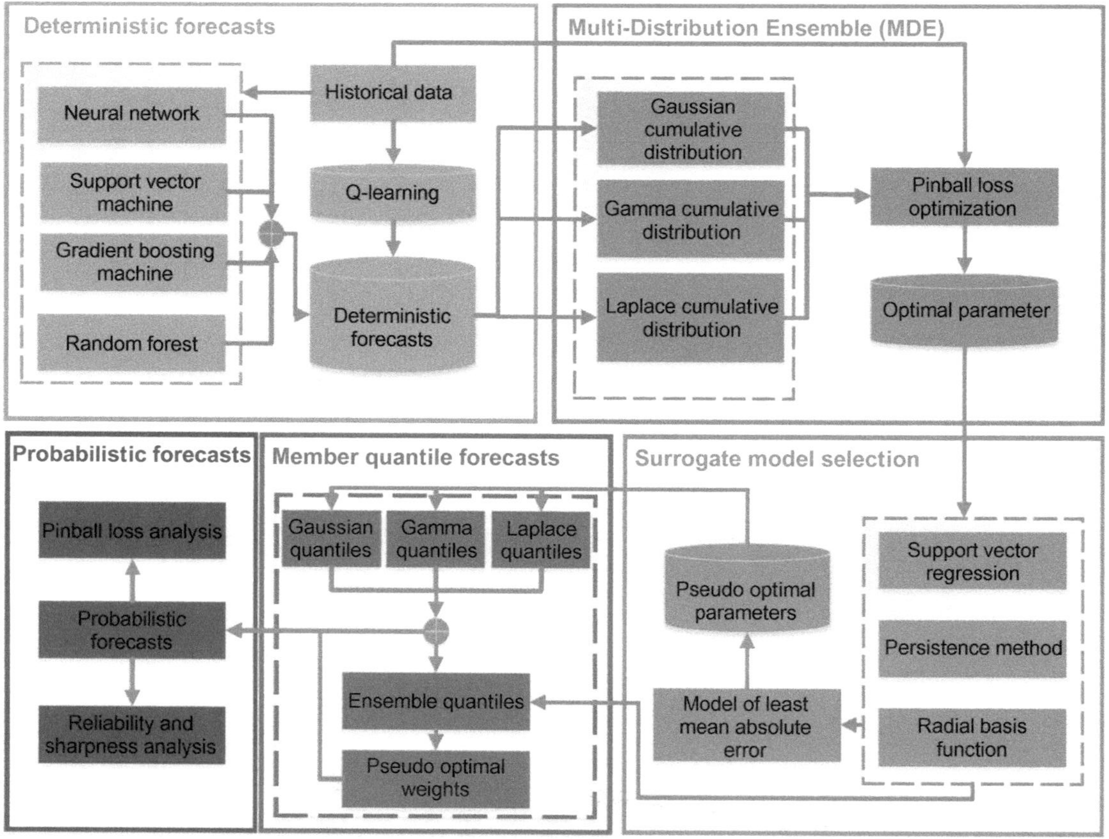

Figure 7.4 The overall framework of the cooperative ensemble probabilistic forecasting.

reward by transferring from the current forecasting model M_i to the next forecasting model M_j in each training step, from which the Q-learning agent will learn the optimal policy of the model selection. Then, this optimal policy will be applied to select the best model for forecasting in the next step based on the current model in the forecasting stage. The dynamic model selection process is expressed as:

$$S = \{s\} = \{s_1, s_2, \ldots, s_I\} \tag{7.12}$$

$$A = \{a\} = a_1, a_2, \ldots, a_I \tag{7.13}$$

$$R^t\left(s_i, a_j\right) = \text{ranking}\left(M_i\right) - \text{ranking}(M_j) \tag{7.14}$$

$$Q^{e+1}(s_e, a_e) = (1-\alpha)Q^e(s^e, a^e) + \alpha\left[R^e(s^e, a^e) + \gamma \max Q^e\left(s^{e+1}, a\right)\right] \tag{7.15}$$

where S, A, R, and Q are state space, action space, reward space, and Q-table in the dynamic model selection Markov Decision Process, respectively. The parameters s and a are possible state and action, respectively. The parameter I is the number of models (M) in the model pool, e is the episode index with the maximum of 100, $\alpha = 0.1$ is the learning rate that controls the aggressiveness of learning, and $\gamma = 0.8$ is a discount factor that weights the future reward. The reward function is defined as the model performance improvement, which ensures the effective and efficient convergence of Q-learning. More details of this Q-learning-based dynamic forecasting model selection can be found in Ref. [58].

In the probabilistic forecasting stage, individual probabilistic forecasts are generated by using each single predictive distribution based on the training dataset [44]. The unknown parameters (i.e., standard deviations) of each predictive distribution are optimized. A weight parameter is assigned to quantile forecasts from each individual model, and these weight parameters are optimized again by minimizing the pinball loss. Then a surrogate model is developed to represent each optimal weight as a function of the deterministic forecast. During online forecasting a set of pseudo-optimal parameters of the ensemble model is estimated by the surrogate model and deterministic forecasts. Finally, the method with the minimum pinball loss is chosen to produce the final ensemble probabilistic forecasts.

7.2.3 Forecasting results

Performance of the single-algorithm and ensemble machine learning models is evaluated in this section for both deterministic and probabilistic wind speed/power forecasting. The ensemble deterministic and multi-model probabilistic forecasting models are tested on wind speed forecasting. The Q-learning–enhanced deterministic and ensemble probabilistic forecasting models are tested on wind power forecasting. Two evaluation metrics are utilized to evaluate the deterministic forecasting accuracy in both case studies: the normalized mean absolute error (nMAE) and the normalized root mean square error (nRMSE):

$$\text{nMAE} = \frac{1}{n}\sum\nolimits_{i=1}^{n} \left|\frac{\hat{x}_i - x_i}{x_{\max}}\right| \tag{7.16}$$

$$\text{nRMSE} = \frac{1}{x_{\max}}\sqrt{\frac{\sum_{i=1}^{n}(\hat{x}_i - x_i)^2}{n}} \tag{7.17}$$

where $\hat{x}_i$ is the forecast value, x_i is the actual value, $x_{\max}$ is the maximum actual value, and n is the sample size.

Pinball loss, one of the most popular metrics, is used to evaluate the performance of probabilistic forecasting in both case studies. Pinball loss is a function of observations and quantiles of a forecast distribution. A smaller pinball loss value indicates better probabilistic forecasting. The pinball loss value of a certain quantile L_m is expressed as:

$$L_m(q_m, x_i) = \begin{cases} \left(1 - \dfrac{m}{100}\right) \times (q_m - x_i), \& x_i < q_m \\ \dfrac{m}{100} \times (x_i - q_m), \& x_i \geq q_m \end{cases} \tag{7.18}$$

where x_i represents the ith observation, m represents a quantile percentage from 1 to 99, and q_m represents the predicted quantile. For a given m percentage the quantile q_m represents the value of a random variable whose CDF is m percentage.

7.2.3.1 Case study I: wind speed forecasting based on competitive ensemble learning

Both the single-algorithm models and ensemble algorithm are applied to the wind speed data collected near hub height with a 1-hour resolution at eight sites listed in Table 7.1. For all the eight locations the first 2/3 of data are used as training data, in which the first 11/12 is used to train ensemble algorithm and the remaining 1/12 of the training data is used to build the surrogate model of the optimal standard deviation. The accuracy of the forecasts is evaluated by the remaining 1/3 of data. The nMAE and nRMSE are compared in Tables 7.2 and 7.3, respectively.

The multi-model framework includes multiple individual models in the first layer and several models in the second layer. Different algorithms are tested in both layers, which include (1) three SVR models with linear (SVR_l), polynomial (SVR_p), and radial base (SVR_r) kernels; (2) five ANN models with different numbers of hidden layers (n_l), neurons in each layer (n_o), and weight decay parameter (n_d) values, and the selected models employ the feed-forward–backpropagation learning function and

Table 7.1 Data duration at selected sites.

Name	Data duration	Height (m)
Boulder_NWTC (C1)	2009-01-02 to 2012-12-31	80
Megler (C2)	2010-11-03 to 2012-11-01	53.3
CedarCreek_H06 (C3)	2009-01-02 to 2012-12-31	69
Goodnoe_Hills (C4)	2007-01-01 to 2009-12-31	59.4
Bovina50 (C5)	2010-10-10 to 2012-10-08	50
Bovina100 (C6)	2010-03-03 to 2012-03-01	100
CapeMay (C7)	2007-09-26 to 2009-09-24	100
Cochran (C8)	2008-06-30 to 2011-06-29	70

Note: The case notations (C1–C8) are different from Sections 7.2.3.2 and 7.4.3.

Table 7.2 The normalized mean absolute error (nMAE) (%) of 1HA forecasts.

Method	C1	C2	C3	C4	C5	C6	C7	C8
SVR_r	5.101	3.114	6.229	3.799	5.145	4.554	3.662	4.014
SVR_l	4.765	2.886	3.927	3.718	4.891	4.466	3.449	3.984
SVR_p	4.772	2.919	4.267	3.734	4.913	4.572	3.553	4.009
ANN1	4.793	2.921	4.155	3.738	4.936	4.671	3.717	4.007
ANN2	4.789	2.938	4.042	3.735	4.939	4.536	3.560	4.012
ANN3	4.817	2.927	4.096	3.738	4.932	4.494	3.502	4.017
ANN4	4.792	2.906	4.022	3.735	4.924	4.481	3.500	4.005
ANN5	4.793	2.902	3.859	3.727	4.899	4.487	3.480	4.006
GBM1	4.822	2.945	4.468	3.739	4.961	4.479	3.562	4.022
GBM2	4.808	2.941	4.474	3.736	4.963	4.478	3.550	4.020
GBM3	4.806	2.936	4.730	3.768	4.969	4.491	3.504	4.002
GBM4	4.845	2.946	4.348	3.754	4.974	4.544	3.554	4.023
RF1	4.965	3.060	4.207	3.883	5.115	4.703	3.715	4.159
RF2	4.920	3.012	4.221	3.852	5.057	4.637	3.659	4.110
M3	**4.623**	**2.712**	**3.731**	**3.654**	**4.683**	**4.256**	**3.223**	**3.871**

Note: The smallest nMAEs among all the models are in bold.

Table 7.3 The normalized root mean square error (nRMSE) (%) of 1HA forecasts.

Method	C1	C2	C3	C4	C5	C6	C7	C8
SVR_r	8.039	4.883	11.769	5.318	7.166	6.363	5.383	5.477
SVR_l	6.954	3.993	5.638	5.095	6.539	6.224	4.813	5.401
SVR_p	6.953	4.011	6.381	5.109	6.545	6.350	4.946	5.423
ANN1	6.907	4.019	6.250	5.123	6.585	6.364	5.065	5.409
ANN2	6.908	4.038	5.974	5.105	6.589	6.260	4.930	5.408
ANN3	6.902	4.026	6.116	5.123	6.579	6.223	4.868	5.402
ANN4	6.930	4.024	5.913	5.124	6.564	6.225	4.888	5.411
ANN5	6.919	4.005	5.483	5.096	6.529	6.207	4.826	5.402
GBM1	6.969	4.105	7.197	5.100	6.605	6.225	4.942	5.413
GBM2	7.016	4.102	7.225	5.103	6.613	6.237	4.936	5.419
GBM3	7.059	4.121	7.886	5.217	6.654	6.258	4.916	5.421
GBM4	7.011	4.116	6.898	5.112	6.620	6.291	4.955	5.426
RF1	7.220	4.238	6.248	5.261	6.788	6.463	5.154	5.585
RF2	7.141	4.159	6.416	5.224	6.715	6.388	5.067	5.526
M3	**6.720**	**3.988**	**5.524**	**4.987**	**6.341**	**6.089**	**4.760**	**5.101**

Note: The smallest nRMSEs among all the models are in bold.

sigmoid activation function; (3) four GBM models based on different loss functions (Gaussian and Laplacian) and parameters, that is, number of trees, learning rate (λ), maximum depth of variable interactions, and minimum number of observations in the terminal nodes; and (4) two RF models with different numbers of variables that are randomly sampled as candidates at each split. In the probabilistic forecasting stage, four widely used predictive distribution types (i.e., Gaussian, Gamma, Laplace, and noncentral *t*) are considered to model the possible shapes of the predictive distribution. The one with the lowest pinball loss is chosen as the final predictive distribution.

It is seen that none of the single models performs better than the ensemble method. The ensemble models have improved the forecasting accuracy of the component models by up to 12.9% based on nMAE and 16.4% based on nRMSE. For the blending algorithms the models with nonlinear blending algorithms have better performance than the models with linear blending algorithms. This shows that the forecasts produced from the first-layer models exhibit a nonlinear relationship with the actual wind speed.

Fig. 7.5 provides an example of the deterministic forecasts along with the confidence intervals in the form of interval plot at the C2 site from 2012-02-01 to 2012-02-04. The colors of the intervals fade with the increasing confidence level, ranging from 10% to 90% in a 10% increment. The intervals are symmetric around the deterministic forecasting curves with a changing width. When the wind speed fluctuates within a small range, the confidence bands are narrow. When there is a significant ramp, the uncertainty of the forecasts is increased and the bands tend to be broader, as shown by hours 50–65. This further proves the necessity of probabilistic forecasting.

The pinball loss values of the eight selected sites with different predictive distributions are summarized in Table 7.4. The sum of pinball loss is averaged over all quantiles from 1% to 99% and normalized by the maximum wind speed at each site. A lower pinball loss score indicates a better probabilistic forecast. Table 7.4 shows that the M3-Laplace with pinball loss optimization has the smallest pinball loss value at all locations. The M3-Laplace model has reduced the pinball loss by up to 35% compared to the M3 forecasts with other predictive distributions [i.e., Gaussian, Gamma, and Noncentral T (nCT)]. Therefore the Laplace distribution is finally chosen to generate probabilistic wind speed forecasts. Note that the models of M3-Gaussian, M3-Gamma, and M3-Laplace perform similarly,

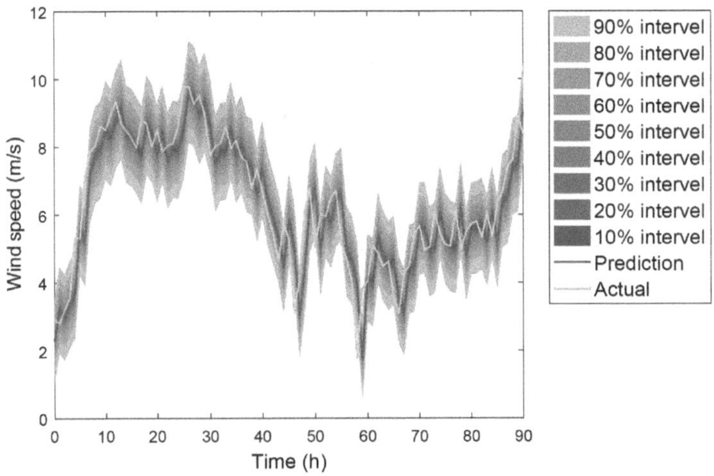

Figure 7.5 Multi-model probabilistic forecasts at the C2 site.

Table 7.4 Normalized averaged sum of pinball loss.

Method	C1	C2	C3	C4	C5	C6	C7	C8
M3-Gaussian	1.74	1.26	1.44	1.36	1.86	1.69	1.28	1.59
M3-Gamma	1.74	1.26	1.44	1.36	1.87	1.69	1.27	1.58
M3-Laplace	**1.72**	**1.25**	**1.43**	**1.35**	**1.85**	**1.63**	**1.26**	**1.57**
M3-nCT	1.74	1.81	2.20	2.21	2.68	3.41	2.56	2.86

Note: The smallest normalized averaged sum of pinball loss at each location is in boldface.

which indicates that the optimization can help achieve better accuracies with different predictive distribution types.

7.2.3.2 Case study II: wind power forecasting based on cooperative ensemble learning

The Q-learning enhanced deterministic forecasting algorithm is applied to the wind power data collected from the WIND Toolkit with 1-hour resolution, which includes seven sites (as shown in Table 7.5) with meteorological information (e.g., wind direction, wind speed, air temperature, surface air pressure, and density at hub height). For all the locations the first 3/4 of the data is used as training data, in which the first 11/12 is used to train the deterministic forecast models and the remaining 1/12 of the training data is used to build the surrogate models of the optimal standard deviations and weight parameters. The effectiveness of the forecasts is validated by the remaining 1/4 of the data. The nMAE and nRMSE of 1HA (hour-ahead) wind power forecasting are listed in Table 7.6. It is shown that the 1HA nMAE and nRMSE from Q-learning model are in the ranges of 5%–8% and 8%–13%, respectively. To show the effectiveness of the Q-learning enhanced deterministic forecasting, the persistence method is used as a baseline. It is seen from Table 7.6 that overall the Q-learning performs better than the persistence method.

Table 7.5 Data summary of the selected seven WIND Toolkit sites.

Site ID (case notation)	Latitude	Longitude	Capacity (MW)	State
4816 (C1)	29.38	− 100.37	80	TX
8979 (C2)	31.53	− 95.62	53.3	TX
10069 (C3)	32.31	− 98.26	69	TX
10526 (C4)	32.44	− 100.55	59.4	TX
1342 (C5)	27.12	− 97.86	50	TX
2061 (C6)	27.95	− 97.40	100	TX
9572 (C7)	31.99	− 100.18	100	TX

Note: The case notations (C1−C7) are different from Sections 7.2.3.1 and 7.4.3.

Table 7.6 The normalized mean absolute error (nMAE) (%) and the normalized root mean square error (nRMSE) [%] of 1HA forecasts.

Model	Metric	C1	C2	C3	C4	C5	C6	C7
Q-learning	nMAE	6.63	6.85	6.70	6.74	6.67	5.38	7.76
	nRMSE	10.55	10.93	11.04	10.66	9.93	8.37	11.93
Persistence	nMAE	7.14	7.18	6.97	7.11	7.12	5.74	8.06
	nRMSE	11.38	11.71	11.82	11.50	10.99	8.93	12.63

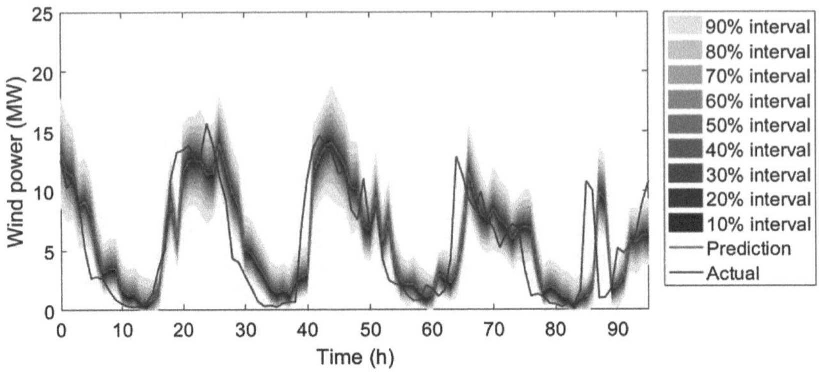

Figure 7.6 Cooperative ensemble probabilistic forecasts at the C2 site.

Table 7.7 Normalized averaged sum of pinball loss.

Method	C1	C2	C3	C4	C5	C6	C7
Q-learning-cooperative	2.23	2.58	2.42	3.01	3.05	1.98	3.67
Q-learning without ensemble	2.46	2.83	2.67	3.38	3.42	2.30	3.94

To better visualize the ensemble forecasting model, Fig. 7.6 provides an example of the deterministic forecasts along with the confidence intervals in the form of interval plot at the C2 site from 2012-07-20 to 2012-07-24. The colors of the intervals fade with the increasing confidence level, ranging from 10% to 90% in a 10% increment. The intervals are symmetric around the deterministic forecasting curves with a changing width. Fig. 7.6 shows that the prediction intervals of the ensemble forecasting model are narrow, which show less uncertainty in the probabilistic forecasts.

The pinball loss values of the seven selected sites with different predictive distributions are summarized in Table 7.7. Results show that the Q-learning-based ensemble probabilistic forecasting method has reduced the pinball loss by up to 16.1% compared with single probabilistic forecasting methods.

7.3 Machine learning–based control and optimization

The powerful learning abilities also enable machine learning models to solve complex control and optimization problems in power systems.

This section reviews the popular applications of machine learning in power system control and optimization. Specifically, a network reconfiguration optimization problem is solved by AI to demonstrate the "learn to optimize" capability of machine learning models.

7.3.1 Prior research work

7.3.1.1 Machine learning–based control

Due to the increasing complexity and uncertainty brought by renewable energy, distributed energy resources, and varying loads, it is of significant importance for the electric power system to employ advanced control strategies to keep the system working reliably and efficiently. Studies have shown that machine learning–based control strategies are capable of addressing the high-dimensional complex nonlinear control challenges, and it could be more effective when properly combined together with conventional mathematical approaches [59,60].

In terms of power system stability, extensive studies have employed neural network to design power system stabilizer (PSS), and controller for AGC or frequency control [61]. For instance, a self-tuning PSS based on ANN was proposed by Segal et al. [62]. In this approach, ANN was introduced for tuning the conventional PSS parameters in real time, showing that the dynamic performance of the system with the ANN-based PSS is robust over a wide range of loading conditions and equivalent reactance. To improve the transient stability of power systems under different operating conditions and parametric uncertainties, Senjyu and Fujita [63] proposed an RNN stabilization controller for both automatic voltage regulators and the governor, where the weights of the controller are adjusted online. The proposed approach was applied to a single-machine infinite-bus system and showed good damping characteristics over a wide range of operating conditions. However, due to the complexity of large-scale power systems, design of intelligent controllers–based ANN requires large training time and a large number of neurons. These drawbacks motivated Chaturvedi et al. [64] to utilize generalized neurons (GNs) to develop a GN-based adaptive PSS. Given the advantages of self-optimizing adaptive control and the quick response of the GN, the proposed GN-based PSS can provide good damping of the power system over a wide operating range and significantly improved the dynamic performance. To stabilize the system after severe disturbances and mitigate the oscillations afterward, Hadidi and Jeyasurya [65] employed a Q-learning RL algorithm to develop a real-time closed-loop wide-area decentralized PSSs. The proposed algorithm proved that the stability boundary of the system could be extended without losing any generator or load area.

Due to the deviation between system load and power generation, there exist frequency fluctuations in the power system, which requires a supplementary control strategy to restore the system frequency to the nominal values. For example, Zeynelgil et al. [66] proposed an ANN controller with the backpropagation through time algorithm to study AGC in a four-area interconnected power system. Chaturvedi et al. [67] proposed a generalized NN-based decentralized controller to control the frequency deviations in power systems. Other modified NN algorithms include dynamic NN [68] and radial base function neural network [69]. Moreover, RL algorithms, that is, model-free Q-learning [70], hierarchically correlated equilibrium Q-learning [71], and deep distributed recurrent Q-networks-action discovery [72], have also been employed to develop frequency controllers for different conditions and purposes.

7.3.1.2 Machine learning–based optimization

For a data-driven nonlinear optimization problem, the first step in general is to establish a mathematical regression model as the objective function in terms of the design variables based on the experimental or collected data. Since the physical or explicit relationships are unable to be observed directly or remain unknown, the modeling process is referred to as black-box. Machine learning (i.e., surrogate prediction model) approaches are widely employed to solve the black-box problem, such as ANN, kriging/Gaussian process (GP) regression, SVM, radial basis functions, and polynomial response surface. By cross-validation the best appropriate models are selected to build up the black-box model. The next step of the optimization process is to apply a gradient-based or heuristic algorithm to solve the problem.

The machine learning–based optimization has also been extensively employed in power system studies. For example, Lucifredi et al. [73] compared the kriging and ANN statistical modeling techniques with a conventional linear multiple regression method for hydroelectric power system maintenance prediction. Pan and Das [74] proposed to use a fractional order control strategy for a microgrid, in which the controller parameters were tuned with a global optimization algorithm to meet system performance specifications. In this research, a kriging-based surrogate modeling technique integrated with GA was employed to alleviate the issue of expensive objective function evaluation for the optimization-based controller tuning. Luh et al. [75] utilized augmented Lagrangian relaxation to form and solve the market clearing prices subproblems by using a surrogate-based optimization framework.

7.3.2 A Machine learning–based network reconfiguration methodology

7.3.2.1 Literature review on network reconfiguration

Over the past decades, power grids have become more prone to be overloaded due to the disproportionate increasing of the load demand and generation units (both dispatchable and nondispatchable units) [76]. Hence, the reconfiguration technique has been proposed to prevent overloading and provide an efficient power dispatch, especially in the emergency conditions when the line current approaches its maximum ampacity [77]. By definition the reconfiguration is the process of changing the topology of the power network by some prelocated sectionalizing and tie switches that can significantly improve the grid reliability [77], voltage profile [78], line loss [79], and load balance [80]. Up to now, many heuristics and mathematical algorithms have been used to solve the reconfigurable power grids, for example, collective decision-based algorithm [77], genetic algorithm [81], and mixed integer linear programming [77,78]. However, fast and effective reconfiguration response for electric service restoration is one of the critical requirements, which can be achieved by using state-of-the-art AI techniques to solve the reconfigurable power grids. Therefore in this section, a deep learning model is applied to solve the reconfigurable power grids.

In this section a new AI technique, known as the deep learning gated recurrent unit (GRU) (DLGRU), is developed to solve the reconfigurable power grids by learning the topological patterns of buses/lines with their physical features. GRU was first designed in Ref. [82] to decrease the complexity of long short-term memory (LSTM) and also enhance the LSTM performance. Same as LSTM, GRU has gates that can control the flow of the information from the input to the output. DLGRU can potentially understand complex nonlinear topological characteristics of the grid. Moreover, as will be shown in the result section, the total operation cost of the power grid by utilizing the DLGRU technique to select optimal switching of the reconfiguration is almost similar to conventional optimization techniques. DLGRU can solve the reconfigurable power grids in real time. However, conventional optimization techniques require a certain amount of time interval to obtain the optimal results by running the optimal power flow. More information regarding the GRU method can be obtained in Ref. [82].

As mentioned, DLGRU is proposed to solve the reconfigurable power grid with the following objective function and constraints through a "learn to optimize" manner. Specifically, the main objective is to minimize the total operation cost as:

$$\min \sum_{i\in\Omega^{DG}} \sum_{t\in\Omega^{T}} [C_i P_{it}^{G} + \mathrm{SU}_{it} + \mathrm{SD}_{it}] + \sum_{k\in\Omega^{S}} N_{\mathrm{RCS},kt} \lambda_{\mathrm{RCS}} \tag{7.19}$$

where C_i is the generation cost of the ith unit, P_{it}^{G} is the generated active power of the ith unit at time t, and $\mathrm{SU}_{\mathrm{it}}$ and $\mathrm{SD}_{\mathrm{it}}$ are the start-up and shutdown costs of the ith unit at time t, respectively. Also, $N_{\mathrm{RCS},kt}$ and λ_{RCS} represent the number of reconfigurable switching actions of kth remote-control switch (RCS) at time t and reconfiguration switching cost, respectively. Moreover, Ω^{DG}, Ω^{T}, and Ω^{S} denote sets of the generation units, switches, and time, respectively. It should be noted that the first and second terms of Eq. (7.19) represent the generation and reconfiguration switching costs of the grid, respectively. The proposed reconfigurable power grid includes some significant constraints in the following paragraphs.

Power balance constraints: the active and reactive power balances of each bus should be constrained as:

$$\sum_{nm\in\Omega^{L}} \left[P_{nm,t}^{L} - R_{nm}\left(I_{nm,t}^{L}\right)^{2}\right] - \sum_{nm\in\Omega^{L}} \left[P_{nm,t}^{L}\right] + P_{it}^{G} = P_{t}^{D} \forall i\in\Omega^{DG}, \forall t\in\Omega^{T} \tag{7.20}$$

$$\sum_{nm\in\Omega^{L}} \left[Q_{nm,t}^{L} - X_{nm}\left(I_{nm,t}^{L}\right)^{2}\right] - \sum_{nm\in\Omega^{L}} \left[Q_{nm,t}^{L}\right] + Q_{it}^{G} = Q_{t}^{D} \forall i\in\Omega^{DG}, \forall t\in\Omega^{T} \tag{7.21}$$

where $P_{nm,t}^{L}$ and $Q_{nm,t}^{L}$ are active and reactive power flow of line nm at time t, respectively. P_{t}^{D} and Q_{t}^{D} are the total active and reactive power demand at time t, respectively. R_{nm} and X_{nm} are resistance and reactance of line nm, respectively. Also, $I_{nm,t}^{L}$ represents the current flow of distribution line nm at time t. It should be noted that Ω^{L} denotes as the set of distribution lines. To apply the KVL to the distribution lines, the following constraints should be considered:

$$(V_{mt})^2 - (V_{nt})^2 = 2\left(R_{nm}P^L_{nm,t} + X_{nm}Q^L_{nm,t}\right) - (Z_{nm})^2\left(I^L_{nm,t}\right)^2 + \Delta V_{nmt}$$
$$\forall nm \in \Omega^L, \forall n, m \in \Omega^N, \forall t \in \Omega^T \tag{7.22}$$

$$(V_{mt})^2\left(I^L_{nm,t}\right)^2 = \left(P^L_{nm,t}\right)^2 + \left(Q^L_{nm,t}\right)^2 \forall nm \in \Omega^L, \forall n, m \in \Omega^N, \forall t \in \Omega^T \tag{7.23}$$

where n and m are defined as the bus indices, and Ω^N denotes the set of buses. Here, V_{mt} represents the voltage of bus m at time t, and ΔV_{nmt} is the auxiliary variable that can be zero if line mn is switched on at time t. Otherwise, it can be positive or negative, which depends on the difference between the voltages of the sending and receiving ends of line nm. Finally, Z_{nm} denotes the impedance of line nm.

Generation units constraints: the active and reactive powers of generation units are constrained to a minimum and maximum capacity as:

$$\underline{P^G_i} I_{it} \leq P^G_{it} \leq \overline{P^G_i} I_{it} \forall i \in \Omega^{DG}, \forall t \in \Omega^T \tag{7.24}$$

$$\underline{Q^G_i} I_{it} \leq Q^G_{it} \leq \overline{Q^G_i} I_{it} \forall i \in \Omega^{DG}, \forall t \in \Omega^T \tag{7.25}$$

where $\underline{P^G_i}$ and $\overline{P^G_i}$ present the minimum and maximum active power capacity of the ith unit, respectively. Similarly, $\underline{Q^G_i}$ and $\overline{Q^G_i}$ present the minimum and maximum active power capacity of the ith unit, respectively. It should be noted that I_{it} is the on/off status of the ith unit at time t, which can be zero (when unit is on) or one (when unit is off).

Bus voltage and angle limits: the voltage and angle of each bus are limited as:

$$\underline{V_n} \leq V_{nt} \leq \overline{V_n} \forall n \in \Omega^N, \forall t \in \Omega^T \tag{7.26}$$

$$-\pi \leq \theta_{nt} \leq \pi \forall n \in \Omega^N, \forall t \in \Omega^T \tag{7.27}$$

where $\underline{V_n}$ and $\overline{V_n}$ represent the minimum and maximum permissible voltage at bus n, respectively. Also, θ_{nt} denotes the angle of bus n at time t.

Reconfiguration constraints: the number of reconfigurable switching per day is constrained as:

$$N_{\mathrm{RCS},k,t} \leq \overline{N_{\mathrm{RCS}}} \forall k \in \Omega^S, \forall t \in \Omega^T \tag{7.28}$$

where $\overline{N_{RCS}}$ denotes the maximum permissible switching per day. Also, the radiality of the network after each reconfiguration process is assured:

$$N_{loop} = N_{branch} - N_{bus} + 1 \tag{7.29}$$

where N_{loop}, N_{branch}, and N_{bus} represent the total number of network main loops, the number of branches, and the number of buses, respectively. More details regarding the reconfigurable power grid as well as the conventional methods for solving the problem can be found in Ref. [77].

Fig. 7.7 illustrates the proposed DLGRU to solve the reconfiguration of the power systems. There are inputs (the load and the generation unit powers), output (the switching numbers of the reconfigurations of the power systems), and hidden GRU layers.

7.3.3 Network reconfiguration results

In this part, as shown in Fig. 7.8, a modified IEEE 33-bus test system is selected to demonstrate the effectiveness of the proposed DLGRU for finding the optimal switching of the reconfigurable power grids. The model includes five tie switches (that are shown by the *dotted lines*), as well as 33 sectionalized switches (that are shown as the *solid lines*). Table 7.8 shows the characteristics of the generation units within the network. It is worth noting that in any time interval, five switches should be open to maintain the radiality of the network. The number of hidden layers in the simulation results is 4, and there are 50 units in each layer.

The reconfiguration switching for both DLGRU and a conventional benchmark method [77] for a 24-hour time horizon is compared in

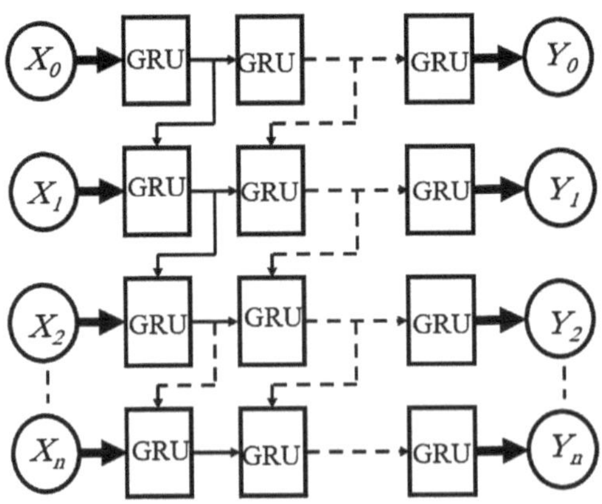

Figure 7.7 DLGRU block diagram. *DLGRU*, Deep learning gated recurrent unit.

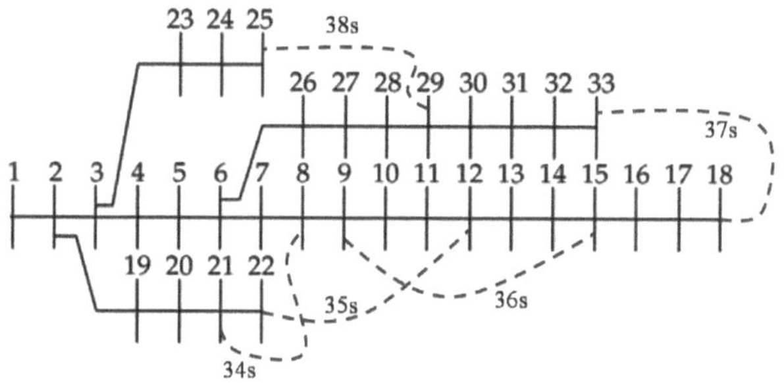

Figure 7.8 Single line diagram of the modified IEEE 33-bus system.

Table 7.8 Generation units features.

Generation type	$\underline{P_i^G}$ (kW)	$\overline{P_i^G}$ (kW)	C_i ($/kWh)	SU/SD ($)
WT1	0	25	1.073	0
WT2	0	25	1.073	0
FC	80	1000	0.294	0.95
MT1	100	1500	0.457	1.65
MT2	100	1500	0.457	1.65

FC, Fuel cell; *MT1*, microturbine 1; *MT2*, microturbine 2; *WT1*, wind turbine 1; *WT2*, wind turbine 2.

Fig. 7.9. It is observed that there is a small difference between the optimal result of the proposed DLGRU method and the conventional technique for generating the switching's of the lines to reconfigure the network. Fig. 7.10 shows the contribution of the generation units for both DLGRU and the conventional technique. The total operation cost of the conventional method and the DLGRU method are $138,017.4 and $138,132.2, respectively. Based on the simulation results, the operation costs of the proposed machine learning technique and the conventional technique are very close. However, the proposed DLGRU technique can find the optimal switching in real time due to its fast convergence speed.

7.4 Advanced artificial intelligence and machine learning applications to building occupancy detection

The occupancy detection is beneficial to improve building energy management and provide important information for demand response.

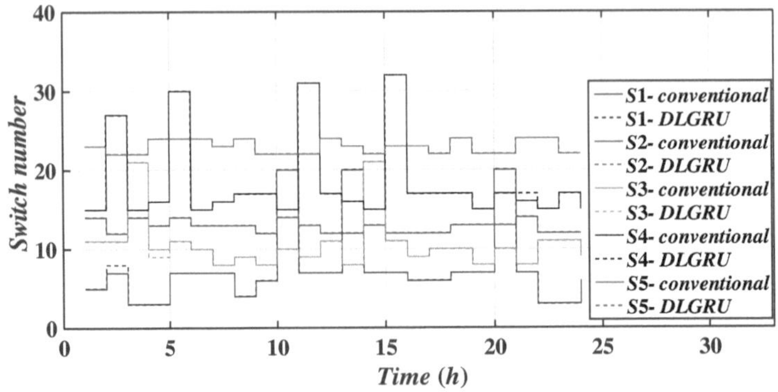

Figure 7.9 Reconfiguration switching of both the conventional and DLGRU techniques. *DLGRU*, Deep learning gated recurrent unit.

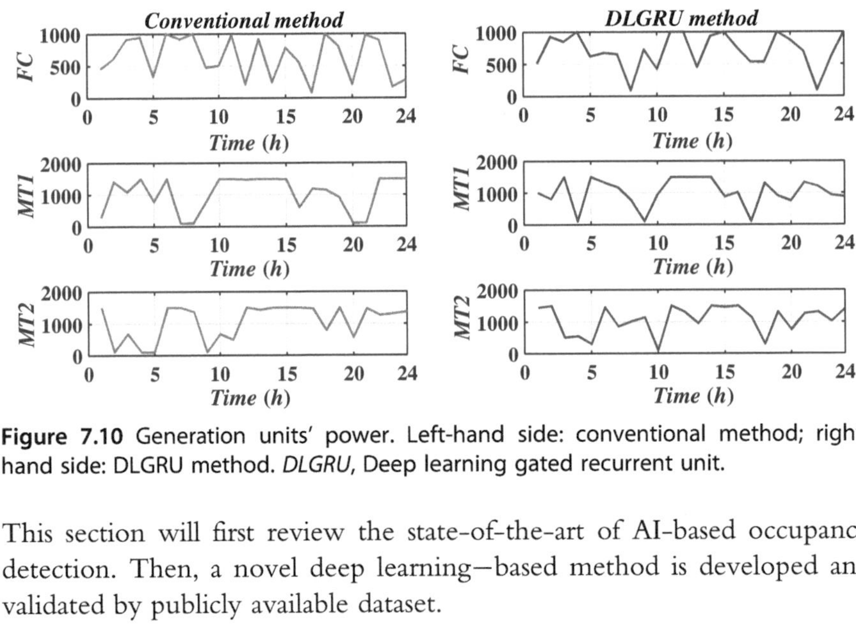

Figure 7.10 Generation units' power. Left-hand side: conventional method; right-hand side: DLGRU method. *DLGRU*, Deep learning gated recurrent unit.

This section will first review the state-of-the-art of AI-based occupancy detection. Then, a novel deep learning–based method is developed and validated by publicly available dataset.

7.4.1 Prior research work

Based on the monitored objects, occupancy detection can be grouped into intrusive and nonintrusive approaches. Intrusive sensors directly measure indoor environments, including motional, acoustic, or climatic parameters [83]. For example, Candanedo and Feldheim [84] compared occupancy detection methods with different indoor climate feature combinations, among which the best model had over 99% accuracy.

Surveillance videos were used for occupancy detection with 95.3% accuracy in Ref. [85]. However, the intrusive occupancy detection is challenging to extensively deploy due to the high installation cost, additional operation requirements (e.g., the illumination condition for cameras), and privacy concerns. Therefore it motivated the development of nonintrusive occupancy detection. The nonintrusive occupancy detection relies on infrastructure sensors that monitor parameters such as Wi-Fi, Bluetooth, or radio-frequency identification (RFID) [86–88]. For example, Wang et al. [89] developed a Wi-Fi–based occupancy detection system with 72.7% accuracy, which helped save 26.4% of energy in cooling and ventilation demands. An RFID-based occupancy detection system was able to track the stationary and mobile occupants with 88% and 62% accuracy, respectively [87]. The nonintrusive methods have fewer privacy issues; however, they still suffer from possibly unsatisfactory accuracies, low infrastructure/device coverage, and extra occupant participation. The limitations of both intrusive and nonintrusive sensors provide an opportunity to infer building occupancy from AMI data. Nevertheless, AMI-based occupancy detection is still limited compared to other sensor-based detection.

Regarding the detection algorithms, data-driven machine learning algorithms are widely applied, due to the growing penetration of sensors. For example, a collection of machine learning models, including a decision tree (DT) model, a SVM, and a Bayes network, were compared in Ref. [90], among which DT was the best model. In addition, SVM beat the hidden Markov model (HMM) and the *k*NN model with 80% accuracy [91]. Moreover, Jin et al. [92] developed an RF model, which outperformed HMM and SVM models with around 90% accuracy. In contrast with the wide deployment of shallow machine learning algorithms, deep learning is far from fully explored in occupancy detection problems. A CNN is the first deep learning model developed for occupancy detection, which provided occupancy information from indoor climate measurements with 95.42% accuracy [93]. Another deep learning architecture utilized an autoencoder long-term recurrent convolutional network, which identified the occupant activity from Wi-Fi–enabled Internet of Things devices [94].

To bridge the gap in occupancy detection using AMI data, a deep learning architecture is proposed in this section. The developed deep learning architecture stacks a CNN and a LSTM network sequentially. The developed CNN–LSTM architecture is expected to capture both spatial and contextual representations of the AMI data. The problem formulation and methodology description are introduced in the rest of this section. The developed model is then compared with state-of-the-art classifiers with a publicly available dataset.

7.4.2 The convolutional neural network–long short-term memory deep learning architecture

7.4.2.1 Occupancy detection problem formulation

The objective of the building occupancy detection is to identify the real-time occupancy condition, $\mathbf{y} \in R^{N \times 1}$, of a house from its AMI data, $\mathbf{X} \in R^{N \times F}$, by using a stacking CNN model and an LSTM model [95]:

$$\hat{\mathbf{y}} = F(\mathbf{X}, \mathbf{W}) = F_R\left\{F_C(\mathbf{X}, \mathbf{W}_C), \mathbf{W}_R\right\} \tag{7.30}$$

where N and F are the sample size and dimension, respectively. $\mathbf{y}$ and $\hat{\mathbf{y}}$ are actual and detected occupancy conditions, respectively. $F(\cdot)$, $F_C(\cdot)$, and $F_R(\cdot)$ indicate the developed CNN–LSTM model and its CNN and LSTM components, respectively. $\mathbf{W}$, $\mathbf{W_C}$, and $\mathbf{W_R}$ are the parameters in the according models. With the condition of either occupied or vacant at every timestamp, that is, $y \in 0, 1$, the occupancy detection is a binary classification problem. Therefore the objective is to minimize the loss function, which is based on the weighted binary cross-entropy, given by:

$$J(\mathbf{W}) = -\frac{1}{N}\sum_{n=1}^{N}\left[\omega y_n \log\left(\hat{y}_n\right) + (1-\omega)(1-y_n)\log\left(1-\hat{y}_n\right)\right] \tag{7.31}$$

where ω is the binary cross-entropy weight, defined as $\omega = P_Y[y = 0 | y \in \mathbf{y}]$.

7.4.2.2 Convolutional neural network

With one or multiple feature learning blocks (FLBs) that consist of a convolutional layer and a max-pooling layer, CNN has powerful feature learning ability. A convolutional layer (indexed by l) contains $D^{(l+1)}$ filters. A convolution operation is performed in each filter to construct a set of feature maps as:

$$\mathbf{Z}^l = \mathbf{W}^l \times \mathbf{X}^l + \mathbf{b}^l \tag{7.32}$$

where $\mathbf{X}^l \in R^{H^l \times D^l}$ is the convolutional layer input. $\mathbf{Z} \in R^{H^{(l+1)} \times D^{(l+1)}}$, $\mathbf{W} \in R^{H^{(l+1)} \times D^l \times D^{(l+1)}}$, and $\mathbf{b} \in R^{H^{(l+1)} \times D^{(l+1)}}$ are the feature map matrix, layer parameter tensor, and bias matrix, respectively. Eq. (7.32) can be expressed in detail as:

$$z^{l}_{h^{(l+1)},d^{(l+1)}} = \sum_{h^l=1}^{H^l} \sum_{d^l=1}^{D^l} \left[w^{l}_{d^l,h^l,d^{(l+1)}} \times x^{l}_{h^l,d^l} + b^{l}_{h^{(l+1)},d^{(l+1)}} \right] \quad (7.33)$$

where (h, d) is a doublet index used to locate the element, x, in $\mathbf{X}$.

To add nonlinearity to the network an element-wise rectified linear unite (ReLU) is used to activate convolution outputs. The ReLU activation function is selected due to its computational efficiency, better convergence, superior performance, and amelioration of vanishing gradients compared to other functions [96]. The ReLU function is expressed as:

$$z^{l}_{h^l,d^l} = \max(0, x^{l}_{h^l,d^l}) \quad (7.34)$$

A max-pooling layer is introduced as the last layer of an FLB to achieve more translation invariance during the spatial representation learning. In this study a unified nonoverlapping moving window is used to subsample the activated convoluted feature maps by a factor of 2. The max-pooling layer is expressed as:

$$z^{l}_{h^{(l+1)},d^l} = \max\left\{ \mathbf{x}^l | \mathbf{x}^l : = x^{l}_{h^l:h^l+\left(H^l/(H^{l+1})\right),d^l} \right\} \quad (7.35)$$

7.4.2.3 Long short-term memory network

RNNs have been effectively applied in time series data analytics due to its capability of capturing temporal correlations. The LSTM is a variant of RNN, which avoids the vanishing gradient problem by gated regulators. A typical LSTM block contains a memory cell, a forget gate, an input gate, and an output gate, shown as Fig. 7.11. The three gates have different functions, where the input gate determines the new information stored in the memory, the forget gate decides the useless old information to exclude, and the output gate exploits useful information to output from the memory cell. The tensor operations in the gates of a forward LSTM are expressed as:

$$\mathbf{f}_t = \sigma(\mathbf{W}_f[\mathbf{h}_{t-1}, \mathbf{X}_t] + \mathbf{b}_f) \quad (7.36)$$

$$\mathbf{i}_t = \sigma(\mathbf{W}_i[\mathbf{h}_{t-1}, \mathbf{X}_t] + \mathbf{b}_i) \quad (7.37)$$

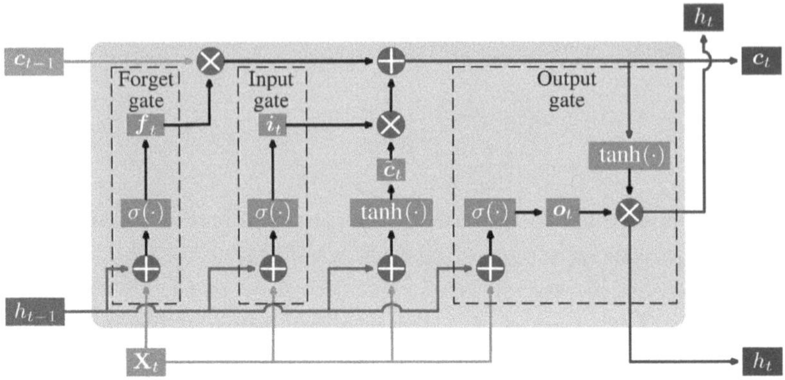

Figure 7.11 The LSTM building block. *LSTM*, Long short-term memory.

$$\tilde{\mathbf{c}}_t = \tanh(\mathbf{W}_c[\mathbf{h}_{t-1}, \mathbf{X}_t] + \mathbf{b}_c) \tag{7.38}$$

$$\mathbf{c}_t = \mathbf{f}_t \times \mathbf{c}_{t-1} + \mathbf{i}_t \times \tilde{\mathbf{c}}_t \tag{7.39}$$

$$\mathbf{o}_t = \sigma(\mathbf{W}_o[\mathbf{h}_{t-1}, \mathbf{X}_t] + \mathbf{b}_o) \tag{7.40}$$

$$\mathbf{h}_t = \mathbf{o}_t \times \tanh(\mathbf{c}_t) \tag{7.41}$$

where $\mathbf{f}$, $\mathbf{i}$, $\mathbf{o}$, and $\mathbf{c}$ are the activation vectors of the forget, input, output gates and the memory cell, respectively. $\mathbf{W}_f$, $\mathbf{W}_i$, $\mathbf{W}_c$, and $\mathbf{W}_o$ are the input matrices of gates or memory cell. $\mathbf{b}_f$, $\mathbf{b}_i$, $\mathbf{b}_c$, and $\mathbf{b}_o$ are the corresponding bias vectors. $\sigma(\cdot)$ and $\tanh(\cdot)$ are the sigmoid and hyperbolic tangent activation functions, respectively. $\mathbf{h}$ is the hidden state and the output of an LSTM hidden layer. $\tilde{\mathbf{c}}$ is the new state candidate vector. The bracket is the concatenation operator. The nodal connections in an LSTM cell are shown in Fig. 7.11.

One typical issue of most recurrent models is that they can only deal with unidirectional dependencies in the data. To capture both forward and backward dependencies in the AMI data a bidirectional LSTM (BiLSTM) layer is included in the LSTM configuration. The nodal connection and the tensor calculations in a BiLSTM are almost the same with those in a unidirectional LSTM, except for the processing directions. The bidirectional operations within a BiLSTM can be expressed as:

$$\vec{\mathbf{f}}_t = \sigma\left(\vec{\mathbf{W}}_f\left[\vec{\mathbf{h}}_{t-1}, \vec{\mathbf{X}}_t\right] + \vec{\mathbf{b}}_f\right) \tag{7.42}$$

$$\overleftarrow{\mathbf{f}}_t = \sigma\left(\overleftarrow{\mathbf{W}}_f\left[\overleftarrow{\mathbf{h}}_{t+1}, \overleftarrow{\mathbf{X}}_t\right] + \overleftarrow{\mathbf{b}}_f\right) \tag{7.43}$$

where $\rightarrow$ and $\leftarrow$ denote the forward and backward operations, respectively. Both $\vec{h}_t$ and $\leftarrow h_t$ are generated and concatenated as the output of the BiLSTM:

$$\mathbf{h}_t = [\overrightarrow{\mathbf{h}}_t, \overleftarrow{\mathbf{h}}_t] \tag{7.44}$$

7.4.2.4 The developed convolutional neural network–long short-term memory architecture

The overall framework of the developed CNN–LSTM and the tensor manipulations are shown in Fig. 7.12. The developed framework contains three stacking components, which are a CNN network, an LSTM, and a dense layer configuration. The AMI data is first convoluted through the CNN network to generate spatial feature maps, as shown in the bottom left dashed box. Inspired by the VGGNet, two FLBs with four layers (i.e., two convolutional layers and two max-pooling layers) are consisted in the CNN. There are 128 and 64 filters in the two FLBs, respectively, which are used to extract the high-level abstract spatial features from the AMI data.

The output of CNN configuration serves as the input to LSTM configuration. To better capture the contextual patterns in the AMI data, we increase the depth of the architecture by stacking three LSTM layers vertically [97]. As shown in Fig. 7.12, the first LSTM layer extracts temporal features from the previous CNN output and generates hidden states. Then, a BiLSTM layer takes both forward and backward dependencies into account, which is combined by the last LSTM layer. The developed BiLSTM configuration is expected to learn hierarchical representations of the convolutional time series of AMI data by operating hidden states at different timescales.

The last component in the CNN–LSTM model is a dense layer configuration with two fully connected layers and a final classification layer. The output of LSTM is first flattened and fed into the first dense layer, whose output is then fed into the second dense layer:

$$\mathbf{Z}^l = \mathbf{W}^l \cdot \mathbf{X}^l + \mathbf{b}^l \tag{7.45}$$

where all the inputs are transmitted to the output. The last layer is a classification layer with sigmoid activation function due to the mutual-exclusive character of occupancy detection results:

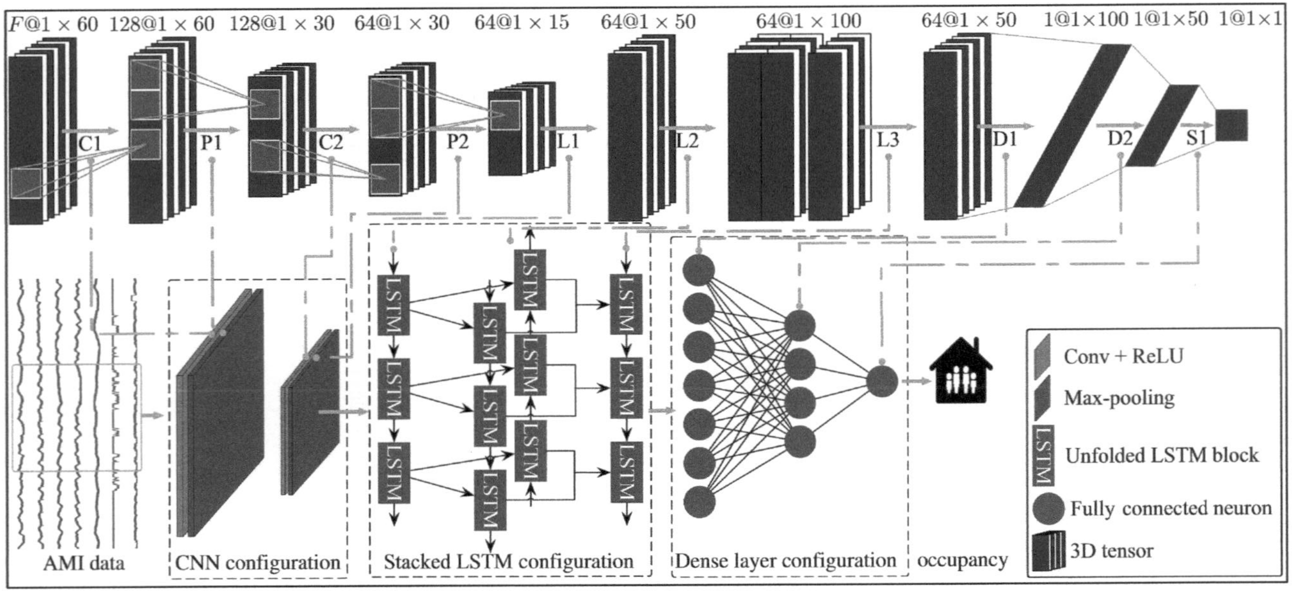

Figure 7.12 The developed end-to-end CNN–LSTM occupancy detection framework. *CNN*, Convolutional neural network; LSTM, long short-term memory.

$$\mathbf{Z}^l = \frac{1}{1 + e^{-\mathbf{W}^l \cdot \mathbf{X}^l + \mathbf{b}^l}} \tag{7.46}$$

where $l = 10$. The occupancy condition could be determined by a applying a threshold to the last-layer output:

$$\hat{y} = H(z - th) \tag{7.47}$$

where $H(\cdot)$ is the Heaviside step function and $th = 0.5$ is the threshold value.

7.4.3 Experiments

The developed CNN–LSTM model is tested with the largest publicly available dataset with both AMI and occupancy data, that is, the Electricity Consumption and Occupancy dataset [98]. The dataset has summer and winter data of five houses with 1-second resolution (aggregated from data with 1 Hz frequency). To avoid the noise and redundancy, basic features were extracted from the AMI data [99]. The minute data is flattened every hour to fit the CNN input format with $W = 60$. The quality-control with two criteria was applied to the dataset: (1) the data length should be more than 900 and (2) both occupancy labels (i.e., occupied and vacant conditions) should be more than 10% of the total occupancy data. As a result, four periods of data from three houses were qualified and selected for case studies, which are summarized in Table 7.9. The ratio of training, validation, and testing data is 3:1:1 for each case study.

Hyperparameters of the developed CNN–LSTM model are determined by a trial-and-error manner with the training and validation dataset and are listed in Table 7.9. The model has a total of 240,489 trainable parameters (2.5 MB), which is a relatively small network, compared to other famous deep learning architectures (e.g., the 16-layer VGGNet has 528 MB weights). Before being trained, parameters in each layer are

Table 7.9 Data summary of case studies.

Case notations	Data dimension (N, W, F)	Label (occupied/vacant)
C1	(935, 60, 30)	774/161
C2	(1103, 60, 30)	831/272
C3	(1079, 60, 24)	771/308
C4	(1367, 60, 29)	1044/323

Note: The case notations (C1–C4) are different from Sections 7.2.3.1 and 7.2.3.2.

Table 7.10 Hyperparameters of layers in the developed convolutional neural network–long short-term memory (LSTM) model.

Layer	Type	Hyperparameter	Trainable parameters
C1	Convolutional	Input size: $60 \times F$ Filter size: 3 Filter number: 128	11,648
P1	Pooling	Window size: 2×128 Stride: 2	0
C2	Convolutional	Input size: 30×128 Filter size: 3 Filter number: 64	34,640
P2	Pooling	Window size: 30×64 Stride: 2	0
L1	LSTM	Output length: 50	23,000
L2	BiLSTM	Output length: 200	120,800
L3	LSTM	Output length: 50	50,200
D1	Dense	Neurons: 100	5100
D2	Dense	Neurons: 50	5050
S2	Classification	Neurons: 1	51

initialized with the Xavier method, where biases are initialized as zeros and weights conform the Gaussian distribution [100]. Twenty percent of neurons in the convolutional, LSTM, and dense layers are randomly dropped to overcome the possible overfitting [101] (Table 7.10).

In this study, mini-batch stochastic gradient descent (SDG) is selected as the optimizer to train the developed CNN–LSTM model. The SDG minimizes the objective function $\boldsymbol{J}(\boldsymbol{W})$ in Eq. (7.31) by updating the parameters in the opposite direction of its gradients. The complete data is passed forward and backward through the network with 100 epochs. The mini-batches ($\boldsymbol{B} := [\boldsymbol{X},\boldsymbol{y}]$), with a batch size of 30, are randomly generated to shuffle the data order in each epoch. Gradients in each iteration are calculated by averaging over the mini-batch. The learning rate scheduling is used to dampen training oscillations. Specifically, the training starts with a learning rate of 0.1 and reduces the learning rate by 50% when a plateau is reached for more than 10 epochs. To reduce the risk of convergence to local minima, momentum is adopted in the training. The weights are updated with the above techniques as:

$$\mathbf{V}_{i+1} = \gamma \mathbf{V}_{\mathrm{i}} + \eta_{i+1} \nabla_{\mathbf{W}} \mathrm{J}(\mathbf{W};\mathbf{B}) \tag{7.48}$$

$$\mathbf{W}_{i+1} = \mathbf{W}_i - \mathbf{V}_{i+1} \tag{7.49}$$

where $\boldsymbol{i}$ is the iteration index, $\boldsymbol{V}$ is the weight update matrix, $\gamma = \boldsymbol{0.9}$ is the momentum, and η is the learning rate.

Six state-of-the-art machine learning–based classifiers are adopted to compare with the developed CNN–LSTM method, which are a *k*NN model, an SVM model with linear kernel, a GP model, an RF, a multilayer perceptron (MLP) classifier, and an adaptive boosting model (AdaBoost). We believe the selected benchmarks are general and representative, since they cover the nonparametric model (*k*NN), kernel-based model (SVM), feedforward NN (MLP), and ensemble learning models (RF and AdaBoost). The hyperparameters and parameters of these models are determined empirically and listed in Table 7.11, including the number of neighbors (`n_neighbors`) in *k*NN, the penalty weight (C) in SVM, the covariance function (`kernel`) of the GP, the number of trees (`n_estimators`), the maximum depth of the tree (`max_depth`), the number of features for the best split (`max_features`) in RF, the number of hidden layer neurons (`neuron`), activation function (`activation`), regularization penalty (`alpha`) in MLP, and the maximum number of estimators (`n_estimators`), and learning rate (`learning_rate`) in AdaBoost.

7.4.4 Results

The case studies are implemented in Python version 3.6 with the `Keras`, `Tensorflow`, and `scikit-learn` libraries. The experiments are repeated 10 times to improve the reproducibility and consistency. All the experiments are conducted on a workstation with an Intel Xeon(R) E5-2603 1.6 GHz CPU and an NVIDIA TITAN V GPU. It took 11 minutes to train a CNN–LSTM model with the experiment setups described above and 3.7 ms to detect the occupancy at a time step. The computational time is applicable for real-time detection.

Table 7.11 Benchmark model hyperparameters and parameters.

Model	Hyperparameters/parameters
*k*NN	`n_neighbors` = **3**
SVM	`C` = 0.025
GP	`kernel` = ***1.0 × RBF(1.0)***
RF	`n_estimators` = 10, `max_depth` = 5, `max_features` = 1
MLP	`neuron` = 100, `activation` = ReLU, `alpha` = 0.01
AdaBoost	`n_estimators` = 50, `learning_rate` = 1.0

RF, random forest; *GP*, Gaussian process; *kNN*, *k*-nearest neighbor; *SVM*, support vector machine; *MLP*, multilayer perceptron classifier.

Being a typical binary classification problem, the occupancy detection can be evaluated based on the binary confusion matrix, which consists of four elements (as shown in Fig. 7.13): (1) true positive (TP), denoting the count that a house is actually occupied and is detected occupied; (2) false positive (FP), denoting the count that a house is actually vacant but is detected occupied; (3) true negative (TN), denoting the count that a house is actually vacant and is detected vacant; and (4) false negative, denoting the count that a house is actually occupied but is detected vacant. Based on the confusion matrix, five metrics are used to quantify the overall performance of the classification models, which are accuracy (ACC), sensitivity (SNS, also known as TP rate or recall), specificity (SPC, also known as TN rate), precision (PRC), and $F1$ score ($F1$):

$$\mathrm{ACC} = \frac{\mathrm{TP} + \mathrm{TN}}{\mathrm{TP} + \mathrm{TN} + \mathrm{FP} + \mathrm{FN}} \tag{7.50}$$

$$\mathrm{SNS} = \frac{\mathrm{TP}}{\mathrm{TP} + \mathrm{FN}} \tag{7.51}$$

$$\mathrm{SPC} = \frac{\mathrm{TN}}{\mathrm{TN} + \mathrm{FP}} \tag{7.52}$$

$$\mathrm{PRC} = \frac{\mathrm{TP}}{\mathrm{TP} + \mathrm{FP}} \tag{7.53}$$

$$F1 = 2\frac{\mathrm{SNS} \times \mathrm{PRC}}{\mathrm{SNS} + \mathrm{PRC}} \tag{7.54}$$

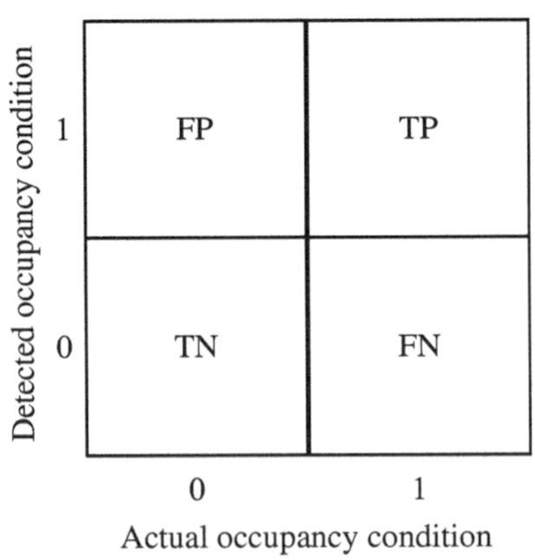

Figure 7.13 A binary confusion matrix.

where ACC indicates the overall accuracy of the detection. SNS measures the proportion of the actual occupied conditions that are correctly detected as such. SPC measures the proportion of the actual vacant conditions that are correctly detected as such. PRC is the success probability of detecting a correct occupied condition. $F1$ combines several metrics in the imbalanced classification problem.

The occupancy detection accuracy is not only affected by the models, but also related to the threshold shown in Eq. (7.47). Therefore in addition to the five overall metrics, another set of metrics are used to assess the goodness of the classifiers over the entire operating range, which are receiver operating characteristic (ROC) curve and the area under the ROC curve (AUC). The former one is a curve of TP rate (SNS) against FP rate ($1 - \text{SPC}$) at various threshold settings, which can also be used to determine the best th. A larger deviation between the ROC curve and the diagonal line represents a better occupancy detection. The AUC value is a comprehensive measurement of the ROC curve, where a larger AUC value (maximum $\text{AUC} = 1$) indicates a better result.

The confusion matrix of one repeat is shown in Fig. 7.14. Results of different models are located in different columns and matrices in different rows indicate results of different cases. Even though the labels are highly imbalanced in the whole dataset (as listed in Table 7.9), the testing data labels could be balanced. For example, the vacant conditions and occupied conditions ratio is 115:106 in the testing data of C. In addition, the diversity of the four cases directly impacts the model performance. For instance, kNN detects more FP than TN in C1 but vice versa in C2 and C4. Different models perform distinctively in the same case. For example, SVM, GP, and MLP accurately detect more occupied conditions; however, AdaBoost is more powerful in detecting the vacant conditions in C2. We can conclude that both the dataset and the benchmarks are diverse and general for the comparisons.

Based on confusion matrices, the evaluation metrics are calculated and listed in Tables 7.12 and 7.13. From Table 7.12, it is observed that every benchmark model has the chance to generate satisfactory occupancy detection, since five out of six benchmark models outperform others in some cases and based on some metrics. Nevertheless, none of the benchmark models can always beat others in all the four cases and all the metrics. In addition, some benchmark models make extremely assertive detections. For example, GP detects occupancy conditions with 100% SNS but 0% SPC in C1, C3, and C4. Some benchmarks are not robust,

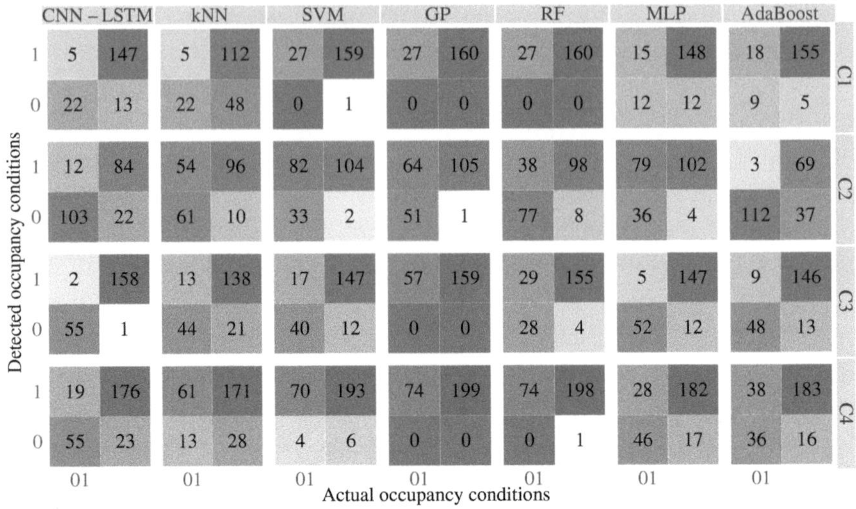

Figure 7.14 Confusion matrices of the occupancy detection results. The darker color indicates a higher frequency, and the gray color means a zero occurrence. Positive diagonal elements indicate right detections, and negative diagonal elements show the wrong detections.

which is revealed by the nonnegligible variance of the metrics over 10 experiment repeats, such as SPC of RF in C2 and SPC of MLP in C4. In contrast, the developed CNN–LSTM model shows encouraging accuracy in both the occupied and vacant conditions. More importantly, the CNN–LSTM model is more accurate than benchmark models in all the four cases, indicated by two overall evaluation metrics, ACC and $F1$. The average ACC and $F1$ over the four cases and 10 experiment repeats are 89.41% and 91.55%, respectively. The robustness of the CNN–LSTM model is not only revealed in diverse cases but also shown in 10 sets of experiments, which is supported by the relatively small variances listed in Table 7.13.

To assess the model performance independent of the choice of th, ROC and AUC are adopted. Specifically, a set of th values, ranging from 0 to 1, are used to determine $\hat{y}$ and calculate TPR and FPR, as shown in Fig. 7.15. The perfect classifier should have the ROC curve straight up the vertical axis then along the horizontal axis. The classifier that randomly generates occupancy detection results sits on the diagonal, and the classifier that detects complete reverse results has a curve in the bottom left part of the ROC space. Therefore the developed CNN–LSTM model has better and more robust occupancy detection capability with

Table 7.12 Mean of evaluation metrics (%) of the 10 occupancy detection experiment repeats.

Case	Metric	Models						
		CNN–LSTM	*k*NN	SVM	GP	RF	NN	AdaBoost
C1	ACC	91.34	71.66	85.03	85.56	85.51	85.29	87.70
	SNS	93.13	70.00	99.38	100.00	99.94	91.06	96.88
	SPC	80.74	81.48	0.00	0.00	0.00	51.11	33.33
	PRC	96.63	95.73	85.48	85.56	85.55	91.74	89.60
	*F*1	94.84	80.87	91.91	92.22	92.19	91.37	93.09
C2	ACC	85.25	71.04	61.99	70.59	80.81	65.43	81.90
	SNS	82.36	90.57	98.11	99.06	89.62	94.43	65.09
	SPC	87.91	53.04	28.70	44.35	72.70	38.70	97.39
	PRC	86.79	64.00	55.91	62.13	76.69	59.08	95.83
	*F*1	84.21	75.00	71.23	76.36	82.12	72.48	77.53
C3	ACC	98.47	84.26	86.57	73.61	84.26	91.76	90.74
	SNS	99.12	86.79	92.45	100.00	95.66	92.45	92.58
	SPC	96.67	77.19	70.18	0.00	52.46	89.82	85.61
	PRC	98.81	91.39	89.63	73.61	85.06	96.23	94.72
	*F*1	98.96	89.03	91.02	84.80	89.99	94.29	93.64
C4	ACC	82.56	67.40	72.16	72.89	73.15	80.04	80.22
	SNS	89.50	85.93	96.98	100.00	99.70	95.33	91.96
	SPC	63.92	17.57	5.41	0.00	1.76	38.92	48.65
	PRC	87.21	73.71	73.38	72.89	73.19	81.02	82.81
	*F*1	88.20	79.35	83.55	84.32	84.41	87.48	87.14

ACC, Accuracy; CNN, convolutional neural network; *GP*, Gaussian process; *kNN*, *k*-nearest neighbor; *LSTM*, long short-term memory; *PRC*, precision; *RF*, random forest; *SNS*, sensitivity; *SPC*, specificity; *SVM*, support vector machine.

different thresholds. The AUC values of all the experiments are shown in Fig. 7.16, where the developed CNN–LSTM outperforms other models consistently.

To dig into the reason of better occupancy detection performance of the developed CNN–LSTM model, the detection probabilities and results of the developed model are compared to those of the three best benchmarks, that is, SVM, MLP, and AdaBoost. The outperformance of the CNN–LSTM model is twofold. First, the developed model is more accurate, revealed by the smaller deviations between detections and the actual conditions. Actually, the CNN–LSTM model only misclassifies the occupancy three times at three discrete hours, which is less harmful to the demand response decisions. In contrast, all the three best competing models generate more and longer period wrong detections. The second advantage is that the CNN–LSTM model is more confident in the detection it

Table 7.13 Standard deviation evaluation metrics (%) of the 10 occupancy detection experiment repeats.

Case	Metric	CNN–LSTM	*k*NN	SVM	GP	RF	NN	AdaBoost
C1	ACC	0.49	0.00	0.00	0.00	0.17	0.85	0.00
	SNS	0.59	0.00	0.00	0.00	0.20	2.52	0.00
	SPC	3.40	0.00	0.00	0.00	0.00	9.85	0.00
	PRC	0.56	0.00	0.00	0.00	0.02	1.34	0.00
	*F*1	0.30	0.00	0.00	0.00	0.10	0.66	0.00
C2	ACC	1.09	0.00	0.00	0.00	6.80	4.50	0.00
	SNS	6.55	0.00	0.00	0.00	4.15	4.41	0.00
	SPC	6.18	0.00	0.00	0.00	15.85	12.50	0.00
	PRC	4.40	0.00	0.00	0.00	9.92	4.11	0.00
	*F*1	1.51	0.00	0.00	0.00	4.67	1.81	0.00
C3	ACC	0.38	0.00	0.00	0.00	4.04	0.57	1.95
	SNS	0.44	0.00	0.00	0.00	1.98	0.89	1.59
	SPC	1.29	0.00	0.00	0.00	15.24	3.86	2.96
	PRC	0.45	0.00	0.00	0.00	3.98	1.33	1.11
	*F*1	0.26	0.00	0.00	0.00	2.32	0.35	1.35
C4	ACC	1.58	0.00	0.00	0.00	0.49	2.56	0.00
	SNS	3.95	0.00	0.00	0.00	0.35	3.03	0.00
	SPC	11.82	0.00	0.00	0.00	2.30	16.68	0.00
	PRC	3.66	0.00	0.00	0.00	0.42	3.86	0.00
	*F*1	1.07	0.00	0.00	0.00	0.23	1.15	0.00

ACC, Accuracy; *CNN*, convolutional neural network; *GP*, Gaussian process; *kNN*, *k*-nearest neighbor; *LSTM*, long short-term memory; *PRC*, precision; *RF*, random forest; *SNS*, sensitivity; *SPC*, specificity; *SVM*, support vector machine.

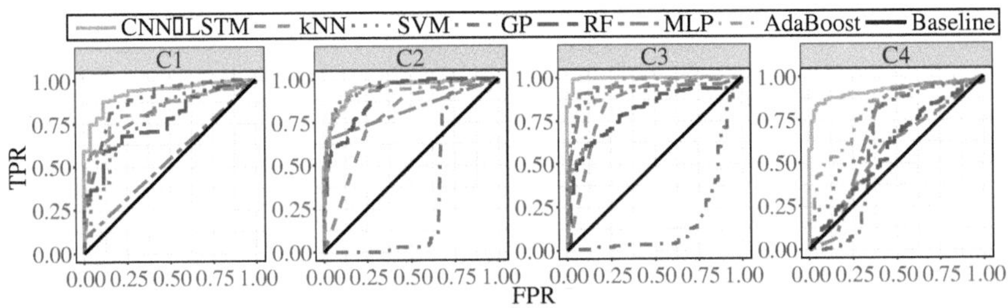

Figure 7.15 ROC curves in randomly selected experiments. *ROC*, Receiver operating characteristic.

makes. This is illustrated by the darker color of the classification probabilities, compared to other models, such as AdaBoost. Therefore the outperformance of the developed model is well-explained (Fig. 7.17).

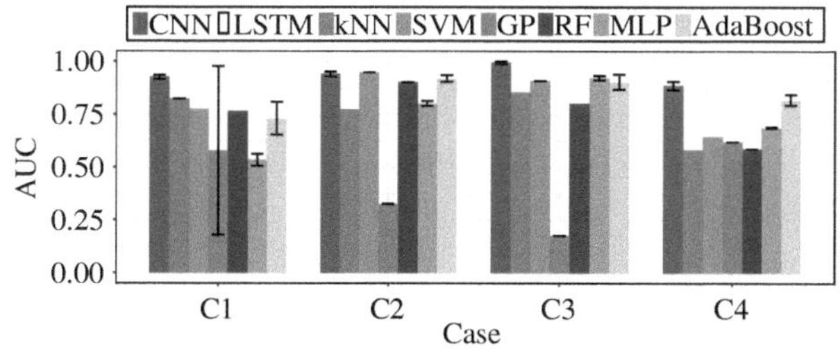

Figure 7.16 The AUC statistics of the 10 experiment repeats. *AUC*, Area under ROC curve.

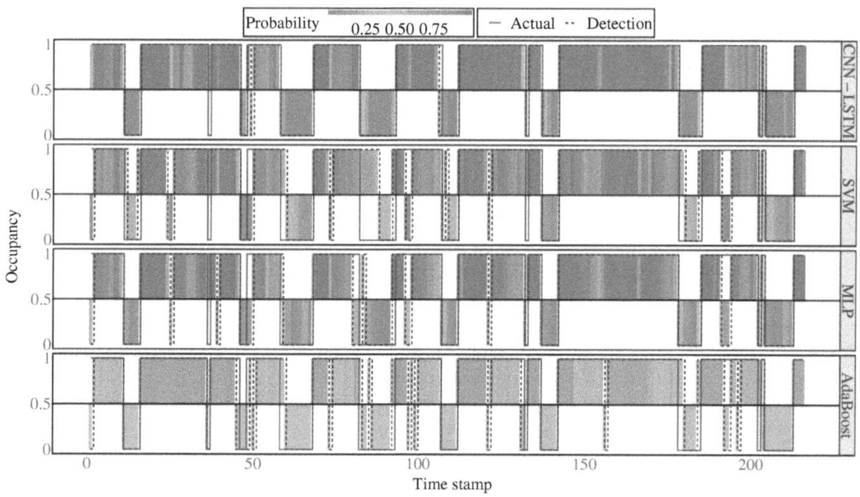

Figure 7.17 The occupancy labels, detection probability, and results of C3.

7.5 Conclusion

AI has been witnessed to have significant impacts on power system operations. Large amounts of data are being collected by smart grid devices, such as the AMIs and the PMUs, which facilitate the wide applications of AI techniques to power systems. In this section, two emerging machine learning subfields were introduced: ensemble learning and deep learning. Three use cases were used to demonstrate their applications in power systems. Specifically, competitive and cooperative ensemble learning models were developed to provide short-term wind forecasts. Both methods included state-of-the-art machine learning models, for example, ANNs, SVR models, GBMs, RF models, and Q-learning models. In addition, a DLGRU was used to solve a network reconfiguration problem through a "learn to optimize" manner. In the last application case, a deep learning configuration with a CNN and a LSTM network was developed to detect real-time occupancy conditions in smart buildings. Three sets of case studies based on publicly available datasets showed that the developed ensemble learning methodologies and deep learning methods outperformed the corresponding benchmarks. The accurate forecasts and detection are valuable to renewable integration, power system reliability, and building integration into the smart grid.

References

[1] Z. Huang, H. Luo, D. Skoda, T. Zhu, Y. Gu, E-sketch: gathering large-scale energy consumption data based on consumption patterns, 2014 IEEE International Conference on Big Data (Big Data), IEEE, 2014, . 656–665.

[2] C. Ramos, C.-C. Liu, AI in power systems and energy markets, IEEE Intell. Syst. 26 (2) (2011) 5–8.

[3] D. Ernst, M. Glavic, L. Wehenkel, Power systems stability control: reinforcement learning framework, IEEE Trans. Power Syst. 19 (1) (2004) 427–435.

[4] T.I. Ahamed, P.N. Rao, P. Sastry, A reinforcement learning approach to automatic generation control, Electr. Power Syst. Res. 63 (1) (2002) 9–26.

[5] T. Yu, J. Liu, K. Chan, J. Wang, Distributed multi-step Q (λ) learning for optimal power flow of large-scale power grids, Int. J. Electr. Power 42 (1) (2012) 614–620.

[6] J.M. Malof, K. Bradbury, L.M. Collins, R.G. Newell, Automatic detection of solar photovoltaic arrays in high resolution aerial imagery, Appl. Energy 183 (2016) 229–240.

[7] Y. He, G.J. Mendis, J. Wei, Real-time detection of false data injection attacks in smart grid: A deep learning-based intelligent mechanism, IEEE Trans. Smart Grid 8 (5) (2017) 2505–2516.

[8] Y. Wang, M. Liu, Z. Bao, Deep learning neural network for power system fault diagnosis, 2016 35th Chinese Control Conference (CCC), IEEE, 2016, . 6678-6683.

[9] L. Cheng, T. Yu, A new generation of AI: A review and perspective on machine learning technologies applied to smart energy and electric power systems, Int. J. Energy Res. 43 (6) (2019) 1928–1973.

[10] C. Feng, J. Zhang, Wind power and ramp forecasting for grid integration, Advanced Wind Turbine Technology, Springer, 2018, pp. 299–315.

[11] C. Feng, M. Sun, M. Cui, E.K. Chartan, B.-M. Hodge, J. Zhang, Characterizing forecastability of wind sites in the United States, Renew. Energy 133 (2019) 1352–1365.

[12] L. Cibulka, M. Brown, L. Miller, A. Meier, User requirements and research needs for renewable generation forecasting tools that will meet the need of the CAISO and utilities for 2020, in: CIEEA White Paper Report, 2012.

[13] M. Henderson *et al.*, ISO New England wind integration study, 2011 IEEE Power and Energy Society General Meeting, IEEE, 2011, . 1–6.

[14] J. Wilczak, et al., The Wind Forecast Improvement Project (WFIP): a public–private partnership addressing wind energy forecast needs, Bull. Am. Meteorological Soc. 96 (10) (2015) 1699–1718.

[15] US Department of Energy, WindView: An Open Platform for Wind Energy Forecast Visualization. Available from: <https://www.energy.gov/eere/wind/renewable-systems-integration>, 2015.

[16] US Department of Energy, Watt-Sun: a multi-scale, multi-model, machine-learning solar forecasting technology. Available from: <https://www.energy.gov/eere/solar/watt-sun-multi-scale-multi-model-machine-learning-solar-forecasting-technology>, 2012.
[17] US Department of Energy, Solar uncertainty management and mitigation for exceptional reliability in grid operations. Available from: <https://www.energy.gov/eere/solar/solar-forecasting-2>, 2017.
[18] L. Zhang, G. Wang, G.B. Giannakis, "Real-time power system state estimation and forecasting via deep neural networks," arXiv preprint arXiv:1811.06146, 2018.
[19] H. Sun, X. Chen, Q. Shi, M. Hong, X. Fu, N.D. Sidiropoulos, Learning to optimize: training deep neural networks for interference management, IEEE Trans. Signal Process. 66 (20) (2018) 5438–5453.
[20] A.S. Zamzam, X. Fu, N.D. Sidiropoulos, Data-driven learning-based optimization for distribution system state estimation, IEEE Trans. Power Syst. (2019).
[21] P.N.P. Barbeiro, J. Krstulovic, H. Teixeira, J. Pereira, F.J. Soares, J.P. Iria, State estimation in distribution smart grids using autoencoders, IEEE Eighth International Power Engineering and Optimization Conference Langkawi, Malaysia, IEEE, 2014, . 358–363.
[22] R.C.B. Hink, J.M. Beaver, M.A. Buckner, T. Morris, U. Adhikari, S. Pan, Machine learning for power system disturbance and cyber-attack discrimination, Seventh International Symposium on Resilient Control Systems (ISRCS), Denver, CO, IEEE, 2014, . 1–8.
[23] M. Esmalifalak, L. Liu, N. Nguyen, R. Zheng, Z. Han, Detecting stealthy false data injection using machine learning in smart grid, IEEE Syst. J. 11 (3) (2014) 1644–1652.
[24] J. Brooks, S. Kumar, S. Goyal, R. Subramany, P. Barooah, Energy-efficient control of under-actuated HVAC zones in commercial buildings, Energy Build. 93 (2015) 160–168.
[25] C.D. Korkas, S. Baldi, I. Michailidis, E.B. Kosmatopoulos, Occupancy-based demand response and thermal comfort optimization in microgrids with renewable energy sources and energy storage, Appl. Energy 163 (2016) 93–104.
[26] A. Albert, R. Rajagopal, Smart meter driven segmentation: What your consumption says about you, IEEE Trans. Power Syst. 28 (4) (2013) 4019–4030.
[27] J. Torriti, Demand side management for the European supergrid: occupancy variances of European single-person households, Energy Policy 44 (2012) 199–206.
[28] J. Zeng, W. Qiao, Support vector machine-based short-term wind power forecasting, 2011 IEEE/PES Power Systems Conference and Exposition, IEEE, 2011, . 1–8.
[29] C. Wan, Z. Xu, P. Pinson, Z.Y. Dong, K.P. Wong, Probabilistic forecasting of wind power generation using extreme learning machine, IEEE Trans. Power Syst. 29 (3) (2013) 1033–1044.
[30] J.W. Taylor, P.E. McSharry, R. Buizza, Wind power density forecasting using ensemble predictions and time series models, IEEE Trans. Energy Convers. 24 (3) (2009) 775–782.
[31] Y. Wang, N. Zhang, Y. Tan, T. Hong, D.S. Kirschen, C. Kang, Combining probabilistic load forecasts, IEEE Trans. Smart Grid (2018).
[32] C. Feng, et al., Short-term global horizontal irradiance forecasting based on sky imaging and pattern recognition, IEEE Power & Energy Society General Meeting, Chicago, IL, IEEE, 2017, . 1–5.
[33] M. Espinoza, J.A. Suykens, B. De Moor, Fixed-size least squares support vector machines: A large scale application in electrical load forecasting, Comput. Manage. Sci. 3 (2) (2006) 113–129.

[34] C. Feng, E.K. Chartan, B.-M. Hodge, J. Zhang, Characterizing time series data diversity for wind forecasting, The Fourth IEEE/ACM International Conference on Big Data Computing, Applications and Technologies, ACM, 2017, . 113–119.
[35] J. Xia, P. Zhao, Y. Dai, Neuro-fuzzy networks for short-term wind power forecasting, 2010 International Conference on Power System Technology, IEEE, 2010, . 1–5.
[36] M. Lei, L. Shiyan, J. Chuanwen, L. Hongling, Z. Yan, A review on the forecasting of wind speed and generated power, Renew. Sustain. Energy Rev. 13 (4) (2009) 915–920.
[37] C. Voyant, et al., Machine learning methods for solar radiation forecasting: a review, Renew. Energy 105 (2017) 569–582.
[38] A.R. Khan, A. Mahmood, A. Safdar, Z.A. Khan, N.A. Khan, Load forecasting, dynamic pricing and DSM in smart grid: a review, Renew. Sustain. Energy Rev. 54 (2016) 1311–1322.
[39] M. Landry, T.P. Erlinger, D. Patschke, C. Varrichio, Probabilistic gradient boosting machines for GEFCom2014 wind forecasting, Int. J. Forecast. 32 (3) (2016) 1061–1066.
[40] M. Sun, C. Feng, J. Zhang, Aggregated probabilistic wind power forecasting based on spatio-temporal correlation, IEEE Power & Energy Society General Meeting, Atlanta, GA, IEEE, 2019, . 1–5.
[41] Y. Zhang, J. Wang, GEFCom2014 probabilistic solar power forecasting based on *k*-nearest neighbor and kernel density estimator, 2015 IEEE Power & Energy Society General Meeting, IEEE, 2015, . 1–5.
[42] S. Lou, D.H. Li, J.C. Lam, W.W. Chan, Prediction of diffuse solar irradiance using machine learning and multivariable regression, Appl. Energy 181 (2016) 367–374.
[43] H.-z Wang, G.-q Li, G.-b Wang, J.-c Peng, H. Jiang, Y.-t Liu, Deep learning based ensemble approach for probabilistic wind power forecasting, Appl. Energy 188 (2017) 56–70.
[44] M. Sun, C. Feng, E.K. Chartan, B.-M. Hodge, J. Zhang, A two-step short-term probabilistic wind forecasting methodology based on predictive distribution optimization, Appl. Energy 238 (2019) 1497–1505.
[45] J. Wu, B. Zhang, K. Wang, Application of AdaBoost-based BP neural network for short-term wind speed forecast,", Power Syst. Technol. 36 (9) (2012) 221–225.
[46] C. Feng, J. Zhang, Hourly-similarity based solar forecasting using multi-model machine learning blending, IEEE Power & Energy Society General Meeting (PESGM), Portland, OR, IEEE, 2018, . 1–5.
[47] C. Feng, M. Cui, B.-M. Hodge, S. Lu, H. Hamann, J. Zhang, Unsupervised clustering-based short-term solar forecasting, IEEE Trans. Sustain. Energy (2018).
[48] C. Feng, J. Zhang, Short-term load forecasting with different aggregation strategies, ASME 2018 International Design Engineering Technical Conferences and Computers and Information in Engineering Conference, American Society of Mechanical Engineers, 2018, pp. 1–10.
[49] H. Liu, H.-q Tian, Y.-f Li, Comparison of two new ARIMA-ANN and ARIMA-Kalman hybrid methods for wind speed prediction, Appl. Energy 98 (2012) 415–424.
[50] A. Lojowska, D. Kurowicka, G. Papaefthymiou, L. van der Sluis, Advantages of ARMA-GARCH wind speed time series modeling, 2010 IEEE 11th International Conference on Probabilistic Methods Applied to Power Systems, IEEE, 2010, . 83–88.
[51] R. Ye, Ensemble Time Series Forecasting With Applications in Renewable Energy (Doctor of Philosophy), Nanyang Technological University, 2016.
[52] S. Wang, N. Zhang, L. Wu, Y. Wang, Wind speed forecasting based on the hybrid ensemble empirical mode decomposition and GA-BP neural network method, Renew. Energy 94 (2016) 629–636.

[53] Y. Lin, M. Yang, C. Wan, J. Wang, Y. Song, A multi-model combination approach for probabilistic wind power forecasting, IEEE Trans. Sustain. Energy 10 (1) (2018) 226–237.
[54] D. Kim, J. Hur, Short-term probabilistic forecasting of wind energy resources using the enhanced ensemble method, Energy 157 (2018) 211–226.
[55] C. Cortes, V. Vapnik, Support-vector networks, Mach. Learn. 20 (3) (1995) 273–297.
[56] C. Feng, M. Cui, B.-M. Hodge, J. Zhang, A data-driven multi-model methodology with deep feature selection for short-term wind forecasting, Appl. Energy 190 (2017) 1245–1257.
[57] G. Ibarra-Berastegi, J. Saénz, G. Esnaola, A. Ezcurra, A. Ulazia, Short-term forecasting of the wave energy flux: Analogues, random forests, and physics-based models, Ocean. Eng. 104 (2015) 530–539.
[58] C. Feng, J. Zhang, Reinforcement learning based dynamic model selection for short-term load forecasting, The 10th Conference on Innovative Smart Grid Technologies, IEEE, 2019, . 1–5.
[59] D. Zhang, X. Han, C. Deng, Review on the research and practice of deep learning and reinforcement learning in smart grids, CSEE J. Power Energy Syst. 4 (3) (2018) 362–370.
[60] M. Glavic, R. Fonteneau, D. Ernst, Reinforcement learning for electric power system decision and control: past considerations and perspectives, IFAC-PapersOnLine 50 (1) (2017) 6918–6927.
[61] L. Cheng, T. Yu, A new generation of AI: a review and perspective on machine learning technologies applied to smart energy and electric power systems, Int. J. Energy Res. (2019).
[62] R. Segal, A. Sharma, M. Kothari, A self-tuning power system stabilizer based on artificial neural network, Int. J. Electr. Power Energy Syst. 26 (6) (2004) 423–430.
[63] T. Senjyu, H. Fujita, Recurrent neural network supplementary stabilization controller for automatic voltage regulator and governor, Electr. Power Compon. Syst. 31 (7) (2003) 693–707.
[64] D. Chaturvedi, O. Malik, P. Kalra, Generalized neuron based power system stabilizer,", Electr. Power Compon. Syst. 32 (5) (2004) 467–490.
[65] R. Hadidi, B. Jeyasurya, Reinforcement learning based real-time wide-area stabilizing control agents to enhance power system stability, IEEE Trans. Smart Grid 4 (1) (2013) 489–497.
[66] H. Zeynelgil, A. Demiroren, N. Sengor, The application of ANN technique to automatic generation control for multi-area power system, Int. J. Electr. Power Energy Syst. 24 (5) (2002) 345–354.
[67] D. Chaturvedi, P. Satsangi, P. Kalra, Load frequency control: a generalised neural network approach, Int. J. Electr. Power Energy Syst. 21 (6) (1999) 405–415.
[68] K. Sabahi, M. Nekoui, M. Teshnehlab, M. Aliyari, M. Mansouri, Load frequency control in interconnected power system using modified dynamic neural networks, 2007 Mediterranean Conference on Control & Automation, IEEE, 2007, . 1–5.
[69] H. Shayeghi, H. Shayanfar, Power system load frequency control using RBF neural networks based on/spl mu/-synthesis theory, IEEE Conference on Cybernetics and Intelligent Systems, 2004, IEEE, vol. 1, 2004, . 93–98.
[70] H. Li, Z. Yan, Application of Q-learning approach with prior knowledge to non-linear AGC system,", Autom. Electr. Power Syst. 23 (2008).
[71] M. Tan, C. Han, X. Zhang, L. Guo, T. Yu, Hierarchically correlated equilibrium Q-learning for multi-area decentralized collaborative reactive power optimization, CSEE J. Power Energy Syst. 2 (3) (2016) 65–72.

[72] L. Xi, J. Chen, Y. Huang, T. Xue, T. Zhang, Y. Zhang, Smart generation control based on deep reinforcement learning with the ability of action self-optimization, Sci. Sin. Inf. 48 (10) (2018) 1430–1449.
[73] A. Lucifredi, C. Mazzieri, M. Rossi, Application of multiregressive linear models, dynamic kriging models and neural network models to predictive maintenance of hydroelectric power systems, Mech. Syst. Signal Process. 14 (3) (2000) 471–494.
[74] I. Pan, S. Das, Kriging based surrogate modeling for fractional order control of microgrids, IEEE Trans. Smart Grid 6 (1) (2014) 36–44.
[75] P.B. Luh, et al., Payment cost minimization auction for deregulated electricity markets using surrogate optimization, IEEE Trans. Power Syst. 21 (2) (2006) 568–578.
[76] M. Parvania, M. Fotuhi-Firuzabad, Demand response scheduling by stochastic SCUC, IEEE Trans. Smart Grid 1 (1) (2010) 89–98.
[77] M. Dabbaghjamanesh, A. Kavousi-Fard, S. Mehraeen, Effective scheduling of reconfigurable microgrids with dynamic thermal line rating, IEEE Trans. Ind. Electron. 66 (2) (2018) 1552–1564.
[78] M. Dabbaghjamanesh, A. Kavousifard, S. Mehraeen, J. Zhang, Z.Y. Dong, Sensitivity analysis of renewable energy integration on stochastic energy management of automated reconfigurable hybrid AC-DC microgrid considering DLR security constraint, IEEE Trans. Ind. Inf. (2019).
[79] D. Shirmohammadi, H.W. Hong, Reconfiguration of electric distribution networks for resistive line losses reduction, IEEE Trans. Power Deliv. 4 (2) (1989) 1492–1498.
[80] M.A. Kashem, V. Ganapathy, G.B. Jasmon, Network reconfiguration for load balancing in distribution networks, IEE Proc.-Gen., Transm. Distrib. 146 (6) (1999) 563–567.
[81] S. Golshannavaz, S. Afsharnia, F. Aminifar, Smart distribution grid: optimal day-ahead scheduling with reconfigurable topology, IEEE Trans. Smart Grid 5 (5) (2014) 2402–2411.
[82] K. Cho, B. Van Merriënboer, D. Bahdanau, Y. Bengio, On the properties of neural machine translation: Encoder-decoder approaches, arXiv preprint arXiv:1409.1259, 2014.
[83] J. Chaney, E.H. Owens, A.D. Peacock, An evidence based approach to determining residential occupancy and its role in demand response management, Energy Build. 125 (2016) 254–266.
[84] L.M. Candanedo, V. Feldheim, Accurate occupancy detection of an office room from light, temperature, humidity and CO_2 measurements using statistical learning models, Energy Build. 112 (2016) 28–39.
[85] J. Zou, Q. Zhao, W. Yang, F. Wang, Occupancy detection in the office by analyzing surveillance videos and its application to building energy conservation, Energy Build. 152 (2017) 385–398.
[86] H. Zou, Y. Zhou, J. Yang, C. Spanos, Device-free occupancy detection and crowd counting in smart buildings with WiFi-enabled IoT, Energy Build. 174 (2018) 309–322.
[87] N. Li, G. Calis, B. Becerik-Gerber, Measuring and monitoring occupancy with an RFID based system for demand-driven HVAC operations, Autom. Constr. 24 (2012) 89–99.
[88] Z. Chen, Q. Zhu, Y.C. Soh, Smartphone inertial sensor-based indoor localization and tracking with iBeacon corrections, IEEE Trans. Ind. Inf. 12 (4) (2016) 1540–1549.
[89] W. Wang, T. Hong, N. Li, R.Q. Wang, J. Chen, Linking energy-cyber-physical systems with occupancy prediction and interpretation through WiFi probe-based ensemble classification, Appl. Energy 236 (2019) 55–69.
[90] J. Zhao, B. Lasternas, K.P. Lam, R. Yun, V. Loftness, Occupant behavior and schedule modeling for building energy simulation through office appliance power consumption data mining, Energy Build. 82 (2014) 341–355.

[91] J.G. Ortega, L. Han, N. Whittacker, N. Bowring, A machine-learning based approach to model user occupancy and activity patterns for energy saving in buildings, 2015 Science and Information Conference (SAI), IEEE, 2015, . 474–482.
[92] M. Jin, R. Jia, C.J. Spanos, Virtual occupancy sensing: using smart meters to indicate your presence, IEEE Trans. Mob. Comput. 16 (11) (2017) 3264–3277.
[93] Z. Chen, R. Zhao, Q. Zhu, M.K. Masood, Y.C. Soh, K. Mao, Building occupancy estimation with environmental sensors via CDBLSTM, IEEE Trans. Ind. Electron. 64 (12) (2017) 9549–9559.
[94] H. Zou, Y. Zhou, J. Yang, C. Spanos, Towards occupant activity driven smart buildings via WiFi-enabled IoT devices and deep learning, Energy Build. 177 (2018) 12–22.
[95] Y. Wang, Q. Chen, D. Gan, J. Yang, D.S. Kirschen, C. Kang, Deep learning-based socio-demographic information identification from smart meter data,", IEEE Trans. Smart Grid (2018).
[96] D.-A. Clevert, T. Unterthiner, S. Hochreiter, Fast and accurate deep network learning by exponential linear units (ELUs), arXiv preprint arXiv:.07289, 2015.
[97] A. Graves, A.-R. Mohamed, G. Hinton, Speech recognition with deep recurrent neural networks, 2013 IEEE International Conference on Acoustics, Speech and Signal Processing, IEEE, 2013, . 6645–6649.
[98] C. Beckel, W. Kleiminger, R. Cicchetti, T. Staake, S. Santini, The ECO data set and the performance of non-intrusive load monitoring algorithms, Proceedings of the 1st ACM Conference on Embedded Systems for Energy-Efficient Buildings, ACM, 2014, . 80–89.
[99] W. Kleiminger, C. Beckel, S. Santini, Household occupancy monitoring using electricity meters, Proceedings of the 2015 ACM International Joint Conference on Pervasive and Ubiquitous Computing, ACM, 2015, . 975–986.
[100] X. Glorot, Y. Bengio, Understanding the difficulty of training deep feedforward neural networks, in: Proceedings of the Thirteenth International Conference on Artificial Intelligence and Statistics, 2010, pp. 249–256.
[101] N. Srivastava, G. Hinton, A. Krizhevsky, I. Sutskever, R. Salakhutdinov, Dropout: a simple way to prevent neural networks from overfitting, J. Mach. Learn. Res. 15 (1) (2014) 1929–1958.

CHAPTER EIGHT

Power system operation with power electronic inverter–dominated microgrids

Yuhua Du[1], Xiaonan Lu[1] and Xiongfei Wang[2]
[1]Temple University College of Engineering, Philadelphia, PA, United States
[2]Department of Energy Technology, Aalborg University, Aalborg, Denmark

Contents

Nomenclature 259
8.1 Power system evolution toward modernization 260
8.2 Networked microgrids with parallel inverters 261
8.2.1 Advanced microgrid structures 261
8.2.2 Concept of dynamic microgrids 263
8.3 Parallel inverter operation in microgrids 264
8.3.1 Parallel inverter operation in the context of dynamic microgrids—steady-state operation 264
8.3.2 Parallel inverters operation in the context of dynamic microgrids—transient-state operation 266
8.4 Conclusion 273
References 274

Nomenclature

DER distributed energy resource
DOE Department of Energy
DG distributed generator
EPS electric power system
GFM grid-forming
IEEE Institute of Electrical and Electronics Engineers
PCC point of common coupling
POI point of interconnection
PV photovoltaic
SSW smart switch

New Technologies for Power System Operation and Analysis.
DOI: https://doi.org/10.1016/B978-0-12-820168-8.00008-0

8.1 Power system evolution toward modernization

Traditional power grids are implemented based on a centralized architecture. It provides limited automation for power system operation and lacks situational awareness. For example, only one-way power flow is allowed and the power generation is regulated following the load. Meanwhile, the whole system has little operational flexibility with respect to new producer participation, agile configuration of system topology, etc.

Nowadays, the legacy power system architecture has been dramatically upgraded with the increasing penetration of renewable energy sources and dynamic and controllable load-side management. The concept of "smart grid" was proposed to identify the significant difference of modern power system compared to the conventional one. Different organizations have proposed the descriptions of smart grids from different aspects.

US Department of Energy (DOE) proposed that [1] a smarter grid applied technologies, tools, and techniques available now to bring knowledge to power—knowledge capable of making the grid work far more efficiently.

The Smart Grids European Technology Platform proposed that [2] a smart grid is an electricity network that can intelligently integrate the actions of all users connected to it (generators, consumers, and those that do both) in order to efficiently deliver sustainable, economic, and secure electricity supplies.

Institute of Electrical and Electronics Engineers (IEEE) proposed that [3] the term "smart grid" represents a vision for a digital upgrade of distribution and transmission grids both to optimize current operations and to open up new markets for alternative energy production.

It is worth mentioning that conventional power grids are dominated by rotational generating units in the generation and transmission systems, and the distribution systems, which are the backbone to connect upstream bulk power system and downstream end users, are mainly comprised of passive feeders. Along with electric grid modernization, versatile grid resources are integrated into the electric grids. Especially in distribution grids, conventional passive distribution networks are implemented with increasing penetration of power electronic devices, which are controllable devices that can actively participate in grid services (e.g., voltage regulation [4] and power quality enhancement [5]). Increasing participation of active sources and loads interfaced with power electronic inverters significantly enhances grid operational flexibility and enables additional functionalities that can help address the issues in today's electric grids. However, it still poses challenges in terms of system stability, resiliency in face of power outages, coordinated operation with conventional grid resources, etc., which should be fully addressed and have been investigated in the past years.

8.2 Networked microgrids with parallel inverters

8.2.1 Advanced microgrid structures

Conventional vision of microgrids generally focuses on a single group of loads and distributed generators (DGs) that form a single electric power system (EPS) with clearly defined electric boundaries and a single point of common coupling (PCC). In case, there are multiple microgrids coexisting in the distribution system; they are treated as independent entities individually and would not be physically interconnected to each other.

However, as microgrids being created and integrated into the grid, a viable and advanced microgrid model that interconnects the microgrid with the utility and additional microgrids is proposed in Ref. [6]: advanced hardware, intelligent power electronic inverters, smart controllers, and compatible communications will be the enabling technology mix used to maximize a microgrid system's economic and operational benefits. Advanced communication interfaces and smart controls will increase the value of the energy provided by these advanced microgrid systems. The reliability, resilience, and interoperable electrical service for conventional and advanced microgrid customers will be vastly improved over the results of today's installed microgrids.

The desire to link multiple microgrids gives rise to a new concept of microgrids, that is, nested microgrids. Nested microgrids, also called interconnected microgrids, microgrid clusters, or aggregated microgrids in different works, refer to the interconnection of multiple adjacent microgrids into one network. Compared to a conventional microgrid that operates independently and only interacts with the external system through a single

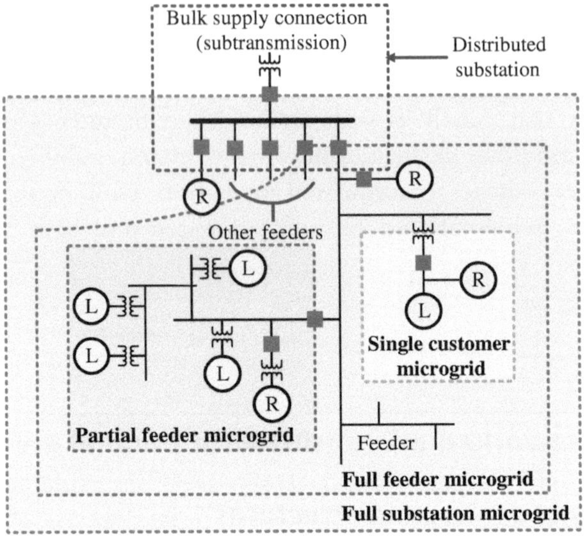

Figure 8.1 Configuration of nested microgrids.

PCC, nested microgrid could have multiple PCCs. A nested microgrid structure envisioned by DOE is presented in Fig. 8.1 [7].

Cooperated operation among nested microgrids is enabled under both grid-connected and islanded mode, which could potentially improve system resiliency, operation efficiency, and economic benefit. The cooperation among microgrids enables the utilization of the most efficient generation methods and has the system run optimally. System resilience could be improved by having areas with deficient generation supported by other nested microgrids that are connected. However, the management of nested microgrids also brings several challenges:

- By connecting multiple microgrids together, system operation and coordination become more complicated with the number of assets increases.
- Managing multiple PCCs results in additional operation complexity (e.g., protection coordination), which has not been exclusively studied.
- Interactions among nested microgrids become critical and require extra control efforts as the nested microgrid network becomes complex.

Besides the constantly increasing interests from academia, there have been several real-world projects that put the concept of nested microgrids into practice. For example, the Bronzeville microgrid is initiated by

Commonwealth Edison to install a microgrid in a location that would allow for the interconnection with a nearby campus microgrid, which has already been implemented by Illinois Institute of Technology. The two microgrids are connected to the grid through different substations; however, they have a tie line allowing them to be connected electrically. Another example is the Olney Town Center microgrid project funded by DOE. The microgrid is arranged into several zones based on the distribution of distributed energy resources (DERs) and critical loads and a microgrid controller is designed such that the system is able to meet the following power quality requirements: system average interruption duration index of less than 2 minutes, reduction in emissions of 20%, and an improvement in efficiency of over 20% [8].

8.2.2 Concept of dynamic microgrids

In addition to the notion of nested microgrids, there is also another type of advanced microgrid structure that has been proposed, which is called dynamic microgrids. A dynamic microgrid can be defined as a microgrid with flexible boundaries that expand or shrink to keep the balance between generation and load. The boundaries of these dynamic microgrids would be the substation breakers and distribution automation reclosers and sectionalizers. Compared to conventional microgrids that have static electric boundaries, the structure of dynamic microgrids could be actively varied as per request to the system operator. For example, the topology of a dynamic microgrid could be varied due to a power balancing request for DER integration or participation of distributed system restoration during natural disasters. In the meantime, since dynamic microgrids could also be dominated by power electronic inverter–dominated resources, flexible control functionalities could be implemented, such as power quality enhancement and advanced networked reconfiguration.

The differences between conventional static microgrids and emerging dynamic microgrids have been briefly presented in Fig. 8.2. A typical multimicrogrid structure is shown in Fig. 8.2. The electric boundary of each static microgrid is predefined. Interconnections among static microgrids are achieved through static breakers at the static PCC. When the breaker is open, the whole microgrid operates as an independent entity with respect to the external system. On the other hand, structure of a

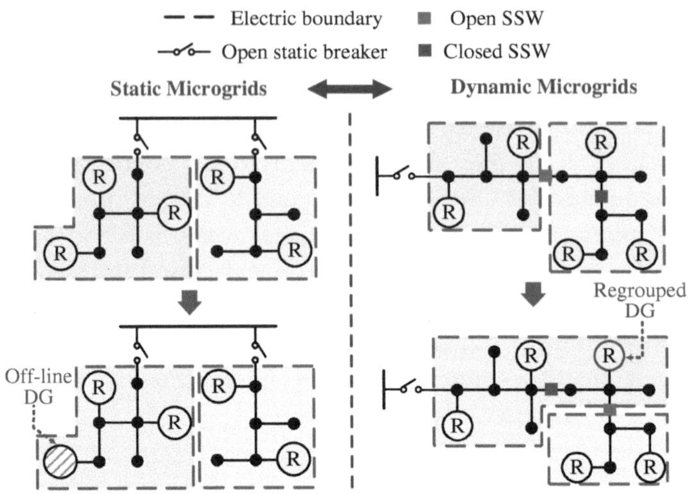

Figure 8.2 Networked microgrids operation in the context of static microgrids, and dynamic microgrids. Static microgrids share the same PCC while dynamic MGs are interconnected through varying POI.

dynamic microgrid is shown schematically in Fig. 8.2. Dynamic microgrids have varying electric boundaries. Interactions between neighboring dynamic microgrids are achieved using conventional or emerging breakers [e.g., smart switches (SSWs)]. As the topology of each dynamic microgrid varies, their points of interconnection (POIs) become dynamic accordingly.

8.3 Parallel inverter operation in microgrids

8.3.1 Parallel inverter operation in the context of dynamic microgrids—steady-state operation

Compared to parallel inverter operation in single microgrids, inverter operation in the context of dynamic microgrids represents more operation challenges on system stability in both steady and transient states.

Inverter operation in single microgrids is designed for microgrids with static electric boundaries. The inverter-based DGs would be properly sized and managed as a group to enable the steady-state operation of a microgrid under both grid-connected and islanded modes. It is noteworthy that DG's "plug and play" operation would not change system electric boundaries and all the online DGs are still managed as a group. However, when dealing with variable system topology, a more vital goal is to find the optimal system operation topology that satisfies the designed objective with consideration of system operational constraints. This problem can be resolved with network reconfiguration techniques. Existing optimization objectives for distribution system reconfiguration are concluded as follows:

- *Transmission line loss minimization:* The network reconfiguration aims to minimize the power loss on transmission lines, and loads are preferable to be supported by nearby generations.
- *System operation reliability maximization:* Such objective is usually designed considering system abnormal operation and the network reconfiguration is used to maximize service restoration.
- *Maximization of renewable penetration (DG hosting capacity):* High penetration of intermittent renewables [e.g., photovoltaics (PVs)] could result in undesired system voltage quality issues, which could be mitigated by online reconfiguration in active distribution networks.

Besides the aforementioned objectives, system operation constraints are also important when deriving the optimization formulation of network reconfiguration. Existing system operational constraints that have been frequently considered are concluded as follows:

- *Operational constraints:* System steady-state operation status (e.g., bus voltage, transmission line current, and voltage unbalance factor at the critical bus) under each topology should be within the acceptable range. Each DG should be operated under their rated capacity.
- *Topology constraints:* Distribution system should always be kept in a radial structure, mainly because (1) limitation of steady-state power flow calculation techniques and (2) challenge on protection setup.
- *Physical constraints:* System topology variation is achieved using controllable switches, whose quantity and position should be optimized to minimize the cost. Another related constraint is the operation frequency of switches.
- *Unbalance constraints:* Inverter-based DGs are more sensitive toward system unbalance and could be forced to trip under extreme conditions. It is suggested that a three-phase inverter-interfaced DG may trip if its voltage unbalance factor is greater than 3% and the unbalanced loading current should be kept within 10%–20%.

The aforementioned system topology optimization problem considers mostly system steady-state operation. Inverter-based DGs in these problems are treated as conventional *PQ*/*PV* buses. Renewable DGs are considered to cause voltage issues and they may be curtailed if needed. Storage is usually considered as a bidirectional power source with bounded power and energy capacity. On one hand, it can provide more flexible reconfiguration solutions; on the other hand, it would make the reconfiguration problem computationally complex.

One special case is when an islanded microgrid is intentionally formed during network reconfiguration. Under such scenario, at least one DG needs to be operated as the slack bus under grid-forming (GFM) mode to stabilize the islanded system voltage and frequency, while the rest DGs could be operated under power control mode, that is, grid-following mode. It is feasible and favorable for parallel inverters to operate under GFM mode simultaneously in an islanded system; however, such scenario has not been extensively discussed in existing works. In addition, cooperative operation among connected DGs regarding system topology variation is usually not covered.

8.3.2 Parallel inverters operation in the context of dynamic microgrids—transient-state operation

8.3.2.1 Inverter dynamic stability during network reconfiguration

Conventional network reconfiguration problems are usually derived under transmission level and consider only system steady-state operation stability while ignoring the transient responses on system frequency and voltage during system topology variation. Such an approach is legit mainly due to the fact that conventional transmission systems are energized using synchronous generators with significantly large inertia and are robust against system transient responses. However, this is not always applicable for distribution system reconfiguration in the context of dynamic microgrid operation. Distribution systems have small line impedances and are usually energized by inverter-based DGs with small inertia when islanded. Transient responses caused by system topology transitions are more significant in such type of system. Meanwhile, inverter-based DGs are also more sensitive toward operational voltage and frequency transients. As per IEEE Std. 1547, the DERs shall cease to energize the area EPS and trip within the respective clearing time. Detailed tripping requirements for different types of DERs (Categories I, II, and III) regarding abnormal frequency and voltage are listed next (Tables 8.1−8.4).

Table 8.1 Distributed energy resource (DER) response to abnormal frequencies for DER of Categories I, II, and III.

Shall trip function	Default settings		Ranges of allowable settings	
	Frequency (Hz)	Clearing time (s)	Frequency (Hz)	Clearing time (s)
OF2	62.0	0.16	61.8–66.0	0.16–1000.0
OF1	61.2	300.0	61.0–66.0	180.0–1000.0
UF1	58.5	300.0	50.0–59.0	180.0–1000.0
UF2	56.5	0.16	50.0–57.0	0.16–1000.0

Table 8.2 Distributed energy resource (DER) response (shall trip) to abnormal voltages for DER of Category I.

Shall trip function	Default settings		Ranges of allowable settings	
	Voltage (p.u.)	Clearing time (s)	Voltage (p.u.)	Clearing time (s)
OV2	1.20	0.16	Fixed at 1.20	Fixed at 0.16
OV1	1.10	2.0	1.10–1.20	1.0–13.0
UV1	0.70	2.0	0.0–0.88	2.0–21.0
UV2	0.45	0.16	0.0–0.50	0.16–2.0

Table 8.3 Distributed energy resource (DER) response (shall trip) to abnormal voltages for DER of Category II.

Shall trip function	Default settings		Ranges of allowable settings	
	Voltage (p.u.)	Clearing time (s)	Voltage (p.u.)	Clearing time (s)
OV2	1.20	0.16	Fixed at 1.20	Fixed at 0.16
OV1	1.10	2.0	1.10–1.20	1.0–13.0
UV1	0.70	10.0	0.0–0.88	2.0–21.0
UV2	0.45	0.16	0.0–0.50	0.16–2.0

Table 8.4 Distributed energy resource (DER) response (shall trip) to abnormal voltages for DER of Category III.

Shall trip function	Default settings		Ranges of allowable settings	
	Voltage (p.u.)	Clearing time (s)	Voltage (p.u.)	Clearing time (s)
OV2	1.20	0.16	Fixed at 1.20	Fixed at 0.16
OV1	1.10	13.0	1.10–1.20	1.0–13.0
UV1	0.88	21.0	0.0–0.88	21.0–50.0
UV2	0.50	2.0	0.0–0.50	2.0–21.0

Electric system topology variation is usually done by changing the on and off status of controllable switches (e.g., SSW). Without proper regulation, undesired transient responses will be introduced to the inverter-dominated system:

- Due to the current continuity in the line impedance, any nonzero current flow through the switch at the moment it opens will result in an undesired system transient response. For islanded systems, such type of operation variations can be modeled as step load current changes and the transient responses are mainly determined by the magnitude of through current, selection of droop control gains, and system admittance matrix.
- Asynchronous voltage phasors on both sides of an opened switch will result in undesired system transient response at the moment it closes. The mismatched voltage phasor can be modeled as an additional voltage source that is connected in series in the system. The transient responses are mainly determined by the magnitude of mismatched voltage phasors on magnitude, phase and frequency, selection of droop control gains, and system admittance matrix.

8.3.2.2 Network reconfiguration with improved inverter operation performance

Based on the aforementioned discussion, it can be concluded that without proper regulation, the transient response during system transits could result in inverter-based DGs ceasing to energization and eventually lead to blackout if the system is not coupled to a strong grid. Such issue is critical during system restoration in the context of dynamic microgrids, as the whole system could be completely islanded and network reconfiguration is supposed to provide grid service restoration. DG dynamic operation constraints should be considered when deriving the optimization problem of system network reconfiguration.

Due to the mathematical difficulties, it is not feasible to observe system transient response using conventional power flow techniques, since it will significantly complicate the computation and makes the process of solving the optimization problem time-consuming. Authors in Ref. [9] proposed an alternative approach to incorporate the stability of microgrids and dynamic performance of DGs during network reconfiguration for system restoration. Dynamic simulations are performed in parallel using simulation software (MATLAB/Simulink) to observe system transient response and evaluate the feasibility of reconfiguration schemes. The search of the optimal shall not end until the system operation constraints are met from both power flow calculation and dynamic simulation. The proposed method has been applied to the Avista distribution system that serves Pullman, including Washington State University. By applying the proposed method, it is observed using dynamic simulation that improper restoration (restoring all the critical loads in one step) will result in out-bounded transient response on system frequency and voltage, while proper switching actions can show an acceptable system dynamic response.

Such approach can effectively avoid generating out-bounded transients during network reconfiguration, which improves system operational resilience. However, it has the following limitations:

- On the one hand, accurate dynamic simulation requires sufficiently detailed system models, and on the other hand, depending on the complexity of system modeling, dynamic simulation could be time-consuming. There is a trade-off between the accuracy of dynamic simulation and the computation time it takes to solve the optimization problem.
- Though such approach can circumvent reconfiguration schemes that result in out-bounded transients, it does not provide any active solutions that help eliminate the undesired transients. Instead, it would further limit the possible reconfiguration options, which could even result in a nonsolvable optimization searching in extreme cases.

8.3.2.3 Seamless network reconfiguration using advanced inverter control

The system operation status can be actively regulated by inverter-based DGs to achieve seamless system topology variations, which can, in turn, ensure the operation stability and resilience of inverter-based DGs. In addition, seamless system topology variations can also enable more reconfiguration options. As previously discussed, to seamlessly open a closed switch, the through current or power needs to be minimized before the corresponding switch changes its status, while to seamlessly reclose an opened switch, the voltage phasors on both sides of the switch need to be synchronized before the switch changes states. Such control objectives can be integrated into the secondary control level. However, conventional secondary control approaches are mainly designed for single microgrids with static topologies and are not applicable for dynamic microgrids operation due to the following reasons:

- Conventional secondary control approaches do not coordinate the interconnections of dynamic microgrids during network reconfiguration. System frequency and voltage regulations can be constantly enabled using inverter-based DGs; however, extra control efforts are required to achieve seamless system transit.

- DGs are managed in a static group in a conventional secondary control level, which could fail when system topology varies. To be more specific, DGs that are connected to the same feeder should regulate system operation states coordinately, while interactions among DGs that belong to isolated microgrids using conventional secondary control could destabilize the entire system.
- Distribution systems are usually unbalanced and conventional secondary control considers system voltage unbalance mitigation at a single critical load bus, which could fail when multiple critical load buses are connected in the context of network reconfiguration with dynamic microgrids. Extra control efforts are required to coordinate voltage unbalance mitigation over multiple critical load buses, which have not been explicitly considered in conventional secondary control diagrams.

Generally speaking, topology variation of dynamic microgrids starts with the reconfiguration request sent from the system operator to the target switch that needs to change status. DGs shall coordinate to minimize the through power at the target switch or minimize the voltage phasor mismatch across the target switch. The target switch shall change status once the system seamless transition criteria are met and the system can transit to a new topology without undesired transients. The time it takes to fulfill seamless reconfiguration is dependent system by system and could be in the range of milliseconds to seconds, as specified in IEEE Std. 2030.7. Proper reconfiguration of dynamic microgrids requires coordinated control among all the DGs and controllable switches (e.g., SSW). A distributed control framework designed for dynamic microgrids operation is introduced [10,11].

As shown in Fig. 8.3, SSWs are used to identify the electric boundaries of dynamic microgrids. Conventional switches only provide fixed a breaking point in the electric system following local protection scheme, while SSW could be equipped with advanced functionalities, such as remote controllability and communication capability. A number of SSWs have been commercialized, and used in the field, such as the IntelliRupter from S&C Electric Company. The concept of minimum microgrid (min-microgrid) is proposed as the building block that constructs dynamic microgrids. A min-microgrid is defined as the smallest set of DGs and SSWs that are able to support local critical loads by forming a microgrid, whose installed generation capacity exceeds the critical load therein (as

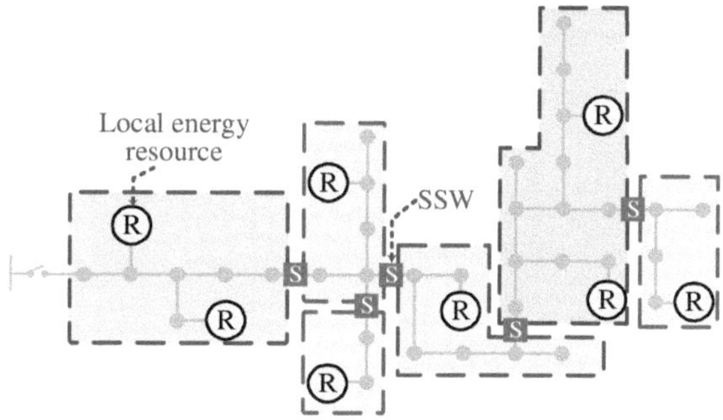

Figure 8.3 Configuration of min-microgrids.distributed feeders divided into minimum-MGs (each block represents a minimum-MG)

per request in IEEE Std. 2030.7). The distribution system is assumed to be divided into several min-microgrids and each microgrid in the dynamic framework is comprised of one or more neighboring min-microgrids. The electric boundary of a min-microgrid is defined by the positions of SSWs and the connectivity among neighboring min-microgrids is determined by the on–off status of SSWs. Following rules apply when locating the SSWs:

- Each min-microgrid has at least one POI that interconnects to the rest of the grid via SSW.
- Two adjacent min-microgrids share only one SSW (this can be easily achieved in a radial distribution system).

In Ref. [11] a modified IEEE 34-bus test feeder is divided into four min-microgrids using three SSWs. Five identical inverter-based DGs are connected to the system and operate under GFM mode. The test feeder is islanded and operates in the context of dynamic microgrids. As shown in Fig. 8.4, the system initially operates under Topology 1 with SSW_{12} and SSW_{23} closed and SSW_{34} open. The system is initially operated with only primary control and secondary control is initiated at t_1 to regulate system frequency and voltage as rated. The system is then requested to transit from Topology 1 to 2 at t_2, and then to Topology 3 at t_4. Without proper regulation, the reconfiguration is done by directly closing SSW_{34} at t_3, and opening SSW_{23} at t_5. As shown in Fig. 8.5, it could be observed that significant transients will be generated each time the system transits ($\Delta f_{max} = 1.6$ Hz and $\Delta V_{max} = 0.2$ p.u.). To achieve seamless system reconfiguration, the aforementioned seamless transit criteria are met before the target SSW changes status. It is observed from Fig. 8.5 that with proper regulation, no significant transients are introduced during system

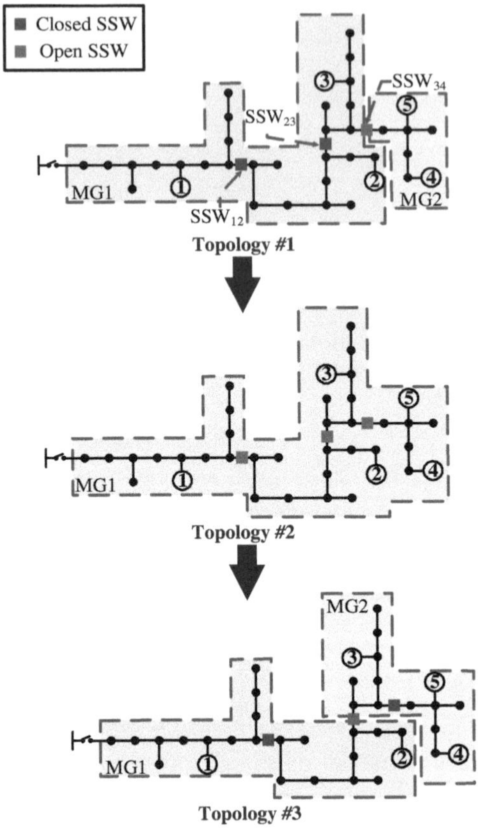

Figure 8.4 Dynamic microgrid reconfiguration operation.

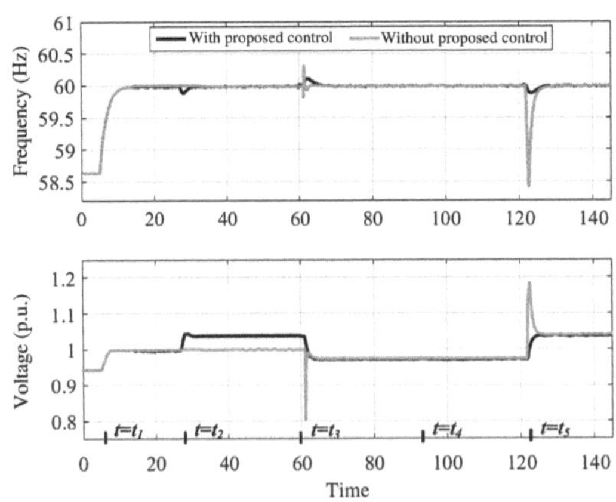

Figure 8.5 Recorded operation states under varying system topology.

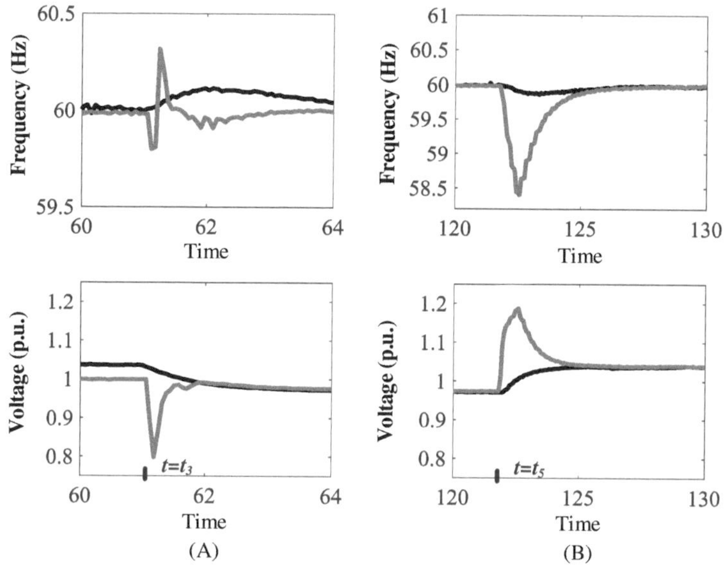

Figure 8.6 Zoomed-in system topology.operation states. (A) SSW_{34} closes and (B) SSW_{23} opens.

reconfiguration. The system transient response during system topology variationreconfiguration are zoomed-in and presented in Fig. 8.6. It can be observed that without proper regulation, undesired transients will be introduced to the system at the instance SSW changes status. Additionally, DGs are grouped and managed dynamically, as the connected DGs share the power consumption proportionally according to the real-time system topology [11].

8.4 Conclusion

Increasing penetration of inverter-interfaced resources has vastly enhanced the controllability and operational flexibility of modern electric grids. Especially in distribution systems, inverter-based DGs can actively participate in grid services and provide additional functionalities to improve grid operation resiliency. However, it is also noteworthy that increasing power-electronics-based devices involve additional challenges. Since conventional rotational generation units are gradually replaced with inverter-based static sources, system inertia could be significantly reduced, which could jeopardize the operational stability and thereby obstruct the deployment of inverter-based resources. Microgrids as local energy systems can effectively organize and coordinate inverter-interfaced devices in a local area. Therefore there are increasing interests in integrating inverter-dominated microgrids in distribution systems. Furthermore, multiple variants of conventional static microgrids, such as nested microgrids and dynamic microgrids, can further enhance the flexibility of microgrids and address the operational issues, for example, active stabilization in both the steady and transient states and voltage unbalance mitigation.

References

[1] U.S. Department of Energy, The smart grid – an introduction, in: Technical Report U.S. Department of Energy, 2008.
[2] The SmartGrids European Technology Platform. [Online] <http://www.smartgrids.eu/ETPSmartGrids>.
[3] M. Olken, Beyond the gridlock, IEEE Power Energ. Mag. 7 (2) (2009) 4–6.
[4] A. Vaccaro, M. Popov, D. Villacci, V. Terzija, An integrated framework for smart microgrids modeling, monitoring, control, communication, and verification, Proc. IEEE 99 (1) (2011) 119–132.
[5] X. Wang, F. Blaabjerg, Harmonic stability in power electronic-based power systems: concept, modeling, and analysis, IEEE Trans. Smart Grid 10 (3) (2019) 2858–2870.
[6] W.I. Bower, D.T. Ton, R. Guttromson, et al., The Advanced Microgrid: Integration and Interoperability, Technical Report Sandia National Laboratory, Albuquerque, NM, 2014.
[7] S. Bossart, DOE perspective on microgrids, in: Workshop on Advanced Microgrid Concepts and Technologies, 2012.
[8] M. Burr, J. Camilleri, D. Lubkeman, Q. Long, Y. Du, Microgrid Optimized Resource Dispatch for Public-Purpose Resiliency and Sustainability, United States, 2017, <https://www.osti.gov/biblio/1415998>.
[9] Y. Xu, C. Liu, K.P. Schneider, F.K. Tuffner, D.T. Ton, Microgrids for service restoration to critical load in a resilient distribution system, IEEE Trans. Smart Grid 9 (1) (2018) 426–437.
[10] Y. Du, X. Lu, J. Wang, S. Lukic, Distributed secondary control strategy for microgrid operation with dynamic boundaries, IEEE Trans. Smart Grid 10 (5) (2019) 5269–5282.
[11] Y. Du, X. Lu, H. Tu, J. Wang, S. Lukic, Dynamic microgrids with self-organized grid-forming inverters in unbalanced distribution feeders, IEEE J. Emerg. Sel. Top. Power Electron., accepted and in press.

CHAPTER NINE

Automated optimal control in energy systems: the reinforcement learning approach

Xiangyu Zhang and Huaiguang Jiang
National Renewable Energy Laboratory, Golden, CO, United States

Contents

9.1 Introduction 276
 9.1.1 Background 276
 9.1.2 Markov decision process and Bellman equations 277
 9.1.3 Solving Markov decision process problems 280
 9.1.4 Value-based reinforcement learning 280
 9.1.5 Policy-based reinforcement learning 283
 9.1.6 Actor—critic reinforcement learning 285
 9.1.7 Summary 285
9.2 Deep reinforcement learning 286
 9.2.1 What is deep reinforcement learning 286
 9.2.2 Introduction to three deep reinforcement learning algorithms 286
 9.2.3 Scalable reinforcement learning frameworks 292
 9.2.4 Curriculum learning 293
 9.2.5 Meta learning 294
 9.2.6 Multiagent system 295
 9.2.7 Summary 296
9.3 Reinforcement learning in energy systems 297
 9.3.1 Advantages of applying reinforcement learning in engineering problems 297
 9.3.2 Training a reinforcement learning controller 298
 9.3.3 Reinforcement learning applications in energy systems 304
 9.3.4 Some interesting research topics 311
 9.3.5 Limitations and challenges 313
 9.3.6 Summary 313
References 314

New Technologies for Power System Operation and Analysis.
DOI: https://doi.org/10.1016/B978-0-12-820168-8.00015-8

9.1 Introduction

In this section the basics of reinforcement learning (RL) will be summarized and reviewed, with a goal to give readers a concise introduction or a refresh of memory on what RL does, what RL components are, and how typical RL problems are usually solved. It is worth noting that this section is not intended to be a thorough explanation of RL preliminary. For detail and more systematic learning of RL, interested readers should refer to Ref. [1] for more thorough explanation.

9.1.1 Background

RL, as one of the three key machine-learning methods (the other two are supervised learning and unsupervised learning), enables computers to solve sequential decision-making optimal control problems in a novel manner without formulating an optimization problem. The computer, also known as an RL agent or decision maker, learns an optimal control strategy from the experience collected by interacting with an environment repeatedly. Typically, problems that can be solved by RL are sequential optimal control problems, namely, they involve multistep decision-making. At each control step, as shown in Fig. 9.1, the RL agent makes a decision on what action to take based on the current state/observation s_t. The chosen action a_t is sent to the environment, which then evolves given the current system state, action, and other factors. In addition, the action will be evaluated by the environment, rendering a scalar reward r_t. By repeatedly making decisions and receiving the r_t over the control horizon, the RL agent obtains feedback on how it performs and uses such experience to adjust its future decision-making process.

However, due to the strong temporal dependency between steps, it is worth noting that the feedback r_t might be delayed, which means a high/low

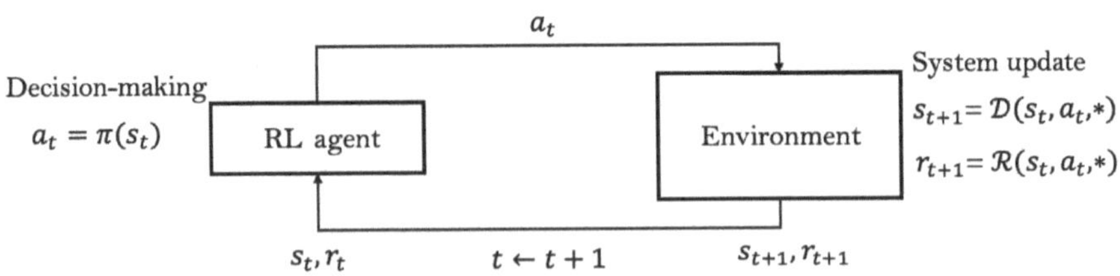

Figure 9.1 RL agent and environment. $\mathcal{D}$ represents the system dynamics of the environment, $\mathcal{R}$ is the reward function and asterisks in $\mathcal{D}$, and $\mathcal{R}$ represent other internal factors of the environment. *RL*, Reinforcement learning.

r_t might not be directly attributed to the immediate action a_t but because of an earlier action a_{t-k} or a combination of actions of many previous steps. That being said, the optimal solution for an RL problem might be far from being greedy at each step; sometimes an action that yields low reward at current step might ultimately lead to a global optima. Therefore the goal of RL is to learn an optimal policy π^* that helps the RL agent achieve the maximum future cumulative reward $G_t = \sum_{k=0}^{T-t} r_{t+k+1}$. To obtain such an optimal policy, many different RL algorithms and techniques are devised to learn from a large amount of experience collected by repeated interaction with the environment. The repetition of experience collection and learning will "reinforce" RL agent's understanding of the environment's behavior and gradually help it acquire a better control policy.

Examples of RL problems

In general, problems that are suitable to be solved using RL should meet the following two important features: (1) they can be formulated as an sequential optimal control problem with an objective to be minimized/maximized and (2) there is strong temporal dependency among the control steps, namely, actions at earlier steps have impact on future states and actions. For a better understanding, refer to the following links for some interesting classic control problems that can be properly formulated as RL problems: https://gym.openai.com/envs/#classic_control. These examples are from OpenAI, an artificial intelligence (AI) research laboratory, who designed a Python implementation of RL problems called gym environment, which we will discuss in more detail in Section 9.3.2.

9.1.2 Markov decision process and Bellman equations

Based on the previous conceptual understanding, now let us look at a more strict mathematical formulation of RL problems. Markov decision process (MDP) is used to formally formulate the environment in an RL problem. In literature, MDP is usually represented using a quintuple $(\mathcal{S}, \mathcal{A}, \mathcal{P}, \mathcal{R}, \gamma)$, representing five key components of MDP.

1. *State set* $\mathcal{S}$ is a set of states, which describes the system status, characterize system history, and can be used to infer the future of the system. The state at step t is represented as $s_t \in \mathcal{S}$, and the transition between states has Markov property (i.e., $P(s_{t+1}|s_t) = P(s_{t+1}|s_1, s_2, \ldots, s_t)$).
2. *Action set* $\mathcal{A}$ is a set of all possible actions that can be taken by the RL agent. The action taken by the decision maker at step t is $a_t \in \mathcal{A}$.

3. *State transition probability set* $\mathcal{P}$ is a set of probability describing the system dynamics. Probability $P(s_{t+1}|s_t, a_t) \in \mathcal{P}$ represents the chance for the system state that moves from s_t to s_{t+1} considering the action a_t taken. Similarly, in classic control theory, the state-space representation (i.e., $X(k+1) = AX(t) + BU(t)$) also describes such state transition relationship, but in a deterministic way.
4. *Reward function* $\mathcal{R}$ is a function describing the environment's immediate response toward the action a_{t+1} taken at state s_t. The environment can be stochastic, meaning the response for a given (s_t, a_t) might be random.
5. *Discount factor* $\gamma \in [0,1]$ is used to discount future reward. Reasons to use a discount factor include avoiding infinite G_t due to infinite horizon or cyclic Markov process. If $T \neq \infty$, $\gamma = 1$ can be used.

The goal to solve a MDP problem is to find a behavior guideline to help the agent/decision maker to achieve the maximum expected future discounted cumulative reward $E(G_t) = E\left(\Sigma_{k=0}^{T-t} \gamma^k \cdot r_{t+k+1}\right)$, where $\forall t \in \mathcal{T}$.

9.1.2.1 Policy

The abovementioned behavior guideline for the RL agent interacting with an MDP problem is called a policy. A policy is a conditional probability distribution, showing the probability of an action being selected under a specific state:

$$\pi(a_t|s_t) = P[a = a_t|s = s_t]$$

In general, a policy can be either deterministic or stochastic. Depending on the specific MDP, one might be better than the other.

9.1.2.2 Value function

Value functions are mapping relationships that project current state s_t or state-action pair $\{s_t, a_t\}$ to the expected sum of future cumulative discounted reward under a specific policy π. Mathematically, the state-value function and action-value function can be represented as follows:

$$\begin{aligned} V_\pi(s_t) &= E_\pi[G_t|s = s_t] \\ Q_\pi(s_t, a_t) &= E_\pi[G_t|s = s_t, \ a = a_t] \end{aligned}$$

Different policies define different agent behavior, and thus even with the same state s_t, values for it under different policies might be different (i.e., $V_{\pi 1}(s_t) \neq V_{\pi 2}(s_t)$).

9.1.2.3 Bellman equations

From the perspective of dynamic programming, Bellman equations show the relationship between the value functions of two adjacent steps:

$$
\begin{aligned}
V_\pi(s_t) &= E_\pi[G_t|s = s_t] \\
&= E_\pi[r_{t+1} + \gamma G_{t+1} \ |s = s_t] \\
&= E_\pi[r_{t+1} + \gamma V_\pi(s_{t+1}) \ |s = s_t]
\end{aligned}
$$

$$
\begin{aligned}
Q_\pi(s_t, a_t) &= E_\pi[G_t|s = s_t, \quad a = a_t] \\
&= E_\pi[r_{t+1} + \gamma G_{t+1} \ |s = s_t, \quad a = a_t] \\
&= E_\pi[r_{t+1} + \gamma V_\pi(s_{t+1}) \ |s = s_t, \quad a = a_t] \\
&= E_\pi[r_{t+1} + \gamma E_a[Q_\pi(s_{t+1}, a_{t+1})] \ |s = s_t, \quad a = a_t]
\end{aligned}
$$

The Bellman equations show a very important dynamic programming concept called *bootstrapping*, namely, use a prediction (e.g., $V_\pi(s_{t+1})$) to update another prediction (e.g., $V_\pi(s_t)$). This concept lays crucial groundwork for the development of RL algorithms later.

9.1.2.4 Bellman expectation equations

The following two equations represent the relationship between state-value functions and action-value functions:

$$
\begin{aligned}
V_\pi(s_t) &= \sum_{a \in \mathcal{A}} \pi(a \mid s_t) \cdot Q_\pi(s_t, a) \\
Q_\pi(s_t, a_t) &= r_{t+1} + \gamma \cdot \sum_{s \in \mathcal{S}} P(s \mid s_t, a_t) \cdot V_\pi(s)
\end{aligned}
$$

Merging these two equations, there will be:

$$
\begin{aligned}
V_\pi(s_t) &= \sum_{a \in \mathcal{A}} \pi(a|s_t) \cdot \left(r_{t+1,a} + \gamma \cdot \sum_{s \in \mathcal{S}} P(s \mid s_t, a_t) \cdot V_\pi(s) \right) \\
Q_\pi(s_t, a_t) &= r_{t+1} + \gamma \cdot \sum_{s \in \mathcal{S}} P(s|s_t, a_t) \cdot \left(\sum_{a \in \mathcal{A}} \pi(a|s) \cdot Q_\pi(s, a) \right)
\end{aligned}
$$

9.1.2.5 Bellman optimal equations

Note that in the Bellman expectation equations, value functions are evaluated for a given policy π. Imaging the following policy that always provides the maximum possible values:

$$
\pi^*(a|s) = \begin{cases} 1 & \text{if } a = \underset{a \in \mathcal{A}, \pi \in \Pi}{\arg\max} Q_\pi(s, a) \\ 0 & \text{else} \end{cases}
$$

Π above is an infinite set of all possible policies. Such a π^* is called an *optimal policy*. With π^*, there will be $V^*(s) = max_\pi V_\pi(s)$ and $Q^*(s, a) = max_\pi Q_\pi(s, a)$. When considering π^*, Bellman expectation equations become Bellman optimality equations:

$$V^*(s_t) = \max_{a \in \mathcal{A}} Q^*(s_t, a)$$

$$Q^*(s_t, a_t) = r_{t+1} + \gamma \cdot \sum_{s \in \mathcal{S}} P(s|s_t, a_t) \cdot V^*(s)$$

$$V^*(s_t) = \max_{a \in \mathcal{A}} \left(r_{t+1,a} + \gamma \cdot \sum_{s \in \mathcal{S}} P(s|s_t, a) \cdot V^*(s) \right)$$

$$Q^*(s_t, a_t) = r_{t+1} + \gamma \cdot \sum_{s \in \mathcal{S}} P(s|s_t, a_t) \cdot \left(\max_{a \in \mathcal{A}} Q^*(s, a) \right)$$

9.1.3 Solving Markov decision process problems

For a known MDP problem (all elements in the quintuple $(\mathcal{S}, \mathcal{A}, \mathcal{P}, \mathcal{R}, \gamma)$ are known), by solving the Bellman optimality equations above, we can obtain the optimal policy π^*. These functions are nonlinear and usually require iterative algorithms to solve (i.e., dynamic programming). Interested readers refer to Sections 4.3 and 4.4 of Ref. [1] for details about these iterative methods such as policy iteration and value iteration.

However, one practical issue is that when MDP is not perfectly known, which is often the case for many real-world MDP problems, we will not be able to directly use $P(s | s_t, a_t)$ and $\mathcal{R}$ to calculate the value functions and solve the Bellman optimality equations. Though state transition and reward information are not directly available, if the agent can learn them or even the optimal policy from interacting experience, this will lead to many different RL algorithms.

9.1.4 Value-based reinforcement learning

In this section the first type of RL methods, known as value-based methods, are explained. Similar to solving MDP using dynamic programming above, value-based methods learns an optimal policy indirectly by learning the value functions. When the optimal value functions are learned, a greedy policy can be implemented by always choosing the action that gives higher value (i.e., $(a_t = \text{argmax}_{a \in \mathcal{A}} Q^*(s_t, a))$).

Depending on how value functions are estimated, there are two approaches called Monte Carlo (MC) and temporal difference (TD) learning, which will be introduced next.

9.1.4.1 Monte Carlo reinforcement learning

MC RL learns the value functions from experience, the idea is straightforward: since the value function reflects the expected return given a state or a state-action pair, we can obtain the expected return by sampling and use the empirical mean to approximate the expected value, given enough experience (reason why this is called MC).

The Monte Carlo learning algorithm for control follows the following steps (Algorithm 9.1):

To guarantee continual exploration, ε-greedy method is used for policy update. Instead of being purely greedy, ε-greedy policy chooses the greedy action ($a = \arg\max_{a \in \mathcal{A}} Q_\pi(s, a)$) with probability $1 - \varepsilon$ (where $\varepsilon \in (0,1)$) and evenly samples all possible actions with probability ε. In Algorithm 9.1, ε is set to be the reciprocal of the iteration number k. By doing this, a larger value of ε encourages exploration at the early learning phases and eventually when the agent has collected enough experience, ε becomes smaller and this gradually reduce the probability of exploring new actions.

One apparent drawback is that MC control requires a complete trajectory to update the Q-value function using G_t, this means policy update is relatively infrequent (can only update till the sequence is completed) and since G_t from a specific sequence is used, the update has higher variance. Also, due to the

Algorithm 9.1 Monte Carlo control [1,2]

Result: Optimal Control Policy π^*
k=0, $N(s_t, a_t) = 0$, $\forall t \in \{0, 1, ..., T-1\}$, randomly initialize the policy π_0 and Q function (Q_0);
while *π_k is not converged* **do**
    Sample experience trajectory by interacting with the environment use current policy π_k:
    $\mathcal{T} = \{s_0, a_0, r_1, ..., s_{T-1}, a_{T-1}, r_T\}$;
    $k \leftarrow k + 1$;
    For all $t \in \{0, 1, ..., T-1\}$, incremental Monte Carlo updates for $Q_k(s_t, a_t)$: $N(s_t, a_t) \leftarrow N(s_t, a_t) + 1$
    $Q_{k+1}(s_t, a_t) \leftarrow Q_k(s_t, a_t) + \frac{1}{N(s_t, a_t)}(G_t - Q_k(s_t, a_t))$;
    Policy improvement using ϵ-greedy method: $\epsilon \leftarrow \frac{1}{k}$,
    $\pi_{k+1} \leftarrow \epsilon\text{-greedy}(Q_{k+1})$
end

requirement for a complete trajectory, MC can only be used for finite step problem ($T \neq \infty$).

9.1.4.2 Temporal difference reinforcement learning

Another type of learning, called TD learning, is introduced to address the issue of MC learning mentioned before. Interestingly, TD learning not only possesses the merit of MC learning (ability to learn from an unknown MDP through experience) but also resembles dynamic programming that updates an estimate using another estimate (i.e., *bootstrapping* mentioned earlier) and thus able to learn from incomplete sequence.

In order to learn from incomplete sequence, instead of using the actual episodic return G_t, TD learning estimates the value of G_t. Depending on how the estimation is made, there comes two different implementations: SARSA learning and Q-learning. Before delving into the difference between them, let us first understand the following two concepts:

1. *On-policy learning* approaches learn to improve the policy π from experience collected using π.
2. *Off-policy learning* approaches learn to improve the policy π from experience collected using another policy π'.

SARSA is an on-policy TD learning algorithm, which follows the steps shown in Algorithm 9.2. According to the algorithm, one difference between SARSA and MC learning is that SARSA updates Q-value function every time step. Some literature also call this *online learning* or

Algorithm 9.2 SARSA learning [1,2]

Result: Optimal Control Policy π^*
k=1, randomly initialize the policy π_0 and Q function (Q_0);
while *π_k is not converged* **do**
  Initialize a new control episode ($t = 0$) and provide s_0, choose action $a_0 = \pi_k(s_0)$;
  while *Episode is not terminated* **do**
    Take action a_t and receive r_t and s_{t+1} from the environment. Choose action $a_{t+1} = \pi_k(s_{t+1})$. Update Q-value function:
    $Q_{k+1}(s_t, a_t) \leftarrow Q_k(s_t, a_t) + \alpha(r_t + \gamma Q_k(s_{t+1}, a_{t+1}) - Q_k(s_t, a_t))$
    Policy improvement using ϵ-greedy method: $\epsilon \leftarrow \frac{1}{k}$, $\pi_{k+1} \leftarrow \epsilon\text{-greedy}(Q_{k+1})$
    $t \leftarrow t+1$, $k \leftarrow k+1$;
  end
end

incremental learning, in which learning happens within control steps, while MC learning is in an offline manner, since learning starts only after all experiences are collected. In addition, an even larger difference between these two learning method is the way they update the value function: in SARSA, by bootstrapping, $r_t + \gamma Q_k(s_{t+1}, a_{t+1})$ is used to estimate G_t. Since the next step action based on current policy is leveraged (i.e., $a_{t+1} = \pi_k(s_{t+1})$), and the algorithm uses the experience it collected using π_k to update π_k related Q_k, it is why SARSA is an on-policy learning method.

Q-learning [3], on the other hand, is an off-policy learning method. Instead of estimating G_t using $r_t + \gamma Q_k(s_{t+1}, a_{t+1})$, it takes a greedy approach and uses $G_t = r_t + \gamma \max_{a \in \mathcal{A}} Q_k(s_{t+t}, a)$ in Q_k update. This greedy update let Q_k directly approximate the optimal value and is able to lead to an early convergence. However, it is worth noting that even though Q_k is updated greedily, during agent's rollout of the control horizon, the action it takes still follows the ε-greedy manner (i.e., $a_k = \pi_k(s_k)$). This is to guarantee the exploration of more actions and states, which leads to a successful convergence of Q.

In summary the value-based RL algorithm above learns the optimal value function values from experience. With optimal value functions the action at each control step is determined by $a = \operatorname{argmax}_{a \in \mathcal{A}} Q_\pi^*(s, a)$ during control.

9.1.4.3 Value function approximation

If the number of RL state and action is limited, a table with dimension of $|\mathcal{S}| \times |\mathcal{A}|$ can be used to store the optimal values. This table is called Q-table and combining Q-table with Q-learning is the learning approach often called *tabular Q-learning*. However, one issue with Q-table is that it does not work if the number of state is large or even infinite (for continuous states). In this case a parameterized function is used to approximate the relationship between state/action values and state/state-action (i.e., $V_w(s)$ for state-value function and $Q_\theta(s, a)$ to action-value function, and w and θ are parameters to be identified).

9.1.5 Policy-based reinforcement learning

As introduced in previous section, value-based RL algorithms first learn values of actions and then choose actions with highest values during control. A policy is implicitly established based on all learned action values. In this section the policy-based RL, which explicitly learns a parameterized policy, is discussed. Notation $\pi_\theta(a_t \mid s_t)$ is used to represent the policy, and $\theta \in \mathbb{R}^m$ is the parameters vector that determines the policy (m is an arbitrary number representing parameter dimension). Essentially, $\pi_\theta(a_t \mid s_t)$ represents a probability of action selection, as shown next:

$$\pi_\theta(a_t \mid s_t) = P(a = a_t \mid s = s_t, \theta)$$

Since the controller's performance measure depends on the control policy, which is parameterized by θ, the performance measure can thus be expressed as a function of θ, as in $J(\theta)$. To train an optimal controller, one can adjust θ to the direction where the performance measure $J(\theta)$ is maximized. Mathematically, controller parameters can be updated in the following gradient ascent manner:

$$\theta_{k+1} = \theta_k + \alpha \nabla J(\theta_k)$$

For episodic learning the performance measure can be represented in the form of the value function: $J(\theta) = V_{\pi_\theta}(s_0)$, where s_0 is the initial state. Using Bellmen equations, the gradient of $J(\theta)$ can be estimated as follows:

$$\nabla_\theta J(\theta) = E_{\pi_\theta}\left[Q^{\pi_\theta}(s, a) \nabla_\theta \log \pi_\theta(a|s)\right]$$

This is called *policy gradient theorem* (deduction process is omitted here; refer to Section 13.2 in Ref. [1] for details). RL algorithms that are based on this theorem are called policy gradient methods. Specifically, an MC version of policy gradient algorithm, also named REINFORCE [4], is

Algorithm 9.3 Monte Carlo policy gradient algorithm [4]

Result: Optimal Policy π^* defined by optimal parameter θ^*
Randomly initialize the policy by setting parameter θ_0;
while θ_k *is not converged* **do**
  Sample experience trajectory by interacting with the environment use current policy π_{θ_k}:
  $\mathcal{T} = \{s_0, a_0, r_1, ..., s_{T-1}, a_{T-1}, r_T\}$;
  For all $t \in \{0, 1, ..., T-1\}$, incremental Monte Carlo updates:
  $\theta_{k+1} \leftarrow \theta_k + \alpha \gamma^t G_t \nabla_\theta \log \pi_\theta(a_t|s_t)$;
  $k \leftarrow k + 1$;
end
$\theta^* = \theta_k$

introduced in Algorithm 9.3. Due to $E_{\pi_\theta}[Q^{\pi_\theta}(s,a)] = E_{\pi_\theta}[G_t|s,a]$, the gradient of $J(\theta)$ is estimated using $E_{\pi_\theta}[G_t \nabla_\theta \log \pi_\theta(a|s)]$ in REINFORCE.

In many other policy gradient algorithms the term $Q^{\pi_\theta}(s,a)$ is replaced with other terms in order to reduce variance level. Interested reader can refer to Section 2 of Ref. [5] for a summary of different $Q^{\pi_\theta}(s,a)$ variants.

9.1.6 Actor–critic reinforcement learning

To reduce variance and accelerate learning the idea of TD learning introduced earlier in value-based learning is again applied in the policy gradient algorithms. In actor–critic methods, not only the policy π_θ but also a parameterized value function $V_w(s_t)$ are learned. The policy learning part is called *actor* since the policy guides the RL agent to act and the value function learning part is called *critic*, which evaluates the value of each state or state-action pair.

As a result, when updating the policy, instead of using G_t as in REINFORCE, bootstrapping is leveraged and thus using $r_{t+1} + \gamma V_w(s_{t+1})$ to represent the reward-to-go. The updating rule is shown next, and the part with underline is called baseline, which is another technique that helps reduce variance during learning.

$$\theta_{k+1} \leftarrow \theta_k + \alpha_\theta(r_{t+1} + \gamma V_w(s_{t+1}) - \underline{V_w(s_t)})\nabla_\theta \log \pi_\theta(a_t \mid s_t)$$

Critic update for the value function parameters is shown as follows:

$$w_{k+1} \leftarrow w_k + \alpha_w(r_{t+1} + \gamma V_w(s_{t+1}) - V_w(s_t))\nabla_w V_w(s_t)$$

9.1.7 Summary

In this section, basic concepts regarding RL are first introduced and the nature of RL problems is described (i.e., sequential optimal decision-making). Then, a mathematical formulation of typical RL problems, the MDP, is presented and Bellman equations are introduced to solve MDP. Next, the need for RL algorithms is introduced in the case when MDP is not fully known and three types of RL algorithms (value-based, policy-based, and actor–critic methods) are discussed. By reading this section, we hope readers can have a preliminary understanding about RL and the logic behind its development. Again, for more in-depth and systematic study of RL, we suggest interested readers to refer to Ref. [1] and many other learning resources.

9.2 Deep reinforcement learning

9.2.1 What is deep reinforcement learning

Deep RL (DRL) is the combination of deep learning and RL; specifically, deep neural network (NN) is used to approximate the RL value function or the control policy. In the 2010s the successful story from DeepMind for using DRL to train a controller to play Atari game marks a significant milestone in RL's development [6]. Their experiment for using a convolutional NN (CNN) for value function approximation and with only raw game video pixels as inputs demonstrates the possibility of having end-to-end RL, which avoids labor-intensive work for crafting features for function approximation and truly shows the potential of artificial intelligence (AI). Moreover, NN can represent more complex input–output relationship than linear functions and it is expected to help solve more complicated decision-making problems.

9.2.2 Introduction to three deep reinforcement learning algorithms

In this part, three DRL algorithms of different types are introduced: a value-based method, a policy gradient/actor–critic method, and evolutionary strategies–based method. These three algorithms are three representatives to a series of RL algorithm categories; as a result, after introducing each of them, other algorithms of the same category are briefly mentioned and interested readers are suggested to read the corresponding papers for more details.

9.2.2.1 Deep Q-network: a value-based approach

Recall the Q-learning discussed in the previous section—it is a value-based algorithm, which learns the action values for different state-action pairs Q (s, a) and then establish an implicit control policy based on the learned Q-values. Based on Q-learning, a deep Q-network (DQN) uses a deep NN for action-value function approximation in contrast to earlier tabular Q-learning or approaches using linear function approximation. Specifically, an NN with state as input and action values as outputs will be learned. There are several reasons for using NN for value function approximation: (1) NN can properly take in high-dimensional input data and map it to the Q-value; (2) unlike linear function approximation, NN does not require handcrafted features as inputs; and

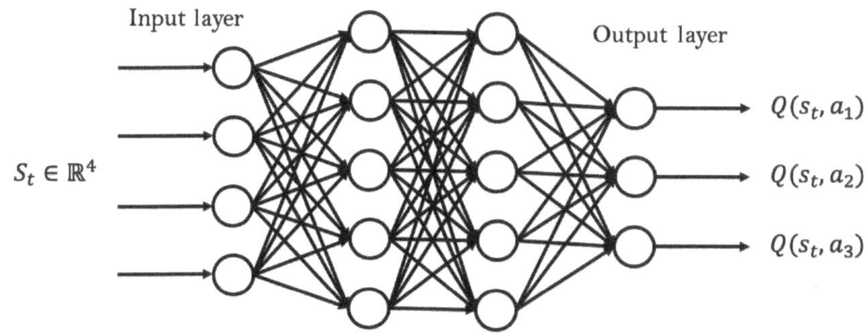

Figure 9.2 Example Q-network for a simple hypothetical RL problem with $s_t \in \mathbb{R}^4$ as the state and three actions from the legal action set. As shown in the figure, s_t is the network input and outputs are action values of three actions $Q\ (s_t, a_t)$, $i \in \{1, 2, 3\}$.

(3) NN is differentiable by back-propagation [7], which can be used in gradient-based method.

Fig. 9.2 shows an example of a DQN ($Q(s, a, w)$, in which w represents the NN parameters): typically the input is s_t and output dimension equals to the number of discrete actions (yes, DQN only works with RL problems with finite actions). It is worth noting that although a fully connected NN is shown in Fig. 9.2 as an example, DQN can be in any NN form such as CNN (as in the original DQN paper, s_t is the raw pixels of game video image) or a recurrent NN (RNN) (if the input contains some time series information such as renewable generation forecasts in the energy-related control problem) or with mixed type of inputs.

In the same vein as training a NN in supervised learning, training a DQN can be done by minimizing the loss between DQN outputs and the ground truth values. However, in RL, there are no ground truth values, but only experience. As a result, one idea is to use bootstrapping to generate the "ground truth" values and thus the loss function can be defined as:

$$\mathcal{L}(w) = E\left[\left(r + \gamma \max_{a \in \mathcal{A}} Q(s', a', w) - Q(s, a, w)\right)^2\right]$$

However, when directly using Q-learning with NN function approximation and bootstrapping, there will be algorithm stability issue and might even cause divergence. As a result, two techniques are proposed in the original DQN paper [6,8] and they are proven to be effective in improving the training stability: first, introducing a target Q-network, whose parameters (w^-) are copied from the learning Q-network (w) periodically and stay fixed for many steps, to generate the "ground truth" values. Therefore the loss function becomes:

$$\mathcal{L}(w) = E\left[(r+\gamma \max_{a\in\mathcal{A}} Q(s', a', w^{-}) - Q(s, a, w))^2\right]$$

The second mechanism used for stabilizing training is the use of experience pool called experience replay. Specifically, the agent's experience at each time step are saved in a replay memory, and during Q-learning update, sample of experience will be drawn at random from the pool. The benefit that this experience replay brings include (1) improving data efficiency as each step of experience might be used in multiple updates and (2) randomly sample experience from the pool can effectively break correlations between samples, which helps reduce variance.

In addition to the original DQN paper, other studies propose some improvement based on the plain DQN. For example, double Q-learning [9,10] is used to overcome the issue that Q-learning sometimes suffer from action-value overestimation and a prioritized replay [11] is introduced to prioritize samples in the experience replay and to learn from the most important experience. Wang et al. introduces a dueling network [12] to improve the learning performance and Hessel et al. combine six different techniques proposed in literature for improving DQN into a RAINBOW DQN [13] and show that it can achieve a better performance.

9.2.2.2 Asynchronous advantage actor–critic: an actor–critic approach

Asynchronous advantage actor–critic (A3C), as its name indicates, is an actor–critic algorithm that focuses on parallel learning by having workers update the learning policy in an asynchronous manner. With A3C, each worker has its own gym environment instance as an action test bed, shown as ① and ② in Fig. 9.3. With the collected experience, each worker calculates gradients regarding to all samples and accumulates them together. Once all samples are processed, the accumulated gradient will be used to update a global shared parameter vector (③). The global parameters, being updated by all workers asynchronously, are then fetched to each worker (④), who will use the updated parameters to start a new iteration of experience collection.

Pseudocode for each A3C worker is shown as in Algorithm 9.4. From the gradient update part, it can be seen that this follows exactly the same rule as we describe in the last section about the actor–critic algorithm.

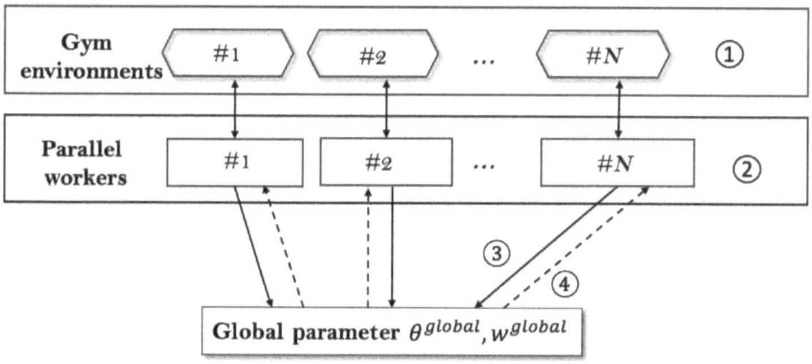

Figure 9.3 A3C illustration. Parallel works collect experiences independently and update a set of global parameter asynchronously. *A3C*, Asynchronous advantage actor–critic.

But by relying on parallel workers and update policy periodically shows an outstanding stabilizing effect on training a deep NN with online RL algorithm. Similar to DQN, which uses an experience replay to decorrelate the sample used for policy update, A3C uses experience collected from parallel workers, which are collected from different environments and thus break the correlation. This parallel learning, in addition, has demonstrated even better learning performance using single machine with multicore processor when compared with the GPU-based algorithm.

A variant of A3C is A2C (advantage actor–critic, removing the A from asynchronous in A3C), which is a synchronous version of A3C. Instead of updating policy asynchronously, A2C waits for every parallel worker and updates the policy synchronously, and then each worker will fetch the same updated global policy for the next iteration. According to the experiment from OpenAI, A2C implementation is "more cost-effective than A3C when using single-GPU machines, and is faster than a CPU-only A3C implementation when using larger policies" [15].

Besides A2C and A3C, there are many other actor–critic algorithms. They all hold the core concept of policy gradient/actor–critic method, but they proposed different mechanisms to stabilize the learning process or accelerate the policy convergence. Some examples are as follows: deep deterministic policy gradient (DDPG) [16] is proposed to learn a deterministic policy for problems with continuous action spaces; trust region policy optimization (TRPO) [17] establishes a trust region using KL divergence when updating policy in order to provide a monotonic performance improvement; similarly, a proximal policy optimization (PPO) algorithm [18] is proposed to simplify the implementation of TRPO but

Algorithm 9.4 Asynchronous advantage actor—critic pseudocode for each parallel worker [14]

Inputs: Initial values for global shared parameters of policy and value functions θ^{global} and w^{global}. Global step counter $T = 0$. Maximum training steps T_{max}.
Result: Optimal Policy π^* defined by optimal parameter θ^*.
Declare local copy of parameters θ and w;
Local step counter $t = 1$;
while $T \leq T_{max}$ **do**
  $d\theta^{global} \leftarrow 0$ and $dw \leftarrow 0$;
  Fetch global parameters $\theta = \theta^{global}$ and $w = w^{global}$ (④ in Fig. 3);
  $t_{start} = t$;
  Get state s_t from environment.;
  while s_t *is not terminal state and* $t - t_{start} < t_{max}$ **do**
    Sample a_t according to current policy $\pi_{\theta'}$;
    Apply a_t to environment and obtain r_t and s_{t+1};
    $t \leftarrow t + 1$ and $T \leftarrow T + 1$
  end
  if s_t *is terminal state* **then**
    $R = 0$
  else
    $R = V(s_t, w)$
  end
  for $i \in t - 1, ..., t_{start}$ **do**
    $R \leftarrow r_i + \gamma R$;
    $d\theta^{global} \leftarrow d\theta^{global} + (R - V_w(s_i))\nabla_\theta \log \pi_\theta(a_i|s_i)$
    $dw^{global} \leftarrow dw^{global} + \nabla_w (R - V_w(s_i))^2$
  end
  Update the global parameter vector (③ in Fig. 3):
  $\theta^{global} \leftarrow \theta^{global} + \alpha_\theta d\theta^{global}$;
  $w^{global} \leftarrow w^{global} + \alpha_w dw^{global}$
end
$\theta^* = \theta_{global}$

remain the learning performance; actor—critic using Kronecker-factored trust region (ACTKR) [19] is proposed that uses Kronecker-factored approximation curvature to update gradient; and more [20,21].

9.2.2.3 Evolution strategies—based reinforcement learning

In the abovementioned algorithms, gradient back-propagation is needed for NN training at each training step. In contrast to this, there have been some efforts to use even less computation for NN training. As a result, evolution strategies (ES), a type of gradient-free black-box optimization, has been utilized for network training in RL (Algorithm 9.5) [22].

Algorithm 9.5 Evolution strategies–based reinforcement learning [22]

Inputs: Learning step size α, noise standard deviation σ, total iteration number K and initial policy parameter θ_0
Result: Optimal parameter θ^*.
Iteration counter $k = 1$;
while $k \leq K$ **do**
  Sample noise $\epsilon_1, ..., \epsilon_n \sim N(0, I)$;
  Obtain rewards using policy determined by $\theta_t + \sigma\epsilon_i$, that is, collect $F_i = F(\theta_t + \sigma\epsilon_i)$ by interacting with the environment ($\forall i \in \{1, ..., n\}$).;
  Update policy parameter $\theta_{t+1} \leftarrow \theta_t + \alpha\frac{1}{n\sigma}\sum_{i=0}^{n} F_i\epsilon_i$
end
$\theta^* = \theta_t$

ES algorithm mimics the evolution process in the nature: starting from a generation with an initial set of parameters (known as genotypes), the ES perturbs such genotypes in various ways and created new genomes (known as mutation), and the fitness of the new genomes will be evaluated by the "nature" and the best one will be recombined as base gene for the next generation. By repeating this mutation and selection iteration a set of optimal parameters are expected to be obtained generation after generation. Similarly, the same algorithm can be used in optimizing RL policy, which is assumed to be parameterized by a parameter vector θ. Assuming by following the policy π_θ, the RL received is $F(\theta)$ and the goal is to find θ that maximize a stochastic objective of $E_{\epsilon\sim N(0,I)}[F(\theta + \sigma\epsilon)]$ (σ as the noise standard deviation is given). This maximization problem can be solved by stochastic gradient ascent and the gradient can be directly approximated by samples [22]:

$$\nabla_\theta E_{\epsilon\sim N(0,I)}[F(\theta + \sigma\epsilon)] = \frac{1}{\sigma}E_{\epsilon\sim N(0,I)}[F(\theta + \sigma\epsilon)\epsilon]$$

According to Ref. [22], because ES does not need back-propagation, it can reduce the computation load by two-thirds when compared with other RL algorithms. Besides, ES-RL can also be scaled conveniently without tremendous communication overhead increase: consider scaling the PPO [18], an actor–critic algorithm, at each iteration, all parallel workers need to share with each other the entire gradient information, thus requiring higher communication bandwidth. In contrast, ES only requires sharing the scalar reward F_i and ϵ_i. Moreover, sharing ϵ_i can be avoided by synchronizing random seeds among workers before optimization, as a result, ϵ_i from other workers can be generated locally using the known seed. The information sharing and policy parameter update part only accounts for a small fraction of the total learning time, according to Ref. [22], even when scaled up to 1440 workers. Another advantage identified is that ES demonstrates better exploration behavior than other policy gradient–based methods, as ES is able to perform actions that are not possible to be learned by other algorithms.

In summary, ES-RL provides an additional perspective on how RL can be trained. The approach of solving RL problems by leveraging black-box optimization is called direct policy search. Another example of this type of algorithm, augmented random search, can be found in Ref. [23]. The authors show that the sample efficiency of direct policy search is not as bad as believed.

9.2.3 Scalable reinforcement learning frameworks

Recall that in RL, the agent is generally repeating two steps iteration after iteration: experience collection and policy update. This pattern applies to many different RL algorithms, value-based or actor–critic, and up until a point, the two steps are conducted sequentially: in DQN the single RL agent collects experience from environment and puts them in experience replay and when a certain amount of experience has been collected, it starts to update the Q-network; in PPO, though multiple workers can be used, they first collect experience together, and until everyone is done, the gradient information are shared among all workers and they update the policy together using the averaged gradient and considering the average KL divergence. In A3C, although each work updates the policy asynchronously, but for a single worker, the two steps are still sequential. Later, researchers start to think: can these two steps be conducted in a parallel manner, namely, to decouple the "acting" and "learning" parts. As a result, a distributed architecture for scalable learning is proposed in Ref. [24].

In this architecture, as shown in Fig. 9.4, there are distributed workers that are specialized in experience collection in a parallel manner with their own environments (shown as ①), and the experience are then pushed into an experience replay pool shared by all actors (shown as ②). The actors are also responsible to compute initial priorities for the data. In parallel with experience collection, there is a learner that samples from the prioritized experience replay and use them for policy update (as shown in ③ and ④). The learner is usually powered by GPU to enable fast gradient computation and high data throughput. The priorities of experience are

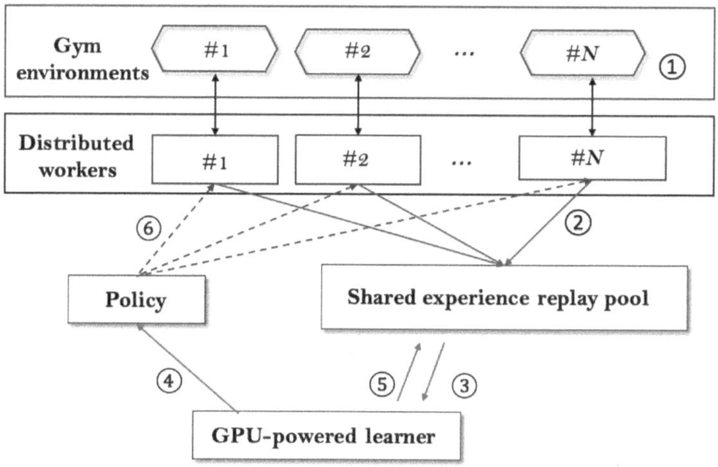

Figure 9.4 Ape-X illustration.

also updated by the learner (shown as ⑤) to enable a biased sampling that helps the learner focus on learning using the most valuable experience. Depending on the actual control problems (whether continuous action space or discrete), the learner can use DQN or DPG (or other off-policy algorithms) for value/policy update. Finally, as shown in ⑥, the updated policy will be periodically fetched by actors and use it to collect more experience.

In experiments, this distributed architecture can achieve state-of-the-art performance on both discrete and continuous tasks, when compared with other algorithms from the perspectives of performance and learning wall time. However, it is worth noting that the highly distributed learning architecture is based on one assumption: large amount of learning experience can be generated in parallel easily. Otherwise, if the data take long time to simulate or the simulator itself cannot be properly run in parallel, Ape-X might not be the optimal option.

Another architecture in a similar vein is the *Importance Weighted Actor-Learner Architecture (IMPALA)* [25].

9.2.4 Curriculum learning

In the previous section and the one before the efforts of enabling DRL and scaling it for accelerated learning are introduced. In this section, work from another perspective for efficient learning is introduced, and that is *curriculum learning*.

Human learners benefit from learning using a curriculum: the knowledge are presented in a meaningful order and the student's understanding to more complex concepts and more intricate problems gradually deepens. The same idea intrigues researchers from cognitive science domain, and they start to think: will machine learning also benefit from a similar learning setting? Bengio et al. [26] conduct some experiments and lead to some results that support this hypothesis, and there are some other works [27,28] that find success when using curriculum learning to solve difficult tasks. Though there are some signs that curriculum learning is effective, one question arises: how to generate a reasonable curriculum and especially generate in automated manner, instead of being handcrafted. To this end, Matiisen et al. [29] propose a teacher—student curriculum learning framework, in which the teacher will automatically generate learning tasks for the student based on its learning performance. An experiment on solving a Minecraft maze problem is conducted, and it shows that the automatic curriculum learning can lead the student to successfully solve the problem by breaking the complicated problem into several subtasks and master them progressive. In contrast, direct RL on the problem cannot lead to a solver.

9.2.5 Meta learning

Meta learning is subarea in machine learning, which widely employed on metadata with its automatic learning algorithms. This algorithm aims to solve the scalability problem, which is defined as the capability of an algorithm to handle an increasing number of works in different applications or conditions. In general machine learning the scalability is a tricky problem and difficult to tune, for example, a designed algorithm performs very well in one domain, but not in other domains or new domains. The reason is a well-trained algorithm fitting its own data domain very well, and this blocks the algorithm to adapt the problem in new domain. In other words, it is difficult for the traditional algorithms to learn the inner connection between the data domains. In addition, the metalearning concept is widely broaden to few-shot learning and one-shot learning.

In general, there are three major types of meta learning to solve this problem: RNNbased (or model-based), metric-based, and optimization-based.

RNN-based: This kind of metalearning contains an RNN [such as long short-term memory (LSTM) and gated recurrent units (GRU)] as a memory to keep the parameters (to be tuned). In the training step the parameters are updated recurrently as both input and output of the RNN. Considering the large computation load, multiple GPU (hundreds or even thousands of level) are employed for training the model concurrently. In the objective designed the performance of the algorithm such as accuracy and residual errors are treated as rewards to tune the policy of the RNN controller. Neural architecture search (NAS) is a famous example in this kind of algorithm [30], which is a critical algorithm in automated machine learning (AutoML).

Metric-based: The nearest neighbor idea is the core for the metric-based metalearning algorithm, which aims to find and measure the distance generated in high-dimensional space by the kernel tricks. Therefore the objective function is based on the relative distance in high-dimensional space. Different NNs and kernel tricks can be seemed as embedding functions to map the original data into high-dimensional functions. In this area, face verification is a typical application, and convolutional Siamese neural network is a famous algorithm [31], which can be widely employed in energy system security.

Optimization-based: The optimization-based algorithm is similar as calculus of variations, which employs a bunch of subfunctions for different subtasks. The metalearning function learns the inner connections between different subtasks and generate new functions to adapt new data domains. Specifically, this process can be explained to learn the prior distribution of the parameters in machine learning. According to the Bayesian law, the posterior distribution is computed with prior distribution and maximum likelihood, which can be received by traditional machine-learning technologies. In this area, two typical algorithms are MAML [32] and Reptile [33], which can be used as basic algorithms for few data state estimation and forecasting algorithms in power system applications.

9.2.6 Multiagent system

The multiagent system is defined as a group of organized and multiinteracting intelligent agents, which provide a more accurate model for self-organized systems such as power system with a lot of distributed energy resources (DERs) and transportation systems [34]. Considering the behaviors and cooperation manner of the agents, the multiagent system can be divided into three types of algorithms: cooperative, competitive, and mixed setting [35].

Cooperative multiagent system: In the cooperative multiagent system the Nash equilibrium can be seemed as global optimum [35], and all the agents contains similar goal and share full or partial information. For example, in Ref. [34], the distributed bidirectional charging and discharging stations and electrical vehicles in a urban area are modeled as multiple agents; they cooperate and share full information to achieve the same global objective in large scale, while also keeping its own objective in small scale. In Refs. [36,37] a distributed algorithm are employed to model the multiagent system in both discrete and continuous domain. In very large scale the mean-field algorithm is used to model a large volume of agents in Ref. [38], which formulates the agents into two equivalent agents.

Competitive multiagent system: According to Game theory, the competitive multiagent systems are usually modeled as zero-sum system, which also indicates this kind of system are usually modeled or formulated into two equivalent players. In this area the algorithms can be divided into value-based algorithms and policy-based algorithms [35]. In value-based algorithms a linear program method is presented in Refs. [35,39]. In policy-based algorithms a global convergence of nested policy gradient is discussed in Ref. [40].

Mixed multiagent system: Compared with the cooperative multiagent system and the competitive multiagent system, the mixed multiagent system is more challenging and less well understood [35]. In this area the major challenging is how to find a Nash equilibrium in different scenarios. Similar as before, this area can also be divided into two major areas—policy-based and value-based, and the detailed information can accessed in Refs. [35,41,42]. In power system areas, considering different applications, these three major models can be combined flexibly and to handle the problems in different scales.

9.2.7 Summary

In this section, DRL is introduced and by presenting three representative DRL algorithms, how to integrate deep NN for RL training is demonstrated. Techniques such as scalable learning platform and curriculum learning are shown as some efforts for achieving more efficient learning. Other topics such as metalearning and multiagent learning are briefly covered. It is worth noting that at this moment, this is a very active and productive research community, and domain scientist from energy engineering domains can closely follow the development in DRL and to see how the state-of-the-art algorithms be leveraged to solve real-life engineering problems, which will be discussed in more detail next.

9.3 Reinforcement learning in energy systems

In this section, how RL is and can be used for solving control problems in energy systems is discussed. This section will be divided into four parts. First, advantages of RL that make it a promising candidate for solving some engineering-related optimal control problems are presented. Second, the tasks and workflow to design an RL learning environment for specific engineering problems are demonstrated. Third, RL applications on several energy-related research topics are reviewed. Finally, limitations and challenges for using RL in real-life applications are discussed.

9.3.1 Advantages of applying reinforcement learning in engineering problems

The need for advance optimal control under sophisticated environments in modern engineering and the rapid development in AI have inspired many emerging research on applying RL in solving engineering problems. Practically, following are the several potential advantages of RL that make it competitive to the state-of-the-art optimal controllers.

1. *Real-time readiness*

   Existing approaches for conducting sequential optimal control are mostly based on optimization approaches, which might need to solve optimization problems at each control interval repeatedly (e.g., model predictive control). Two issues arise in some cases: (1) it does not work for some complicated control problems that involve optimization problems take long time to be solved; (2) it requires powerful computers for the on-demand computation. In contrast, an RL controller can be trained/optimized before deployment in an offline manner and during control, the decision-making only involves a situation evaluation process (i.e., evaluating value or policy network, which requires very light computation). Therefore RL controllers are suitable for problems that prefer fast responses and would like to avoid heavy computation load on-the-fly.

2. *Model adaptability*

   Due to the nature of optimization-based approaches, systems being controlled need to be properly modeled into mathematical representations to be put in optimization problems. However, formulating accurate models in some domains might be labor-intensive: for instance, to model hundreds of houses in order to implement smart home control or to model all intersections in the city for intelligent traffic light control. In contrast, RL does not place constraints on system models and a controller can be trained on models that do not even have explicit mathematical expressions. As a result, many data-driven models in the form of various supervised learning models can be used for RL training. These models are much cheaper to obtain and the acquisition can be fully automated, especially considering massive deployment of optimal controllers.

3. *Handling nonlinearity*

   Besides using a powerful computer to increase the efficiency of solving optimization problems, an alternative that is frequently used in real-life applications is model simplification. Though transforming the original problem to a similar, simpler problem (e.g., linear programming) can reduce the computing time, the simplification might lead to a suboptimal control performance on a real system. In contrast, RL can learn nonlinear control problem directly (nonlinearity from either controlled system or the reward structure).

4. *Modeling uncertainty*

   Many control problems involves uncertainty, on one hand, disregarding it and use deterministic optimization will inevitably cause suboptimal control, while on the other hand, leveraging stochastic programming requires explicit effort for estimating probability distribution. In contrast, RL can directly learn from data about the uncertainty in an implicit way, making considering uncertainty in control a much easier effort (which might be valuable in some applications). Of course, whether RL can achieve or even exceed the performance of a stochastic programming—based controller regarding the control performance needs further investigation.

9.3.2 Training a reinforcement learning controller

9.3.2.1 Typical workflow

Though training an RL controller on an actual system is possible, it is very unlikely to do so in reality in many cases for two reasons: first, at the beginning of the learning phase, an RL controller needs to adequately explore a variety of actions and some of them might be dangerous for a real system (e.g., testing different and random control signals in a normally operating power grid); second, sample efficiency in RL learning is limited to real-time while experience collection with software simulations plus parallel learning is much faster (e.g., it might take 3 years in real life to collect enough experience for an RL controller to out-compete a rule-based controller [43]). As a result, in most cases,

RL controllers are trained in a simulated environment first until there is a well-trained policy, and only then will it be deployed on a real system for continual learning and controlling. Admittedly, in areas other than engineering, where simulators are not as common and easily accessible or simply because wrong decisions during exploration are not as harmful, directly learning on the actual system might not be as undesired (design an RL-based recommender system [44]).

Fig. 9.5 shows the typical workflow for training an RL controller for an actual physical control problem. First, starting from upper left corner, the control subject is modeled in a software simulation. Then, the system simulator is wrapped by the OpenAI Gym framework [45] (introduced in detail in the next part). Next, by utilizing a proper RL algorithm, an RL controller is trained based on the simulator wrapped in a gym environment. Finally, after adequate examination and testing, the trained controller is deployed to the actual system for control and continual learning.

9.3.2.2 Building a reinforcement learning environment

In this section, how to design an RL learning environment for specific control problems is explained. By reading this part, readers should be able to tell the basic components of an RL learning environment and understand how to design and build one.

For the purpose of efficient benchmarking, increasing code reusability, and facilitating domain researches, the RL community has standardized an interfacing framework, the OpenAI Gym [45], for RL algorithms to collect control experiences from a simulator and learn to have a better performance. Such an interface, which defines several key function calls and

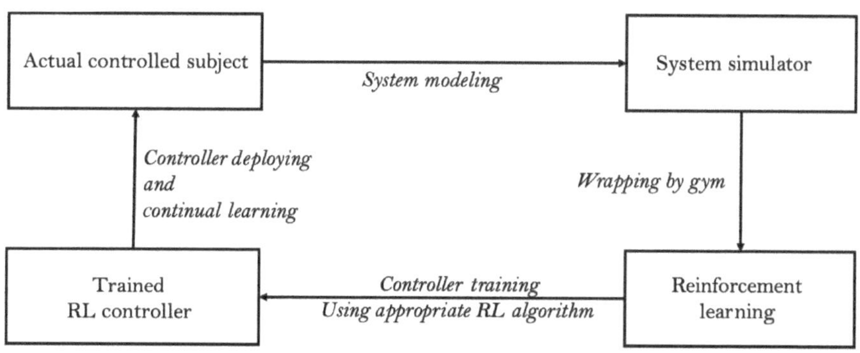

Figure 9.5 Typical workflow for using RL for a real-world control problem. *RL*, Reinforcement learning.

corresponding data structures, largely accelerates the development needed by both algorithm developers and domain scientists who use the algorithm to solve a problem of interest.

Fig. 9.6 shows the interaction between an RL agent and a specific domain problem written in the OpenAI Gym format (referred to as a gym environment hereinafter). Below are some discussions based on three aspects: (1) controlled subject simulator, (2) environment key attributes, and (3) interfacing functions.

1. *Simulator*: A gym environment wraps a simulator of the controlled subject, and the simulator determines the system dynamics and can realistically reflect the system next state given current state and RL agent's action. Typically, the following three types of simulators are commonly used in the gym environment (i.e., the "core simulator" shown in Fig. 9.6).
   a. *Simplified physics models*: Some simple mathematical models that describe the system dynamics such as state-space model (which are commonly used in classic control), power flow equations for power system control, and building resistance−capacitance (RC) thermal models [46] for building engineering. Such simulators have fast computation and can

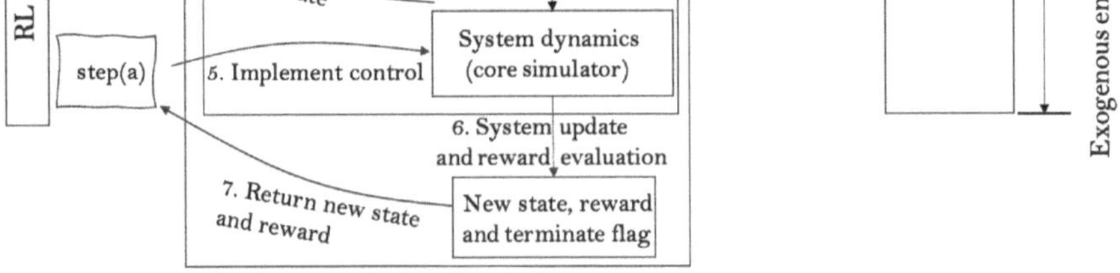

Figure 9.6 Interaction between an RL agent and a gym environment: (1) Agent calls the reset function to start a new control episode. (2) Episode initialized with exogenous data (e.g., for wind turbine yaw control problem, wind data of speed and direction for a period of time are sampled and loaded). (3) and (4) Control episode initialized and return the initial state so to the RL agent. (5) Agent implements control at each interval. (6) and (7) The controlled system updated, reward r_t is evaluated and returned with the new state s_{t+1}. Steps (5)−(7) are repeated until this control episode is terminated and the agent will call the reset function again to start learning in a new episode. *RL*, Reinforcement learning.

be easily scaled but are also comparatively less accurate and require domain expertise for the system identification.

b. *Domain-specific simulators*: Unlike traditional optimization-based approaches, RL does not place any limitation on system model, which can be nonlinear, nonconvex, stochastic, and even time-variant. As a result, many use domain-specific, first principle simulators to gain high modeling accuracy (e.g., OpenDSS [47] for power system engineering and EnergyPlus [48] for building control). Of course, domain expertise is required to build such model case by case, and despite the capability for precise modeling, the underlying computation for some simulators is heavier, which affect learning efficiency directly or indirectly. In addition, learning scalability largely depend on the scalability of the simulator (e.g., unable to use scalable RL algorithms for learning acceleration if the simulator itself cannot/is hard to run in parallel).

c. *Data-driven models*: Recently, with the development of machine-learning/deep-learning techniques and advanced sensing technologies, many data-driven models are proposed, aiming at combining fast computation, preserving modeling accuracy, and reducing modeling cost. The biggest advantage of this type of simulators based on data-driven models is their low modeling cost (since data-driven models can be learned in a fully automated manner, and thus this avoids labor modeling cost), which is of great importance in some applications that requires case by case modeling extensively. Another merit of the data-driven model is that it can adapt to model drift by periodically retrain the model.

In summary, for a specific control problem, researchers need to identify the best simulator for their problems, considering the model accuracy, readiness, cost, and scalability.

2. *Environment key attributes*: There are four key attributes for a gym environment: action space, observation, reward structure, and termination condition. Proper design of these attributes is of great importance and will determine if an effective controller can be successfully trained.

a. *Action space* defines the number and range of the control variable(s). Gym environment allows action space for both discrete and continuous actions. Action space needs to be declared in the gym class's constructor function (i.e., the "_ _ init _ _" function in the gym class). This information is indispensable since RL algorithms require it to construct the outputs of the policy network/value network.

Discussion A: Choosing an appropriate action space

Imaging a single battery optimal control problem in which the power is being controlled. When formulating this problem into a gym environment, there are many choices for the action space: (1) discrete: $a \in \{charging, discharging\}$, (2) continuous w/o constraints: $a \in [-P_{ch}, P_{dis}]$, and (3) continuous w/ ramping constraints: $a \in [-\Delta P, \Delta P]$. All three action spaces can be used to train a battery controller, but it is up to the researchers to determine which one suits the best for their specific control problem.

b. *Observation/state space* defines the number and range of the observable system state. In implementation the observation can either be a vector or a tensor (e.g., an image) or a combination of both. Unlike action space design, choosing an appropriate observation space is not as straightforward: researchers need to decide which parameters to be included, and the decision, in many cases, will impact the training results. Similarly, the observation space is defined in the gym class's constructor function as well and will be used by RL algorithms when building the input of the policy network/value network.

Discussion B: Making observation space inclusive

Considering the battery control example in Discussion A, if we choose to use action space (3), it is of great importance to include current power output P into the observation space since the optimal action ΔP will P the RL controller needs to be aware of. In contrast, if action space (2) is used, P can be excluded. In summary, when designing the observation space, researchers need to include all possible information that may influence the optimal action and meanwhile try to avoid variables that cannot be observed or predicted, though their existence might help improve the controller performance.

c. *Reward structure* defines how the RL agent will be rewarded or penalized at each step given its action taken. Defining an appropriate reward is the key for solving a problem using RL: in general, the researcher needs to design a reward structure that perfectly aligned with the control objective. Reward can be dependent on exogenous data (e.g., energy cost is dependent on electricity price), current state (e.g., penalty if voltage being controlled breaches boundaries) or states of past few steps (e.g., penalizing improper use of a device: turning a device off before its minimum operating time).

d. *Termination condition* defines when the control episode will be terminated. Typically, there are two choices: first, all control episodes terminate at maximum control step (i.e., step count is the only terminating criteria). Second, control episode terminates with some specific conditions. For instance, when training a controller to bring a faulty system back to normal (e.g., power system black start), the episode can terminate when normalcy is obtained, even before the maximum step allowed.

3. *Two key interfacing functions*: Two key functions are repeatedly called by the RL algorithms, "reset" and "step." As mentioned in Section 9.1, the objective of the sequential optimal control is to maximize expected reward over a specific control horizon, called an episode. The "reset" function is called to start a new episode of simulation by reinitializing the environment, when an episode is terminated. The "step" function is called by an RL agent to implement a one-step control. It triggers the simulator inside the gym environment to simulate one step and return the reward and observation at the next step.

reset( ) and step(a)

- *reset( )* does not take in any argument and it returns the initial observation of the system. For instance, in a gym environment related to battery optimal control problem. There will be a code snippet in the "reset( )" to reinitialize the battery SOC.
- *step(a)* contains the following four steps. First, preprocessing the action if it is normalized, for example, $a = 0.5$ is given for the battery control problem, then the actual power output from battery is $p = a \times \overline{P}_{ch}$. Second, pass the processed action into the simulator to simulate the dynamics of this control interval and receive the next state from the simulator and provide the action evaluation signal, namely, a scalar reward. Then, tick the step by plus one ($t \leftarrow t + 1$) and finally determine if the episode is terminated by examining some termination conditions. The step function will return a tuple of (state, reward, if_terminated, info), in which "info" is a Python dictionary that either is empty (i.e., info = {}) or contains additional information.

9.3.2.3 Selecting the right algorithm

With the gym environment for the specific control problem properly formulated, the next question to think about is which RL algorithm should be used given so many available options out there? Below are three perspectives for considerations:

First, some algorithms have limitations: DQN and its variants can only be used for problems with discrete actions and DDPG-related algorithms only work for continuous control problems. As a result, algorithm should be chosen according to the nature of the specific problem.

Second, some algorithms have been tested better than others in RL literature using some benchmark test cases; thus, consider them first is generally a good option. For example, as mentioned in Section 9.2.2, TRPO considers a KL divergence constraint during policy update and its performance is better than vanilla policy gradient (VPG) algorithm. Again, PPO overcomes the TRPO's complicated implementation issue and achieves similar performance by using a clipped surrogate objective. Therefore training an RL controller using PPO might lead to more reliable and faster learning than using VPG or TRPO.

Third, factors such as hardware availability or learning scalability should be taken into consideration as well: If GPU is not available for training, A3C can be considered since it can be parallelized on single machine with multicore CPU with performance similar or better than GPU-based algorithms in some cases [14]. In the case where scalability is important and the simulator run in parallel, evolutionary strategies RL (ES-RL) is preferred than PPO, since parallel workers in ES-RL only need to share a scalar [22], while workers in PPO need to share all gradient information, and this makes ES-RL learn faster than PPO from wall-time perspective.

9.3.3 Reinforcement learning applications in energy systems

In this section, RL applications on several energy system topics are discussed, and some representative studies are presented.

9.3.3.1 Smart buildings

According to data from the U.S. Energy Information Administration (EIA), in 2018, around 40% of total energy consumption in the United States are from buildings, both residential and commercial [49]. Proper building control is expected to maximize occupancy comfort, optimize building energy usage, and provide potential grid services to power system. Currently, the state-of-the-art intelligent building control are mostly based on model predictive control (MPC) [50], which optimize a specific control objective over a period of control horizon, while considering several operation constraints. Of course, successful stories of MPC in many domains have proven it to be an outstanding controller. However, when applying MPC for smart building/smart home control, there will be several practical issues [51,52] as discussed earlier in Section 9.3.1: the requirement for on-demand computation and modeling effort required to obtain an accurate but simple mathematical model for the controlled subject.

In contrast, RL, as an alternative approach for sequential optimal control, does not have these issues and can be a potential powerful candidate for building control, especially considering massive deployment in many buildings. First, as introduced in Section 9.3.2, RL controller can be trained offline before it is deployed in a real building, and only limited computation is required for the policy evaluation during actual control. Thus an RL controller can be implemented using cost-effective embedded system to reduce implementation cost. Second, RL controllers do not place any constraint on the format of building models, and these models do not necessarily need to have an explicit mathematical formulation. As a result, RL can adapt to more model types, including some data-driven building models, which are accurate but much cheaper to obtain. In all, RL controllers seems to be very promising to solve smart building control problems in a low-cost but effective way and thus inspired many research.

Before 2015, most building RL research uses value-based RL (mostly Q-learning and some other TD learning). For instance, in as early as 1997, Anderson et al. [53] started using Q-learning for a heat coil temperature set point tracking task. Henze and Schoenmann [54] investigated using tabular Q-learning for building thermal energy storage system control under the influence of time-variant electricity price (i.e., time-of-use price and real-time price) in a simulated environment. Dalamagkidis et al. [55] uses TD(λ) with linear function approximation to solve a building optimal control problem that minimizes the energy consumption for achieving indoor comfort. In these studies, experiments show that RL controller can achieve good performance if the control task is relatively simple (e.g., in the set point tracking task [53]); however, if control problems are not as straightforward (e.g., when multiple control objectives are considered), RL controllers' performance is not as good. For instance, it takes 4 simulated years for RL controller to approximate the performance of a Fuzzy-PD controller in Ref. [55] and the RL controller cannot match a predictive optimal controller in Ref. [54]. Admittedly, building control is a complicated problem and it might not be easy for tabular Q-learning or linear function approximated RL to learn. However, as of today, will the DRL be a game changer? Many researchers have started their investigation on how DRL can be used for modern building control. For example, Mocanu et al. [56] leverage deep Q-learning and deep policy gradient to perform online scheduling for building energy management system and showed that they can out perform a tabular Q-learning baseline (an expected result). In order to achieve an overall optimum and encourage device coordination, Ding et al. [57] tested a DRL-based holistic building controller in simulation, with a goal to optimize the trade-off between four different building subsystems. Besides these two examples, there are many other studies which focus on this emerging study, and the interested readers can refer to Ref. [58] for a more comprehensive and up-to-date literature review. According to Ref. [58], although as of today, the most used DRL algorithm in building control research is DQN, and more studies are expected to test and compare other algorithms including actor—critic methods, as researchers gradually move from discrete control to more accurate continuous control.

Future studies should consider two questions before an RL controller can be widely used for building control in real-life applications:

1. *What simulator should be used for building RL controller training?*

   As discussed earlier, directly training building RL controller in a real building environment suffers from low sample efficiency (i.e., real-time) and might cause building operation issue. In this case, what simulator should be used is a good question. To fully take advantage of RL's capability of handling nonlinear system, Wei et al. [59] utilized the high fidelity EnergyPlus model to train a control policy for the building air-conditioning system. However, considering that such a model might be costly to obtain, using a crude EnergyPlus model composed during the building design phase to pretrain an RL agent and then deploy it in a real building for continuing learning is proposed in Ref. [60]. But this still does not work for existing buildings, which accounts for the majority, and not all newly constructed buildings have an EnergyPlus model during planning. As a result, Zhang et al. [61] point out by deploying low-cost Internet of Things (IoT) devices and leveraging machine-learning methods, data-driven building models can be automatically learned from building operation data, and thus accurate building specific model can be obtained with much lower cost. With more studies and field tests on data-driven building models, using these models for building RL controller training is a practical and promising solution for future real-life applications.

2. *How to widely deploy RL controllers in buildings with low cost*

   Intelligent building control is not only beneficial to both building owners but also favorable to the power system as it provides responsive loads to the grid. The advantage grows with the number of smart buildings. Reducing the cost and increasing the return-on-investment is important to encourage more building owners to commission such a control system. As discussed earlier, RL controllers do not require an on-site powerful computer to solve optimization problems (reduces both hardware and software cost) and can also use a data-driven building model for training (reduces the modeling cost). Another approach worth future research study is transfer learning, which transfer the knowledge from an already trained controller to a new controller for another but similar building. Typically, the transfer is done by warm-start the NN training by loading parameters from another controller's NN. By doing this the training time needed for the new controller is decreased (comparing with training the controller from scratch) and thus the training cost is further reduced. But it is worth noting that transfer learning only works with similar buildings (at least with the same state and action space); to this end, there should be some studies that investigate how to quantitatively evaluate the similarity between buildings and how the similarity is related to the decrease of training time when transfer learning is applied.

9.3.3.2 Demand response

Demand response (DR) is to encourage electricity customers to change their power consumption in certain periods in response to a signal received from utility company and in exchange of financial benefits [62]. With proper implementation and adequate customer adoption rate, DR can have a vital role in improving power system reliability and reducing the risk for having major blackouts. As a result, DR is considered to be a significant component of future smart grid.

Usually, before and during DR events, DR program organizers and participants need to make sequential optimal decisions of their own. For instance, on one hand, utility companies, as decision makers, need to dynamically determine the real-time price over a period of time, considering the wholesale price, weather, and customer's behavior, in order to properly guide the end users for load reduction, while satisfying other objectives and constraints. On the other hand, customers are also decision makers and need to optimally operate their houses, businesses, or industrial plants during the DR events to maximize benefit, while minimizing the cost from discomfort or production discontinuance. And in many cases, uncertainty is involved in such decision-making processes (e.g., price uncertainty for customers and the uncertainty of customer behavior for utilities). Therefore many studies investigate using RL for these decision-making processes. We will look into some examples using a bottom-to-top sequence, namely, DR problems at the device level, the customer level, the aggregator level, and finally the utility level.

First, device level DR control problems are the most well-studied ones among the four-level problems mentioned above, and they account for the most number of publications as well. In the literature, according to this review paper [63], devices being control during DR events are typically of two types: (1) thermostatically controlled devices (TCL, i.e., air-conditioning systems and water heaters) and (2) storage-based devices (e.g., batteries and electric vehicles). These two types of devices introduce strong temporal dependency to those control problems and thus made them suitable for RL to solve. Typical control problem solved in these studies is to minimize electricity cost while satisfying some operation constraints, considering a deterministic or stochastic time-variant price signal (e.g., real-time price in price-based DR programs). To add more complexity, uncertainty from user behavior or local renewable generation can be considered. For instance, an RL controller based on fitted Q-iteration is proposed in Ref. [64] to help water heaters to participate in DR programs. Two highlights are presented in this paper: (1) an auto-encoder network is used for feature dimension reduction which helps RL for faster convergence and (2) a lab experiment is conducted using an actual water heater and validated the proposed method. For a similar problem the authors of Ref. [65] solve it using a model-based RL approach.

Second, the difference between problems at the customer level and those at the device level is the involvement of more devices and potentially other customers. Thus there is a higher requirement for those controllers aiming at solving customer level problems since they need to control multiple devices of a single or multiple customer(s) in a coordinated manner (increased complexity when compared with single device control). Bahrami et al. [66] propose an online appliance scheduler based on an actor–critic RL method to reduce both cost and peak demand for multiple residential customers without knowing the electricity price ahead of time. Another example of customer level RL controller for DR programs can be found in Ref. [67], where a fully automated energy management system is proposed.

At the third level, some studies investigate how to use RL to properly control customers' devices at the aggregator level (i.e., optimal control for device clusters). Nowadays, to facilitate customers to participate in DR programs, many aggregators emerge, thanks to the development in the information and communication technologies (especially the IoT). Some aggregators are device manufacturer themselves (e.g., smart thermostats from Google Nest and Honeywell) while some others just support BYOD (Bring Your Own Devices) aggregation. During a DR event the aggregator needs to optimally control a large amount of heterogeneous devices to achieve the goal of load reduction, while meeting the operating constraints of each enrolled customer. One major difficulty is that the system dynamics and operating constraints are not known exactly. In Ref. [68] the authors propose a control scheme for managing a water heater cluster using a model free batch RL algorithm. A simulation on a water heater cluster of 100 devices shows that after 40−45 days, the controller is able to learn to reduce the daily electricity usage.

Finally, at the utility level, a dynamic pricing mechanism should be properly devised in order to guide price-responsive loads to adjust power consumption during grid peak hours. Problems at this level are in general not as well studied when compared with problems from previous three levels. The main reason is the modeling complexity: in previous problems on customer loads control at three levels, hypothetical price signals can be used to model the DR market, and they can be generated to train the corresponding controller; in contrast, when training the dynamic pricing scheme, customer models, and responding dynamics needs to be modeled, which is harder than price generation. Although there are studies [69−71] that propose RL-based decision-making framework that do not assume the availability of users' response functions, these approaches of directly learning customer's response by trial-and-error on the actual system might not be practical to be used in real-life applications.

9.3.3.3 Grid operation

According to Ref. [72], power system condition can be described by five operating states: normal, alert, emergency, in extremis, and restorative state. In this section, some examples using RL to facilitate the decision-making process in some of these states or during states transition are discussed.

First, during system normal operation, due to the increasing penetration of renewable generation and electric vehicles at the user end, grid control is challenged in many ways. For better voltage stability, Duan et al. [73] propose using RL to train a controller for autonomous voltage control: by using real-time measurements from the supervisory control and data acquisition system or wide area measurement system, the trained RL controller can support grid operators with effective and timely control actions. In Ref. [74] a two-timescale voltage regulation scheme based on RL is proposed to coordinate smart inverters on a faster timescale and shunt capacitors on a slower timescale. Similarly, Xu et al. [75] use RL to set the optimal tap position for voltage regulation transformers. In addition to the voltage regulation, there are also works on the frequency regulation: as early as 2002, Imthias Ahamed et al. [76] use RL to design automatic generation control (AGC) systems, which limits area control error excursions or restore the balance between generation and load by monitoring the deviation in system frequency and tie line power flows. Later in Ref. [77], RL and artificial emotion is combined to solve the AGC control problem.

Second, when the system is under the alert state (e.g., under $N-1$ or $N-2$ contingencies), grid operators might need to take adequate actions to keep the system operating in the stable region. For instance, the 2003 north American blackout was initiated by a series of line outages due to overloading [78]; to prevent similar cascading failure events, Zarrabian et al. [79] proposed using Q-learning to implement a real-time optimal control (adjusting generators' output power) to relieve congestion of transmission lines and thus prevent cascading line outages and blackouts. Another important aspect of blackout prevention is to conduct vulnerability analysis; a Q-learning–based scheme is designed in Ref. [80] to identify the most devastating sequential topological attacks. Similarly, Zhang et al. [81] propose an online search method to precisely identify "representative risky fault chains" and use such information for further fault blocking control.

Third, emergency control is of great importance as a successful control can stop system slide to the direction of system failure/in extremis state. At the emergency state, grid operators/automated controllers are required to take a sequence of fast, correct, and coordinated actions to mitigate/minimize the damage. As a result, an RL controller, due to its real-time action readiness, is a great candidate for such type of control. In Ref. [82], RL-based control schemes for generator dynamic braking and undervoltage load shedding are investigated.

Rapid system restoration after failure is also of great importance to improve system resiliency. Wu et al. [83] investigate an RL-based framework to identify the optimal restoration sequence after a cascading failure. RL should also be suitable to solve power system black-start problems, especially if considering using renewable energy sources as black-start units. However, no RL-based black-start sequence designing tool has been investigated in the literature at the time of writing.

9.3.3.4 Renewable generation and battery control

Renewable generation, due to the nature of the intermittent primary energy, also involves stochastic optimal control. One example is the wind turbine yaw control: to maximize power output, wind turbine yaw system needs to track the wind direction and adjust the turbine orientation accordingly. However, in some cases, greedily chasing the volatile wind direction might not be an optimal strategy because yawing is not instantaneous due to the mechanical process involved: by the time turbine turns to a desired direction, the optimal direction has already changed. As a result, some studies use RL to learn from wind historical data about the optimal yaw control strategy that yields the maximum power output over a period of time. Graf et al. [84] combined RL with alternating direction method of multipliers (ADMA!) and proposed an ADMM-RL algorithm that can be used for wind turbine yaw control. In Ref. [85] a knowledge-assisted DDPG algorithm is proposed for cooperative wind farm control; the proposed algorithm uses prior engineering knowledge to ensure safety during learning as well as accelerating the learning process. An RL-based maximum power point tracking (MPPT) control is proposed in Ref. [86] to determine the optimal rotor speed.

Grid-connected battery control is often formulated as an RL control problem. This is because battery introduces strong temporal dependency in the controlled system and as mentioned earlier the temporal dependency is indispensable for an RL problem. Typical battery-related RL problems also include a source of uncertainty: for example, how to optimally charging and discharging the battery to minimize the grid electricity usage under stochastic PV generation or random real-time price. In Refs. [87–89], battery system is leverage to participate in energy arbitrage market in order to maximize financial gain by buying and selling energy. Similarly, optimal decisions are made for EV charging in order to minimize energy cost [90].

9.3.4 Some interesting research topics

Although we have seen that engineers started using RL to solve some real-life engineering problems, there are still some pending and interesting research topics that need to be addressed. Several examples of these topics are discussed in the following:

Proper RL/MDP formulation: To successfully leverage RL to solve real-world engineering problems, properly formulating the MDP is of great importance. But for some problems, the definition of MDP state space is not as straightforward: for instance, to design an RL controller for grid control, the best way to define the state space that can accurately reflect the network topology, power flow distribution and device status is not clear. Similarly, some research efforts are needed to study and summarize the best reward structures for some typical engineering problems (e.g., signal tracking/set point following and cost minimization) that helps them for a more effective learning and faster convergence.

Controller training and evaluation: An RL policy helps achieve the maximum expected reward under certain distribution. For example, that means an RL wind turbine yaw controller trained with winter wind data might not be optimal for control in summer due to the different distributions of wind speed and direction in these two seasons. Therefore there should be a quantitative evaluation on how to design the training environment that shares the same distribution with the actual environment this controller will be used in. Following the same example, to train an RL-based turbine yaw controller to be used in August, an automated and quantitative analysis should help the researcher to choose which period of wind data should be used for training.

Combining prior knowledge with RL: An RL agent typically learns from scratch by trial-and-error. However, engineers and scientists usually have deep understanding (i.e., prior knowledge) about the system to be controlled. How can the prior knowledge to be leveraged to help improve the learning efficiency, guarantee a safer exploration, and increase RL controller's interpretability is an interesting research topic.

Transfer learning: Instead of learning from scratch, an RL agent can take advantage of a previously trained RL controller so that the training time can be largely reduced. This is very helpful in applications such as wind farms turbine control and building optimal control, in which there are many different but similar instances that require RL controllers for the same purpose. The transfer learning can help an efficient and scalable deployment of RL controllers in a great number of instances. But research is necessary to quantify the similarity between two control scenarios and determine if knowledge is transferable between them.

Multiagent RL: Oftentimes, there are multiple decision makers involved in a single engineering problem, and they worked either in a cooperative way or a competitive manner. Multiagent RL is suitable for such scenario; however, the training is much harder due to the nonstationary environment (i.e., for one agent, the other agents are part of the environment, and due to the learning process of other agents, the environment to this agent is constantly changing). Research is needed to investigate the effectiveness and stability of such learning framework.

9.3.5 Limitations and challenges

Admittedly, in addition to the potential advantages that RL might bring in engineering, using RL might also suffer from some limitations and challenges, which are listed below to help the readers to gain a more comprehensive understanding:

1. In RL formulation, forming strict constraints (equality or inequality) is not as convenient as the traditional optimization approach. The workaround is to add large penalty if a constraint is violated. However, if the number of constraints is large, this workaround is not a good solution.
2. The behavior of an RL agent cannot be quantitatively described upon the completion of RL training. To be specific, when using chance constraint stochastic optimization to solve a problem, some risks can be quantitatively limited, but for RL, such information is absent.
3. A trained RL controller only provides optimal control performance if the actual environment is used for stays the same as the one it is trained on. That is to say, an RL controller needs to be retrained once the environment has changed.
4. Because RL optimizes to achieve the maximum expected reward given the distribution of some environment's parameters (e.g., distribution of exogenous data such as outdoor temperature), as a result, its performance on some rare events (e.g., extreme hot days) might not be optimal.

9.3.6 Summary

In this section, we first discussed about some potential advantages of using RL for engineering problems, comparing with existing approaches. Then, the typical workflow to build an RL training environment is explained and demonstrated. Next, some examples for using RL in different domains are presented to show some existing efforts in this area. Finally, some potential interesting research topics and the limitations for using RL are also discussed. In all, we hope by reading this chapter, readers can have a better understanding on (1) the basic concepts of RL, (2) its potential to solve the engineering problems, (3) what types of problems are suitable for RL to solve, (4) how to use RL to solve real-life applications, and (5) what still need to be done in this area.

References

[1] R.S. Sutton, A.G. Barto, Reinforcement Learning: An Introduction, MIT Press, 2018.

[2] D. Silver, Model Free Control. [Online]. Available from: <https://www.davidsilver.uk/wp-content/uploads/2020/03/control.pdf> (accessed 03.07.20).

[3] C.J.C.H. Watkins, P. Dayan, Q-learning, Mach. Learn. 8 (3–4) (1992) 279–292.

[4] R.J. Williams, Simple statistical gradient-following algorithms for connectionist reinforcement learning, Mach. Learn. 8 (3–4) (1992) 229–256.

[5] J. Schulman, P. Moritz, S. Levine, M. Jordan, P. Abbeel, High-dimensional continuous control using generalized advantage estimation, arXiv preprint arXiv:1506.02438, 2015.

[6] V. Mnih, K. Kavukcuoglu, D. Silver, A.A. Rusu, J. Veness, M.G. Bellemare, et al., Human-level control through deep reinforcement learning, Nature 518 (7540) (2015) 529–533.

[7] D.E. Rumelhart, G.E. Hinton, R.J. Williams, Learning representations by back-propagating errors, Nature 323 (6088) (1986) 533–536.

[8] V. Mnih, K. Kavukcuoglu, D. Silver, A. Graves, I. Antonoglou, D. Wierstra, et al., Playing atari with deep reinforcement learning, arXiv preprint arXiv:1312.5602, 2013.

[9] V. Hado, Hasselt. Double q-learning, in: Advances in Neural Information Processing Systems, 2010, pp. 2613–2621.

[10] H.V. Hasselt, A. Guez, D. Silver, Deep reinforcement learning with double q-learning, in: 30th AAAI Conference on Artificial Intelligence, 2016.

[11] T. Schaul, J. Quan, I. Antonoglou, D. Silver, Prioritized experience replay, arXiv preprint arXiv:1511.05952, 2015.

[12] Z. Wang, T. Schaul, M. Hessel, H. Hasselt, M. Lanctot, N. Freitas, Dueling network architectures for deep reinforcement learning, in: International Conference on Machine Learning, 2016, pp. 1995–2003.

[13] M. Hessel, J. Modayil, H.V. Hasselt, T. Schaul, G. Ostrovski, W. Dabney, et al., Rainbow: combining improvements in deep reinforcement learning, in: 32nd AAAI Conference on Artificial Intelligence, 2018.

[14] V. Mnih, A.P. Badia, M. Mirza, A. Graves, T. Lillicrap, T. Harley, et al., Asynchronous methods for deep reinforcement learning, in: International Conference on Machine Learning, 2016, pp. 1928–1937.

[15] OpenAI, OpenAI Baselines: ACKTR & A2C. [Online]. Available from: <https://openai.com/blog/baselines-acktr-a2c/>, 2017 (accessed 03.07.20).

[16] T.P. Lillicrap, J.J. Hunt, A. Pritzel, N. Heess, T. Erez, Y. Tassa, et al., Continuous control with deep reinforcement learning, arXiv preprint arXiv:1509.02971, 2015.

[17] J. Schulman, S. Levine, P. Abbeel, M. Jordan, P. Moritz, Trust region policy optimization, in: International Conference on Machine Learning, 2015, pp. 1889–1897.

[18] J. Schulman, F. Wolski, P. Dhariwal, A. Radford, O. Klimov, Proximal policy optimization algorithms, arXiv preprint arXiv:1707.06347, 2017.
[19] Y. Wu, E. Mansimov, R.B. Grosse, S. Liao, J. Ba, Scalable trust-region method for deep reinforcement learning using Kronecker-factored approximation, in: Advances in Neural Information Processing Systems, 2017, pp. 5279−5288.
[20] T. Haarnoja, A. Zhou, P. Abbeel, S. Levine, Soft actor-critic: off-policy maximum entropy deep reinforcement learning with a stochastic actor, arXiv preprint arXiv:1801.01290, 2018.
[21] S. Fujimoto, H.V. Hoof, D. Meger, Addressing function approximation error in actor-critic methods, arXiv preprint arXiv:1802.09477, 2018.
[22] T. Salimans, J. Ho, X. Chen, S. Sidor, I. Sutskever, Evolution strategies as a scalable alternative to reinforcement learning, arXiv preprint arXiv:1703.03864, 2017.
[23] H. Mania, A. Guy, B. Recht, Simple random search provides a competitive approach to reinforcement learning, arXiv preprint arXiv:1803.07055, 2018.
[24] D. Horgan, J. Quan, D. Budden, G. Barth-Maron, M. Hessel, H.V. Hasselt, et al., Distributed prioritized experience replay, arXiv preprint arXiv:1803.00933, 2018.
[25] L. Espeholt, H. Soyer, R. Munos, K. Simonyan, V. Mnih, T. Ward, et al., Impala: scalable distributed deep-rl with importance weighted actor-learner architectures, arXiv preprint arXiv:1802.01561, 2018.
[26] Y. Bengio, J. Louradour, R. Collobert, J. Weston. Curriculum learning, in: Proceedings of the 26th Annual International Conference on Machine Learning, 2009, pp. 41−48.
[27] W. Zaremba, I. Sutskever, Learning to execute, arXiv preprint arXiv:1410.4615, 2014.
[28] A. Graves, G. Wayne, M. Reynolds, T. Harley, I. Danihelka, A. Grabska-Barwińska, et al., Hybrid computing using a neural network with dynamic external memory, Nature 538 (7626) (2016) 471−476.
[29] T. Matiisen, A. Oliver, T. Cohen, J. Schulman, Teacher-student curriculum learning, IEEE Trans. Neural Netw. Learn. Syst. (2019).
[30] B. Zoph, Q.V. Le, Neural architecture search with reinforcement learning, arXiv preprint arXiv:1611.01578, 2016.
[31] L. Bertinetto, J. Valmadre, J.F. Henriques, A. Vedaldi, P.H.S. Torr, Fully-convolutional Siamese networks for object tracking, European Conference on Computer Vision, Springer, 2016, pp. 850−865.
[32] C. Finn, P. Abbeel, S. Levine, Model-agnostic meta-learning for fast adaptation of deep networks, arXiv preprint arXiv:1703.03400, 2017.
[33] A. Nichol, J. Achiam, J. Schulman, On first-order meta-learning algorithms, arXiv preprint arXiv:1803.02999, 2018.
[34] H. Jiang, Y. Zhang, Y. Chen, C. Zhao, J. Tan, Power-traffic coordinated operation for bi-peak shaving and bi-ramp smoothing—a hierarchical data-driven approach, Appl. Energy 229 (2018) 756−766.
[35] K. Zhang, Z. Yang, T. Başar. Multi-agent reinforcement learning: a selective overview of theories and algorithms, arXiv preprint arXiv:1911.10635, 2019.
[36] K. Zhang, Z. Yang, H. Liu, T. Zhang, T. Başar, Fully decentralized multi-agent reinforcement learning with networked agents, arXiv preprint arXiv:1802.08757, 2018.
[37] K. Zhang, Z. Yang, T. Basar, Networked multi-agent reinforcement learning in continuous spaces, in: 2018 IEEE Conference on Decision and Control (CDC). IEEE, 2018, pp. 2771−2776.
[38] Y. Yang, R. Luo, M. Li, M. Zhou, W. Zhang, J. Wang, Mean field multi-agent reinforcement learning, arXiv preprint arXiv:l 802.05438, 2018.
[39] R.J. Vanderbei, Linear Programming: Foundations and Extensions, vol. 285, Springer Nature, 2020.

[40] K. Zhang, Z. Yang, T. Basar, Policy optimization provably converges to Nash equilibria in zero-sum linear quadratic games, in: Advances in Neural Information Processing Systems, 2019, pp. 11602–11614.
[41] K. Zhang, Z. Yang, H. Liu, T. Zhang, T. Basar, Finite-sample analyses for fully decentralized multi-agent reinforcement learning, arXiv preprint arXiv:1812.02783, 2018.
[42] L. Busoniu, R. Babuska, B.D. Schutter, A comprehensive survey of multiagent reinforcement learning, IEEE Trans. Syst. Man Cybernetics, C: Appl. Rev. 38 (2) (2008) 156–172.
[43] L. Yang, Z. Nagy, P. Goffin, A. Schlueter, Reinforcement learning for optimal control of low exergy buildings, Appl. Energy 156 (2015) 577–586.
[44] T. Mahmood, F. Ricci, Improving recommender systems with adaptive conversational strategies, in: Proceedings of the 20th ACM Conference on Hypertext and Hypermedia, 2009, pp. 73–82.
[45] G. Brockman, V. Cheung, L. Pettersson, J. Schneider, J. Schulman, J. Tang, et al., OpenAI Gym, arXiv preprint arXiv:1606.01540, 2016.
[46] G. Fraisse, C. Viardot, O. Lafabrie, G. Achard, Development of a simplified and accurate building model based on electrical analogy, Energy Build. 34 (10) (2002) 1017–1031.
[47] U.S. Electric Power Research Institute. OpenDss. [Online]. Available from: <https://www.epri.com/pages/sa/opendss> (accessed 03.07.20).
[48] D.B. Crawley, L.K. Lawrie, F.C. Winkelmann, W.F. Buhl, Y.J. Huang, C.O. Pedersen, et al., EnergyPlus: creating a new-generation building energy simulation program, Energy Build. 33 (4) (2001) 319–331.
[49] U.S. Energy Information Administration. How Much Energy Is Consumed in U.S. Residential and Commercial Buildings? [Online]. Available from: <https://www.eia.gov/tools/faqs/faq.php?id = 86&t = 1>, 2019 (accessed 03.12.19).
[50] Y. Ma, F. Borrelli, B. Hencey, B. Coffey, S. Bengea, P. Haves, Model predictive control for the operation of building cooling systems, IEEE Trans. Control. Syst. Technol. 20 (3) (2011) 796–803.
[51] A. Alessio, A. Bemporad, A survey on explicit model predictive control, Nonlinear Model Predictive Control, Springer, 2009, pp. 345–369.
[52] X.I. Yu-Geng, L.I. De-Wei, L. Shu, Model predictive control—status and challenges, Acta Autom. Sin. 39 (3) (2013) 222–236.
[53] C.W. Anderson, D.C. Hittle, A.D. Katz, R. Matt Kretchmar, Synthesis of reinforcement learning, neural networks and pi control applied to a simulated heating coil, Artif. Intell. Eng. 11 (4) (1997) 421–429.
[54] G.P. Henze, J. Schoenmann, Evaluation of reinforcement learning control for thermal energy storage systems, HVAC&R Res. 9 (3) (2003) 259–275.
[55] K. Dalamagkidis, D. Kolokotsa, K. Kalaitzakis, G.S. Stavrakakis, Reinforcement learning for energy conservation and comfort in buildings, Build. Environ. 42 (7) (2007) 2686–2698.
[56] E. Mocanu, D.C. Mocanu, P.H. Nguyen, A. Liotta, M.E. Webber, M. Gibescu, et al., On-line building energy optimization using deep reinforcement learning, IEEE Trans. Smart Grid 10 (4) (2018) 3698–3708.
[57] X. Ding, W. Du, A. Cerpa, Octopus: deep reinforcement learning for holistic smart building control, in: Proceedings of the 6th ACM International Conference on Systems for Energy-Efficient Buildings, Cities, and Transportation, 2019, pp. 326–335.
[58] W. Zhe, T. Hong, Reinforcement learning for building controls: the opportunities and challenges, Appl. Energy 269 (2020) 115036.

[59] T. Wei, Y. Wang, Q. Zhu, Deep reinforcement learning for building HVAC control, in: Proceedings of the 54th Annual Design Automation Conference 2017, ACM, 2017, p. 22.
[60] Z. Zhang, A. Chong, Y. Pan, C. Zhang, K.P. Lam, Whole building energy model for HVAC optimal control: a practical framework based on deep reinforcement learning, Energy Build. 199 (2019) 472–490.
[61] X. Zhang, M. Pipattanasomporn, T. Chen, S. Rahman, An IoT-based thermal model learning framework for smart buildings, IEEE Internet Things J. 199 (2019) 472–490.
[62] M.H. Albadi, E.F. El-Saadany, A summary of demand response in electricity markets, Electr. Power Syst. Res. 78 (11) (2008) 1989–1996.
[63] J.R. Vázquez-Canteli, Z. Nagy, Reinforcement learning for demand response: a review of algorithms and modeling techniques, Appl. Energy 235 (2019) 1072–1089.
[64] F. Ruelens, B.J. Claessens, S. Quaiyum, B.D. Schutter, R. Babuska, R. Belmans, Reinforcement learning applied to an electric water heater: from theory to practice, IEEE Trans. Smart Grid 9 (4) (2016) 3792–3800.
[65] O.D. Somer, A. Soares, K. Vanthournout, F. Spiessens, T. Kuijpers, K. Vossen, Using reinforcement learning for demand response of domestic hot water buffers: a real-life demonstration, in: 2017 IEEE PES Innovative Smart Grid Technologies Conference Europe (ISGT-Europe), IEEE, 2017, pp. 1–7.
[66] S. Bahrami, V.W.S. Wong, J. Huang, An online learning algorithm for demand response in smart grid, IEEE Trans. Smart Grid 9 (5) (2017) 4712–4725.
[67] Z. Wen, D. O'Neill, H. Maei, Optimal demand response using device-based reinforcement learning, IEEE Trans. Smart Grid 6 (5) (2015) 2312–2324.
[68] F. Ruelens, B.J. Claessens, S. Vandael, S. Iacovella, P. Vinger-hoets, R. Belmans, Demand response of a heterogeneous cluster of electric water heaters using batch reinforcement learning, in: 2014 Power Systems Computation Conference, IEEE, 2014, pp. 1–7.
[69] R. Lu, S.H. Hong, X. Zhang, A dynamic pricing demand response algorithm for smart grid: reinforcement learning approach, Appl. Energy 220 (2018) 220–230.
[70] A. Ghasemkhani, L. Yang, Reinforcement learning based pricing for demand response, in: 2018 IEEE International Conference on Communications Workshops (ICC Workshops), IEEE, 2018, pp. 1–6.
[71] B.-G. Kim, Y. Zhang, M. Van Der Schaar, J.-W. Lee, Dynamic pricing and energy consumption scheduling with reinforcement learning, IEEE Trans. Smart Grid 7 (5) (2015) 2187–2198.
[72] L.H. Fink, K. Carlsen, Operating under stress and strain [electrical power systems control under emergency conditions], IEEE Spectr. 15 (3) (1978) 48–53.
[73] J. Duan, D. Shi, R. Diao, H. Li, Z. Wang, B. Zhang, et al., Deep-reinforcement-learning-based autonomous voltage control for power grid operations, IEEE Trans. Power Syst. 35 (1) (2019) 814–817.
[74] Q. Yang, G. Wang, A. Sadeghi, G.B. Giannakis, J. Sun, Two-timescale voltage control in distribution grids using deep reinforcement learning, IEEE Trans. Smart Grid 11 (3) (2019) 2313–2323.
[75] H. Xu, A.D. Dommguez-García, P.W. Sauer, Optimal tap setting of voltage regulation transformers using batch reinforcement learning, IEEE Trans. Power Syst. 35 (3) (2019) 1990–2001.
[76] T.P. Imthias Ahamed, P.S. Nagendra Rao, P.S. Sastry, A reinforcement learning approach to automatic generation control, Electr. Power Syst. Res. 63 (1) (2002) 9–26.

[77] L. Yin, T. Yu, L. Zhou, L. Huang, X. Zhang, B. Zheng, Artificial emotional reinforcement learning for automatic generation control of large-scale interconnected power grids, IET Gener. Transm. Distrib. 11 (9) (2017) 2305–2313.
[78] G. Andersson, P. Donalek, R. Farmer, N. Hatziargyriou, I. Kamwa, P. Kundur, et al., Causes of the 2003 major grid blackouts in North America and Europe, and recommended means to improve system dynamic performance, IEEE Trans. Power Syst. 20 (4) (2005) 1922–1928.
[79] S. Zarrabian, R. Belkacemi, A.A. Babalola, Reinforcement learning approach for congestion management and cascading failure prevention with experimental application, Electr. Power Syst. Res. 141 (2016) 179–190.
[80] J. Yan, H. He, X. Zhong, Y. Tang, Q-learning-based vulnerability analysis of smart grid against sequential topology attacks, IEEE Trans. Inf. Forensics Security 12 (1) (2016) 200–210.
[81] Z. Zhang, R. Yao, S. Huang, Y. Chen, S. Mei, K. Sun, An online search method for representative risky fault chains based on reinforcement learning and knowledge transfer, IEEE Trans. Power Syst. 35 (3) (2019) 1856–1867.
[82] Q. Huang, R. Huang, W. Hao, J. Tan, R. Fan, Z. Huang, Adaptive power system emergency control using deep reinforcement learning, IEEE Trans. Smart Grid 11 (2) (2019) 1171–1182.
[83] J. Wu, B. Fang, J. Fang, X. Chen, K.T. Chi, Sequential topology recovery of complex power systems based on reinforcement learning, Phys. A: Stat. Mech. Appl. 535 (2019) 122487.
[84] P. Graf, J. Annoni, C. Bay, D. Biagioni, D. Sigler, M. Lunacek, et al., Distributed reinforcement learning with ADMM-RL, in: 2019 American Control Conference (ACC), IEEE, 2019, pp. 4159–4166.
[85] H. Zhao, J. Zhao, J. Qiu, G. Liang, Z.Y. Dong, Cooperative wind farm control with deep reinforcement learning and knowledge assisted learning, IEEE Trans. Ind. Inform. 16 (11) (2020).
[86] C. Wei, Z. Zhang, W. Qiao, L. Qu, An adaptive network-based reinforcement learning method for MPPT control of PMSG wind energy conversion systems, IEEE Trans. Power Electron. 31 (11) (2016) 7837–7848.
[87] J. Cao, D. Harrold, Z. Fan, T. Morstyn, D. Healey, K. Li, Deep reinforcement learning based energy storage arbitrage with accurate lithium-ion battery degradation model, IEEE Trans. Smart Grid 11 (5) (2020).
[88] H. Xu, X. Li, X. Zhang, J. Zhang. Arbitrage of energy storage in electricity markets with deep reinforcement learning, arXiv preprint arXiv:190Jh 12232, 2019.
[89] H. Wang, B. Zhang. Energy storage arbitrage in real-time markets via reinforcement learning, in: 2018 IEEE Power & Energy Society General Meeting (PESGM), IEEE, 2018, pp. 1–5.
[90] A. Chiş, J. Lundén, V. Koivunen, Reinforcement learning-based plug-in electric vehicle charging with forecasted price, IEEE Trans. Veh. Technol. 66 (5) (2016) 3674–3684.

CHAPTER TEN

Power, buildings, and other critical networks: Integrated multisystem operation

Kyri Baker
University of Colorado Boulder, Boulder, CO, United States

Contents

10.1 Introduction 320
10.1.1 Aging infrastructure and climate-related impacts 321
10.1.2 Increasing electrification in the built environment 322
10.1.3 Increasing connectivity in power, water, and gas networks 324
10.2 Grid-interactive buildings 327
10.2.1 Distributed energy resources 329
10.2.2 Demand response 330
10.2.3 Emerging considerations for demand response 333
10.2.4 Climate and environment 334
10.2.5 Cybersecurity and privacy 336
10.3 Interdependent critical networks 338
10.3.1 Water and energy 338
10.3.2 Power and gas 344
10.3.3 Combined heat and power 345
10.4 Electrification of the transportation sector 347
10.4.1 Consumer vehicles 347
10.4.2 Public transportation 349
10.4.3 Rideshare services and emerging methods of transportation 350
10.4.4 Vehicle-to-grid 351
10.5 Considerations for future power systems 352
10.5.1 Physical considerations 352
10.5.2 Market and organizational considerations 353
10.5.3 Cyber considerations 353
Acknowledgments 354
References 354

New Technologies for Power System Operation and Analysis.
DOI: https://doi.org/10.1016/B978-0-12-820168-8.00009-2

10.1 Introduction

Electrification is considered the greatest engineering achievement of the 20th century by the National Academy of Engineering and is an essential component of many parts of the world today. Without electricity, water pumping stations would not be able to deliver drinking water to a community; electricity-powered gas compressor stations could not provide natural gas for heating and cooking; cell towers could not receive and transmit signals; subway cars would cease to move, potentially stranding commuters in a dark underground tunnel without power for ventilation.

Power, and reliable access to the electric power grid, is essential for the operation of critical networks such as water, gas, telecommunication, and transportation networks, and for ensuring adequate shelter, heating, cooling, and medical facilities. However, multiple emerging factors are challenging the ability of the grid to reliably deliver power to these critical infrastructure sectors. Due to increased electrification at the building level, low-voltage distribution networks may face challenges in the future without increases in distribution capacity or distributed energy resources (DERs). In addition, the effects of extreme weather and climate change have impacted the grid's ability to reliably deliver power, with weather-related blackouts increasing in frequency in recent decades. Many of the recent developments in the electric power grid that are affected by these changes occur in the low-voltage, distribution networks that provide pathways for consumers to receive and, in some cases, deliver power to the main grid; however, changes in distribution networks are affecting high-voltage networks much more rapidly than grid operators had originally anticipated. In this chapter, we will discuss the impact of these changes in the distribution network and other interdependent networks, such as within buildings, water, gas, and communication networks. In particular, we will address the questions: *How will the increasing interconnectedness of critical infrastructure sectors impact the power sector?* and *What are some of the needs of future power systems that will help us adapt to and be cognizant of these challenges?* Toward this, we first take a look at some of the factors that continue to impact and influence the operation of the grid and its dependent networks.

10.1.1 Aging infrastructure and climate-related impacts

The American Society of Civil Engineers (ASCE) stated in their 2017 report, "Without greater attention to aging equipment, capacity bottlenecks, and increased demand, as well as increasing storm and climate impacts, Americans will likely experience longer and more frequent power interruptions." From the 1950s to the 1980s, large-scale power outages in the grid were rare, averaging roughly five per year or less [1]. However, due to aging grid components and the growing uncertainty in load, generation, and weather, the grid has been increasingly stressed: from 2003 to 2012, roughly 679 outages occurred due to weather, with severe weather being the leading cause of outages in the United States. The Department of Energy estimates that the average cost to the US economy during this time period due to weather-related outages is \$18–\$33 billion [2], with numbers even higher during years with major storms such as Superstorm Sandy in 2012 (\$27–\$52 billion). Due to the reliance of many other critical networks (water, gas, transportation, communications, emergency services, and more) on electricity, the loss of this asset can prove detrimental to components beyond the immediate grid, putting lives at risk. Building loads and peak demands are also expected to increase during certain seasons, resulting in increased stress in power infrastructure [3].

It is important for grid planners and operators to be aware of the impacts on the power grid that go beyond just the power grid itself and design plans for grid resiliency that prioritize and ensure power delivery to critical networks and loads, coordinating dependent networks when necessary to prevent power delivery issues. Recent research works focus on the immediate impact of climate-related disasters and extreme weather events on the grid, offering frameworks for grid resiliency upgrades to preserve the operation of these critical networks [4,5]. In Ref. [4] the focus is placed on improving the resiliency of distribution networks to extreme weather, which are often impacted more heavily than transmission networks from these events. They consider four types of upgrades to improve resiliency: (1) hardening existing power lines (e.g., replacing power poles with stronger ones and adding guy wires, which is generally much cheaper than burying lines underground); (2) building new lines, adding redundancy in the network; (3) installing controllable devices that add operational flexibility, such as switches; and (4) adding more DERs, which reduces reliance on power provided by the main grid. Some public

Table 10.1 Average lifespan and capacity-weighted age of coal, natural gas, and nuclear fueled operating power plants in the United States.

Generator fuel type	Average lifespan (years)	Capacity-weighted average age (years)
Coal	50–60	39
Natural gas	40–50	22
Nuclear	30–40	36

Source: Data from EIA and A. Mills, R. Wiser, J. Seel, Power Plant Retirements: Trends and Possible Drivers, Lawrence Berkeley National Laboratory, Nov. 2017 [9].

service commissions, utilities, and system operators have recognized the impact of climate change and extreme weather events on the operation of the grid. For example, Consolidated Edison in New York has included climate predictions into their grid planning models [6].

These impacts are compounded by the fact that most power system infrastructure, including lines, transformers, and generators, are becoming less efficient with age, and in some cases nearing their end-of-life.

According to Federal Energy Regulatory Commission (FERC), distribution utilities across the country continue to invest thousands of millions of dollars a year in distribution infrastructure, with this number increasing over time. However, due to a lack of visibility and sensor networks within distribution networks, equipment is often oversized, or sized for worst case scenarios. Despite these efforts in 2017, ASCE gave the US's energy infrastructure the grade of a D + , indicating our infrastructure is "poor, at risk" and "requires attention." This is not surprising, as 70% of transmission lines and transformers are over 25 years old [7]; wooden power poles tend to start to rot around 40 years old, depending on where they are located; steel poles and transformers corrode over time; and distribution and transmission lines are given roughly a 50-year expectancy [8] (but transmission lines can last up to 100 years, if kept well maintained). Power generation sources are also aging. In Table 10.1 the average lifespan of coal, natural gas, and nuclear plants in the United States is shown along with the current (as of 2017) capacity-weighted average age of these plants. It is evident that other than natural gas plants, which have experienced a growth in recent years, many traditional forms of generation are beginning to age.

10.1.2 Increasing electrification in the built environment

Many aspects of the built environment, that is, the human-made structures and networks that surround us, are trending from other energy sources to the use of electrical energy. Electrification has many benefits: improved demand-side capabilities to provide grid services, air quality improvements, greenhouse gas reductions, and more. Adoption of heat pumps for water heating and space heating and cooling can also result in increase in the efficiency of heating and cooling systems within homes. If the *full* electrification potential in transportation, industrial processes, and buildings sectors is achieved, a potential 72% reduction in fossil fuel combustion for these end uses could be achieved by 2050 [10], with the transportation sector accounting for the majority of this reduction. Some processes within the industry and buildings sectors face additional challenges; some processes such as cement manufacturing require high temperatures that are difficult to achieve with electric sources of heat alone, and consumer acceptance may hinder the adoption of electric stoves, for instance.

The electrification of the transportation sector is occurring in a rapid fashion, with the adoption of electric vehicles (EVs) increasing dramatically worldwide. By 2030 the International Energy Agency (IEA) predicts that the number of EVs on the road will increase from roughly 3 to 125 million [11]. As another example, in Colorado, the percentage of EVs has increased by over 150% since 2011. As indicated in Fig. 10.1, the industry, buildings, and transportation sectors contain a wide array of electrification opportunities, from space and water heating to transit electrification. The transportation sector in particular currently has much room for electrification; according to the Energy Information Administration (EIA), in 2018, petroleum products accounted for approximately 92% of the total US transportation sector energy use. Electricity was less than 1% of the total energy use in this sector, despite transportation's significant share of overall energy consumption, indicating there is significant potential for future electrification.

Despite the benefits from electrification, strains on aging generation resources and power infrastructure are exacerbated by the resulting increases in electric loads. Within the United States the National Renewable Energy Laboratory estimates that the overall electricity consumption will grow from approximately 3860 TWh in 2016 to 5260 TWh in 2050. The use of electric heating in the winter would increase winter peak demands, spurring utilities and grid planners to plan for shifting load curves and both summer and winter peaks. However, the impacts are hard to measure exactly: energy efficiency improvements in both building construction and electrical appliances and grid-interactive loads and other demand-side management (DSM) strategies will influence

Industry/manufacturing	Residential/commercial	Transportation
Process heating	Space heating	Commercially available consumer vehicles
Space heating	Water heating	Transit buses/trolleys/ delivery and work trucks
Combined heat and power/ cogeneration processes	Clothes drying	Passenger and freight trains
Conventional boilers	Cooking equipment	Off-road vehicles (forklifts, tractors, cargo loaders, etc.)

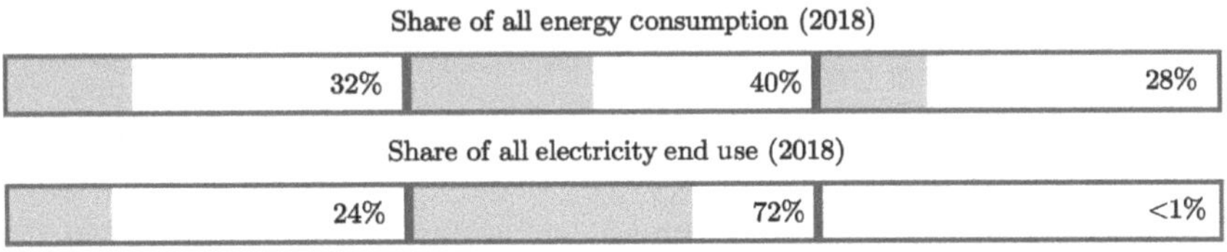

Figure 10.1 Current opportunities for electrification in the industry, buildings, and transportation sectors, along with the distribution of end use energy and electricity in these sectors. Source: *Data from the EIA Monthly Energy Report.*

the overall impact on both peak load and overall energy consumption from electrification. In this chapter, we will focus on the electrification of the transportation sector and the interaction between power and transportation, which has the highest potential of impacting the operation of the grid.

10.1.3 Increasing connectivity in power, water, and gas networks

A distinguishing characteristic of the smart grid as compared to traditional power grids is the proliferation of sensor networks and increased wireless communication and control capabilities, with a two-way communication between the grid and consumers. With increase in phasor measurement units (PMUs) and machine-to-machine communication, the lack of physical wired communication infrastructure that has resulted in wireless communication is becoming ubiquitous throughout the smart grid. Increased visibility into distribution system operations, while not considered as important prior to the smart grid era, is now becoming essential to ensuring grid reliability and stability. PMUs, for instance, were typically only installed at the transmission level but are beginning to emerge at the distribution level, giving grid operators additional insight into the status of the low-voltage networks [12]. While wired connections are generally more reliable and secure and are far more costly than wireless networks.

Various wireless communication technologies are currently used throughout the grid: ZigBee (e.g., smart meters), Bluetooth (e.g., connected appliances), cellular networks (e.g., Global System for Mobile Communications—GSM), and many more [13].

The link between communication networks and power networks enhances the capabilities of the smart grid to solicit demand response (DR) and provide situational awareness to grid operators. Despite the aging of infrastructure and conventional generation sources, rapid additions of distributed solar, storage, and smart thermostats, to name a few, are assisting these conventional controllable generation sources by providing additional resources for the grid. On the demand side, where grid interactivity is increasing with the deployment of DR and dynamic pricing programs, communication between behind-the-meter devices and appliances is essential to continued success. The electricity meter, invented in the late 19th century, made it possible for consumers to be billed for electricity. However, traditional analog electricity meters throughout the country are being rapidly replaced with digital meters to facilitate the efficiency of meter readings as well as to enable grid-responsive buildings.

Automatic meter reading (AMR) systems, for instance, emerged in the mid-1980s and enable a one-way communication between building to grid through telephone protocols (both wired and wireless), radio frequency (RF, by far the most common), or power-line communication, where data is carried over existing power lines. Many residences in the United States today are still fitted with AMR meters. Advanced metering infrastructure (AMI), which is generally what many refer to when they mention "smart meters," can communicate bidirectionally between building and grid on faster timescales, reducing costs from meter readings and enabling faster outage pinpointing and restoration. For example, smart meters helped grid operators in Houston, Texas locate power outages during Hurricane Harvey, fixing damages before power loss spread. While AMR infrastructure installations have stalled since the early 2010s, AMI installations continue to increase (Fig. 10.2), with many states recognizing the benefits of increased communication capabilities between buildings and the grid. AMI meters use both home area networks to communicate to connected devices within a building and wide area networks, generally over cellular networks or RF mesh networks, to communicate with energy providers.

Typically, electricity consumption is recorded in the meter's memory in 5-, 15-, 30-, or 60-minute intervals, with 15 minutes readings being

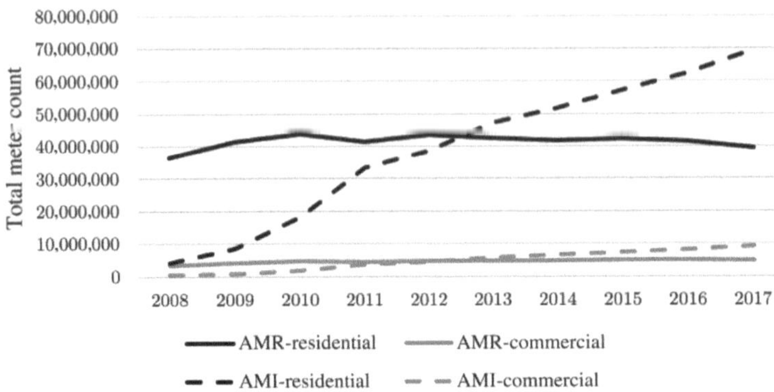

Figure 10.2 Total installed electricity meters on residential and commercial buildings in the United States from 2008 to 2017. The increases in AMI infrastructure is promising for improved demand-side capabilities of providing grid services. *AMI*, Advanced metering infrastructure. Source: *Data from the EIA.*

one of the most common frequencies, and typically read by the utility anywhere from monthly to 15-minute intervals. XCEL Energy, for example, reads some electricity and natural gas meters every 15-minute in order to gather data to help predict system demand. Increased communication capabilities throughout the power and power-dependent networks are not limited to electrical devices, and AMI systems are gaining traction in smart natural gas, water, and even transportation networks. ABI Research estimates that the number of smart meters deployed throughout the world will reach 780 million for electricity, 150 million for gas, and 90 million for water by 2020, meaning that the control and monitoring of water, gas, and power networks will have increased capabilities of working together to improve overall system efficiency and resiliency. Smart water meters began emerging in the 2000s, and harness the two-way communication to inform infrastructure upgrades, improve billing accuracy, perform remote upgrades, and reduce or shut off water supply to customers who have not paid their bills on time. Unlike smart electricity meters, smart water meters are generally battery powered (see Fig. 10.3) as electrical infrastructure to these devices was initially not needed and do not receive power for transmission from the electric grid. This means that the batteries must be in a waterproof casing, have a long-lasting lifespan, and that the communication system cannot make transmissions with the utility too frequently without risking draining the battery.

However, the use of wireless communications throughout the smart grid has its own suite of downsides. For example, the use of GSM

Figure 10.3 The casing covering an AMR water meter in Boulder, CO, which allows water utilities to record consumption remotely, multiple times throughout the day. The battery supply (right) should last for over a decade under present day communication frequency. *AMR*, Automatic meter reading.

networks for smart grid communication is subject to the same pitfalls as the cellular networks themselves, such as being sensitive to network congestion during peak hours. ZigBee technology is versatile, low-power, and allows power, water, and gas utilities to send dynamic pricing signals and consumers to monitor their real-time usage but is susceptible to interference effects from other technologies such as Wi-Fi and Bluetooth [14]. Wireless communication also introduces increased risks over wired networks; cybersecurity risks abound [15–17], and cyberattacks targeted toward vulnerable devices within a network could have the potential of propagating throughout the network to other assets. Public perception about the dangers and health effects of smart meters is also a nonnegligible concern for water and energy utilities. Coordination between these systems, enabled by two-way communication technologies, will be discussed further on in this chapter.

10.2 Grid-interactive buildings

In the United States, buildings consume roughly 40% of the total energy consumption and over 70% of the electricity generated. Traditionally, power flowed throughout distribution networks in a unidirectional fashion—from substation down to building electricity meters. However, with the increase in DERs, including distributed generation and storage, reverse power flows are now occurring, resulting in voltage issues [18] and wear-and-tear on transformers and other substation equipment [19].

Despite these emerging issues, increases in DER penetration, when coupled with enhanced communication and control capabilities, have great potential of increasing the flexibility of distribution networks [20]. The communication capabilities of many smart electricity meters, for example, allow for the ease of information exchange between buildings and utility companies on granular timescales, helping utilities request services from buildings or pinpoint outages during a disaster. This opens up an array of valuable resources for both building owner and utility: automated, price-responsive building loads, better grid-level forecasts of consumption, heightened consumer awareness, and more. In particular, the term "building-to-grid" or "grid-interactive buildings" encapsulates this two-way communication structure that allows for the ease of communication from built environment to the electric power grid.

Case study: 2017 solar eclipse

On August 21, 2017 a total solar eclipse occurred that spanned the contiguous US grid operators, utilities, and other organizations invested significant amounts of time and money to determine how the eclipse would affect the operation of the power grid; the predicted generation drop would require rapidly ramping up dispatchable generation such as gas-fired power plants. The California Public Utilities Commission urged their consumers to reduce consumption by 3500 MW [21]. Smart thermostat manufacturer Nest also prepared for the effects of the solar eclipse. Customers that joined Nest's Solar Eclipse Rush Hour program had their temperatures automatically adjusted during this time, lowering load by about 700 MW (roughly the equivalent of seven gas peaker plants [22]). The success of demand-side flexibility was evident; partially thanks to coordinating building loads with the grid, no eclipse-related outages occurred.

Increased interaction between buildings and the grid cannot be only beneficial for improving system efficiency but can also be a necessity during times of need to prevent blackouts or cascading failures. This is not a new concept: grid operators have been leveraging direct load control (DLC) opportunities from industrial facilities for decades, for instance, requesting load curtailment during times of high grid stress. Industrial facilities and large commercial buildings have also traditionally been subjected to demand charges, where an additional electricity charge is added to their bill based on the 15-minute period with their highest consumption throughout the month. This charge can be a significant portion of their overall bill, as the sizing of distribution lines and other equipment is often driven by the demand peaks of these large loads. Many other DR programs for residential building-to-grid have gained traction in recent years, which will be further discussed in this section.

10.2.1 Distributed energy resources

DERs are assets such as rooftop solar, stationary energy storage, grid-responsive EVs, controllable heating, ventilation, and air conditioning (HVAC) systems, combined heat and power (CHP) plants, and other resources that are connected to low-voltage (distribution) grids and who usually have the potential of providing grid services. DERs are important for DR, which will be discussed in more detail in the following section. Distributed generation comprises one large aspect of DERs and refers to localized, low-voltage generation resources such as rooftop solar and small diesel generators. With DERs, large heat dissipation losses from delivering power generated by large, central power plants, down to end use loads, is significantly lowered due to more localized sources of energy.

For example, Fig. 10.4 shows EIA's estimation for total transmission and distribution losses in the United States as a percentage of overall power generated from 1950 to 2016. Note that for the past decade or so, transmission and distribution losses were around 6%–7%, which amounts to millions of kilowatt hour lost during power delivery. With increased proliferation of generation that is closer to the demand, some of these significant losses could be avoided.

In addition to efficiency benefits, DERs can also improve system resiliency. *Resilience* in the context of power grids is usually defined as the

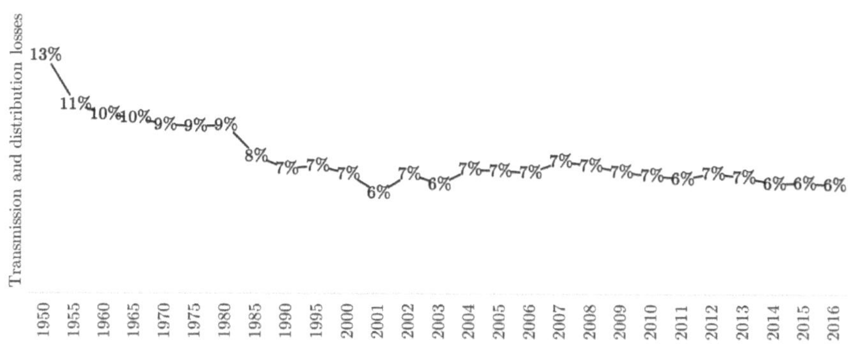

Figure 10.4 Estimated percent power losses due to transmission and distribution in the United States. Source: *Data from the EIA.*

ability to withstand and rapidly recover from power outages and other grid contingencies [23]. When a natural disaster or failure in the transmission grid occurs, DERs can provide resources to communities during a time of need through local generation or energy storage. This differs from grid *reliability*, which refers to the ability of the power system to deliver power to meet the required load. This is not to say DERs do not improve system reliability—certainly, with a greater diversity of assets and geographically spread out resources, less reliance on transmission infrastructure can prove to be very helpful for reliably delivering power to loads.

10.2.2 Demand response

Demand response is an important and essential component of modern power systems. Navigant research estimates that the revenue to residential, commercial, and industrial customers from DR will reach $1.3 billion by the year 2026 [24]. The FERC's official definition of DR is "Changes in electric usage by end use customers from their normal consumption patterns in response to changes in the price of electricity over time, or to incentive payments designed to induce lower electricity use at times of high wholesale market prices or when system reliability is jeopardized." Verbosity aside, DR can be defined as a term that refers to an end user changing their electricity consumption in response to a signal from the utility (usually with some monetary compensation). This was mentioned earlier with DLC of industrial facilities but expands far beyond the curtailment of large industrial loads. Broadly speaking, the different types of DR span two main categories, seen in Fig. 10.5: price-based programs and incentive-based programs. The availability of these programs varies widely by region, utility, and grid needs in that area. For example, utility company Commonwealth Edison is one of the only utilities in the United States to offer residential real-time pricing, which gives customers the option of having their electricity price vary hourly and reflect actual transmission-level marginal prices. Time-of-use (TOU) prices, on the other hand, are found throughout the country and offer users predictable prices across blocks of multiple hours, usually only changing seasonally with seasonal load variations. An example of a real TOU price from XCEL Energy is shown in Table 10.2 for the summer season, showing higher prices during times of peak system loading (2:00–4:00 p.m.) to incentivize users to shift when they consume power.

Price-based DR	Incentive-based DR
Time-of-use pricing: The price of electricity changes during pre-determined different blocks of time	**Direct load control:** The utility and end-user have an agreement allowing the utility control over specific appliances or devices to reduce peak load, line congestion, etc.
Real-time pricing: The price of electricity changes throughout the day, usually hourly, in response to changes in wholesale price	**Emergency DR:** When the grid is in severe trouble, the utility or system operator can send requests to participants to shed (reduce) their load or turn on any on-site generation
Critical-peak pricing: The price of electricity becomes really high for a limited number of hours	**Demand bidding:** When large consumers bid in a load reduction amount at a particular price

Figure 10.5 Demand response can be considered in two categories: price-based and incentive-based. The availability of different programs varies depending on utility, geographical region, and load/building type.

Table 10.2 Summer time of use costs (opt-in) for residential consumers served by XCEL Energy. Last accessed 6/2019.

XCEL Energy's TOU rates	Off-peak price (9:00 p.m. to 9:00 a.m.)	Shoulder price (9:00 a.m. to 2:00 p.m.) (6:00 p.m. to 9:00 p.m.)	On-peak price (2:00–4:00 p.m.)
Summer ($/kWh) (except holidays)	0.08	0.13	0.18

TOU, Time-of-use.

We pursue the balance between demand and generation in the grid at all times. Rather than having the grid handle generation capacity shortages, frequency fluctuations transmission congestion, and voltage control, why not let the demand side help? In some cases, this is more economical than installing additional capacity or turning on another power plant on the grid side—and, in some cases, there are limits to what the supply side can do, and the demand side *must* participate to avoid incurring outages. This was originally done with requesting load sheds at industrial facilities, but DR programs are rapidly emerging at the commercial and residential levels. Because the term "demand response" is so often confused with other concepts, a few examples are provided below to illustrate the concept of DR.

Is it demand response?

Determine if each of the following three examples are considered demand response, or something else.

Replacing all of your light bulbs with energy-efficient LED lamps. This is *not* demand response. This can be considered as an instance of demand-side management (DSM), which is often used interchangeably with demand response but actually is a larger term encompassing both demand response and implementing measures to improve energy efficiency. Here, replacing your light bulbs with LED lamps is considered DSM, but not DR.

Participating in Nest's Rush Hour Rewards. The Rush Hour Rewards program states that "your energy company will reward you to reduce load on the electrical grid during Rush Hours (times when demand for energy is high) like a heat wave or cold snap." In this case, you would be changing your heating/cooling demand in response to a varying price signal from the utility, so this *would be* considered demand response.

Enrolling in XCEL Energy's Windsource Program. For an extra charge on your utility bill each month, you can have a portion (or all) of your electricity consumption generated from wind power. This is *not* demand response. This is considered a renewable energy credit (REC) and does not incentivize the consumer to change their electricity use in any way.

Enrolling in FPL's on call program. Florida Power and Light (FPL) will pay you to let them install switches on some of your appliances, such as your air conditioner, central heater, electric water heater, or pool pump. Then, they are allowed to turn off these appliances in times of need when the grid is heavily loaded. This *does* count as DR; you are allowing your electricity consumption to be changed in response to a grid signal.

Lawrence Berkeley National Laboratory has classified DR into four categories (from longer timescale services to faster timescale services): shape, shift, shed, and shimmy [25]. In addition to being catchy, these terms have provided a common reference when discussing potential DR services that loads can provide. *Shape*, for instance, refers to DR that reshapes load profiles through dynamic pricing or through behavioral campaigns, on the scale of days to years. This includes DSM strategies that improve energy efficiency. As one example, TOU rates, while held constant for multiple months, need to be updated across the years to reflect long-term changes in load. *Shift*, as the name suggests, refers to DR services that occur on the daily to hourly timescale, shifting loads to shave peaks, fill valleys, and better coincide with renewable generation. Overlapping somewhat but providing minute-level responses is *Shed*, which describes loads that can be curtailed when needed. This can happen during either emergency or contingency events within transmission or distribution networks or to relieve line congestion or make up for a lack of generation capacity. And finally, *Shimmy* refers to seconds to minutes level adjustments in load for frequency regulation and short-term ramping. Shimmy can be achieved with energy storage charging/discharging, advanced lighting controls, and variable frequency drives, for instance.

10.2.3 Emerging considerations for demand response

If too many buildings or loads respond to DR requests, a "rebound effect" could be observed. Presently, due to lack of consumer awareness, participation/enrollment apathy due to the relative magnitude of electricity costs, and a shortage of connected devices and buildings, this is generally not a problem, but under a high penetration of demand-responsive loads [26], enough buildings can shift their electricity consumption through techniques such as precooling before high price periods that the resulting peak demand is even greater than without TOU pricing. During June 2016 in California, the demand within the Southern California Edison (SCE) territory was nearing all-time peak levels. SCE requested DR and received a very sudden and sharp 500 MW drop in load within 7 minutes [27]—more and more rapidly than they were expecting.

Another emerging issue is that consumer perception and behavioral patterns stand in the way of large-scale successful deployment of DR. Consumers are not used to changing when they use energy, and often the savings are not significant enough for them to change their behavior. To address this, utilities are emphasizing other positive effects of enrolling in DR programs, such as sustainability and environmental impact, but a high level of automation [28] and "set it and forget it" strategies will prove useful for engaging consumers in DR programs. Lastly, equipment lifetime and maintenance can be impacted by participating in DR. With particular focus on providing ancillary services, loads can be rapidly turned on or off and cycled, which is likely different than the original design specifications for those loads. This could incur early replacement or increased maintenance frequency costs; however, there may be financial benefits from proving DR services, and a full life cycle cost analysis must be performed to determine the overall impact of participating in DR [29]. Long-term planning strategies must also be taken into account by utilities and grid operators to ensure that effects such as the rebound effect do not impact the efficacy of DR as it becomes more widespread.

10.2.4 Climate and environment

Extreme weather events, long-term changes in building loads, and other environmental factors are making it more challenging to reliably deliver power throughout the country. Between 2003 and 2012, weather accounted for 80% of all power outages in the United States; by 2012 the average annual weather-related power outages had doubled since 2003 [30]. In Table 10.3 the total number of severe weather-related major power outages in the United States between 2013 and 2018 is shown in addition to the number of customers affected in the case of the largest outage. Weather-related outages accounted for the majority of major outages in each of the considered years, and the severity of the worst case outages appears to be getting worse.

However, large-scale outages are not always due to ice storms, hurricanes, or thunderstorms; sometimes, extreme temperatures are to blame. In 2003 a heat wave in the northeastern United States resulted in 50 million people losing power within a mere 8 minutes, and the lack of needed generation due to extreme cold weather left 32,000 customers in New Mexico and Texas without heating in 2011. Caused by a heat wave in 2007, the Brown's Ferry nuclear plant in Alabama was forced to shut down due to high river water temperatures. There is a clear increasing trend in the outages due to climate and extreme weather events, and grid operators recognize this; for example, Hurricane Sandy resulted in 8.5 million people losing power across 21 states and spurred investments in grid resiliency upgrades.

One of the US's largest power outages occurred in 2017. On September 20, 2017, Hurricane Maria hit the island of Puerto Rico. By

Table 10.3 Total number of major weather-related power outages and disturbances and number of customers affected in the largest outage.

Year	2013	2014	2015	2016	2017	2018
Major weather-related outages	40	77	51	39	62	80
Largest outage (no. customers affected)	400,000	715,000	500,000	1,200,000	3,500,000	4,200,000

Source: EIA Electric Power Monthly report, Table B.1. *Courtesy of the author.*

8:00 p.m., nearly 100% residents were without power (with the exception of those with backup generators). The total damage to the territory exceeded the previous most costly hurricane by an order of magnitude. Six months after the hurricane, 7% of residents were still without power, and the resulting lack of basic services caused over a quarter million people to leave the island. Schools in rural areas were still without access to electricity; to cope, schools shifted class times earlier, shortening them to 4-hour days due to fuel costs for backup generation. Finally, 11 months after the hurricane occurred, power was fully restored on the island. As a result of this disaster, the perception of power and the grid has changed in Puerto Rico: many consumers have opted for off-grid solar and battery solutions, with innovative hurricane-resistant additions such as storm-shutter–like protections for solar arrays and concrete pads to raise energy storage systems above ground level. The government has also set forth their own goals of power system resiliency; in March 2019 the Puerto Rico Energy Public Policy Act was passed—which will set the island on a path to 100% renewable energy by 2050. A general move toward microgrids for increased energy resiliency is recognized throughout the country in many communities, which challenges and redefines how we think of the grid as a whole.

Other disasters, less commonly thought of as threats to power infrastructure, are also increasing in frequency. The recent increase in wildfires and heatwaves [31], for instance, may only seem like a threat to power infrastructure through burning power poles and substation fires, but one of the greatest risks arises from smoke and particulate matter, which can ionize the air and create unintended and dangerous electrical pathways for current to flow [32]. California utility PG&E rolled out a "wildfire mitigation plan" in early 2019, proposing to remove 375,000 additional trees, burying many existing power lines, and wrapping exposed wire in insulation. When high winds and low humidity conditions are present, the utility cuts power to customers in high-risk areas in attempts to prevent fires from starting.

Looking outside of direct grid impacts from extreme weather and natural disasters, the impact of climate change and increasing global temperatures on building demand will continue to increase stress on the power grid [33–35], meaning that coordination between buildings and the grid and increases in localized resources will become even more essential to avoid further line congestion and transformer overloading. One study found that higher peak electricity demand as a result of climate change—related temperature increases will require an additional 34 GW of generation capacity in the western United States alone by 2050, which is roughly equivalent to more than 100 new power plants [36]. Further, there are additional infrastructure upgrades that may be required for this increase in load, such as higher capacity power lines, or larger transformers. While in some areas of the country, the overall annual electricity demand may decrease over time [37], overall, electricity demand will increase, with peak summer demand increasing almost everywhere. The increases in peaks, driving costs of infrastructure, and generation upgrades further motivate increases in demand-responsive loads and heightened building-to-grid interaction.

10.2.5 Cybersecurity and privacy

The number of devices connected to the internet is estimated to be 30 billion by 2020 [38], and the number of connected devices within commercial buildings and homes is increasing rapidly. While convenient to consumers and potentially relevant for providing grid services, many devices have hit the market before the corresponding cybersecurity measures have been thoroughly tested. For example, in 2014, 100,000 devices, including routers, televisions, and fridges, were taken over by a hacker and used to send spam emails. In 2016 a huge portion of the internet (including sites such as Netflix, Reddit, Spotify, and Twitter) experienced overloaded servers due to a distributed denial-of-service (DDoS) attack and shut down, resulting in a large outcry from consumers. The DDoS attack overwhelmed servers with requests from tens of millions of Wi-Fi—connected smart devices. These devices usually have simple privacy interfaces and passwords and were turned into a botnet capable of repeatedly bombarding the servers of many popular website.

AMR technologies, which a significant number of buildings use to communicate with the grid, do not communicate encrypted data. Using off the shelf tools, researchers were able to intercept AMR smart meter packets from up to 300 m away, fast enough to be able to determine the average power consumption of a house and start to deduce when someone is home [39]. Further, by using nonintrusive load monitoring, which takes an aggregate smart meter signal and disaggregates it into end use loads, even more private data can be recovered. For instance, researchers in Germany were able to uncover what television show a resident was watching from smart meter data and lighting energy patterns, stating

"Our test results show that two 5-minute chunks of consecutive viewing without major interference by other appliances is sufficient to identify the content [40]." The Federal Bureau of Investigation (FBI) released a statement in 2010 in response to exploited smart meters in Puerto Rico, where meters were hacked to underreport consumption data, resulting in millions of dollars of revenue lost for the local utility [41]. The FBI stated, "as Smart Grid use continues to spread throughout the country, this type of fraud will also spread because of the ease of intrusion and the economic benefit to both the hacker and the electric consumer."

Further concerns with consumer privacy arise when considering that the data gleaned from smart meters could potentially be used for targeted advertising and data collection [42]. The concerns do not necessarily stop at passive invasions; however, connected appliances and electricity meters equipped with AMI, due to their two-way communication capabilities, could potentially have capabilities to remotely control smart appliances and devices within a building. On a grid-wide scale, intercepting smart meter signals and manipulating the actual power consumption can result in utilities receiving false information about energy usage, charging the consumer more or less than necessary, or in the long-term, determining planning and operational decisions that do not actually reflect the real state of the grid. Manipulation of the true energy consumption of a building, for example, can make system operators think a potential contingency situation might exist, or falsely lead them to change the operation of grid elements such as switches, capacitor banks, and transformers, threatening the security and resilience of grid operations.

Due to the complexity and interconnectedness of the power grid, even a slight deviation from a stable operating state could cause small problems to affect large regions. The first large-scale power utility cyberattack caused widespread blackouts in Ukraine in 2016. The Russian hacker group "Sandworm" remotely opened breaker switches at substations, which had to be manually closed. In one case, they took over a remote desktop help tool that grid operators used and hijacked the computer mouse of grid engineers, clicking through circuit breakers and turning them off. In 2017 the US power grid was also subject to a cyberattack; a group called Dragonfly 2.0 targeted power companies, gaining access to grid controls, including circuit breakers which have the potential of causing a widespread blackout. Researchers conjecture that the group did not exercise any grid actions during this time, instead exploring the capabilities of potential attacks and possibly waiting for a time that would be more strategic. Efforts to mitigate grid security risks are underway and an active area of research within industry, academia, and government labs.

10.3 Interdependent critical networks

As power engineers, it is important to consider the grid not as an isolated system but as an interconnection and pathway spanning multiple energy systems. Holding myopic views of energy networks when designing operational strategies or planning for power grid upgrades can result in overall system inefficiencies and unexpected behavior. In addition to the buildings and transportation sector, many other energy-dependent networks and systems, such as water and natural gas networks, are highly interconnected with the power system. Concepts such as energy hubs [43] aim to address the optimal control of these multienergy systems, providing multiobjective considerations and joint control decisions that benefit the overall system from economic and environmental perspectives. However, there is much room for improved coordination between power, water, and gas systems, and current practice often lags far behind what is found in the research literature. In this section, we explore these interactions and shed light onto how water and gas systems impact the planning and operation of grid resources. Table 10.4 shows an overview of the difference between these three networks, with common objectives and constraints that are considered in their (independent) optimization.

10.3.1 Water and energy

Dubbed the "water—energy nexus," the interactions between water resources, water networks, and energy production and consumption have been a topic of much discussion and research in recent years [47,48]. Focusing on electrical energy within the water—energy nexus, we see that electricity is used to transport and treat water, while water is needed in many cases to generate electricity (such as from thermoelectric or hydroelectric power plants). In 2014 the US DOE cited the need for joint water—energy policies and national security concerns around climate change impacts on the nexus. The intersection between water use and power production, for example, is complex. Broadly speaking, there are three main ways that water-related issues can impact power generation:

Table 10.4 Brief comparison of power, water, and gas networks, and associated operational goals/constraints.

	Power networks	**Water distribution networks**	**Natural gas networks**
Approximate number of customers connected	**145 million**	**300 million**	**75 million**
Bidirectional flows	Yes, often in transmission (high-voltage) networks and more recently in distribution	Rarely	Sometimes, since 2013, the number of pipes modified for bidirectional flows has grown to account for seasonal variations in gas demand
Storage elements	Transmission and distribution scale energy storage (i.e., batteries, flywheels, and pumped hydropower); electric vehicles as storage	Water tanks, towers, and reservoirs	Underground storage facilities such as salt caverns, aquifers, and depleted natural gas or oil fields; gas can also be liquefied (LNG) and stored in tanks
Common cost and operational objectives	Deliver power to loads in the most economical way and maximize renewable energy use/minimize curtailment	Minimize energy costs for pumping, losses due to pipe friction	Minimize energy costs for gas compressors and maximizing line pack (volume of gas stored in the pipeline)
Common operational requirements and network constraints	Voltage magnitude ranges; frequency regulation, $N-1$ criteria at the transmission level, and power flow dynamics	Water pressure limits, maximum/minimum water levels in storage tanks, water flow dynamics	Pressure and flow requirements, temperature requirements, compressor station constraints, and gas flow dynamics

LNG, Liquefied natural gas.
Source: Data from [44–46].

- lack of water for cooling (e.g., nuclear and coal) or generation (e.g., hydropower);
- high intake water temperatures; and
- high discharge water temperatures.

Thermoelectric power plants are often defined by their fuel source ("coal plants," "natural gas plants," etc.) and typically use steam turbines to generate electricity. This is performed by using the fuel source to boil water, producing high-pressure steam which rotates the turbine, enabling the generator to produce electricity due to the principles of electromagnetic induction. In order to continue this process the steam must be condensed back into water. This is performed by the power plant cooling system.

By far, the two most common cooling systems for thermoelectric power plants in the United States are "once-through" systems and "closed-loop" or "recirculating" systems. As the name suggests, once-through cooling systems draw water from nearby sources (e.g., oceans, aquifers, and rivers) and circulate it in pipes that run throughout the plant. As the water in the pipes is cooler than the surrounding hot steam, the steam is condensed into water and can be used by the steam turbine to generate electricity. After the water absorbs the heat the warmer water is discharged back into the water supply. If the intake water is too warm, the plant may struggle to condense the steam and thus struggle to effectively generate electricity. If the discharge water is too warm, plants and animals in the water supply can face harmful environmental effects (which is why there are typically strict regulations on discharge water temperatures for power plants). However, in dire circumstances where power must be generated, these restrictions on water temperature have been known to have been waived in order to keep generators running—during 2012 in Chicago, for instance, discharge temperatures were temporarily relaxed to 97° from the typical limit of 90° in order to maintain enough power generation during high summer temperatures [49]. There can be resulting ecological impacts from once-through cooling systems, especially when discharge regulations are relaxed to ensure power generation.

Closed-loop, or recirculating cooling systems use less water overall because they reuse the majority of the water they withdraw. Closed-loop systems typically use cooling towers to expose the hot water to air after it has been circulated through the system, allowing the air to cool the water before it is recirculated back through the system. While some of the water is lost to evaporation, much of it can be reused for cooling. Closed-loop cooling systems are more common in newer thermoelectric power plants and in the western part of the country where water resources are more scarce, while once-through systems are more common in the eastern part of the United States.

Water used for thermoelectric power generation represents over 40% of US freshwater withdrawals (Fig. 10.6, [51]), and many power plants throughout the United States have had to shut down or reduce output in recent years due to water-related issues. Droughts continue to impact the water supply, heavily affecting the generation capabilities of power plants, particularly nuclear, hydroelectric, and coal [52]. Hydroelectricity, a source of renewable generation, is one clear instance of the water–energy nexus. According to the National Hydropower Association, hydroelectric generation provides over half of the nation's renewable energy, and 7% of all generation in the United States. Consider the Hoover Dam as one example. The Hoover Dam was constructed in between 1931 and 1935, primarily to regulate the flow of the Colorado River. The Colorado River marks the border between Nevada and Arizona and feeds into the Lake Mead reservoir, a man-made lake that represents the largest reservoir in the country in terms of water capacity. A hydroelectric power plant was constructed in 1936 to harness the energy of the flowing water, providing power to nearly 8 million people. However, after nearly two

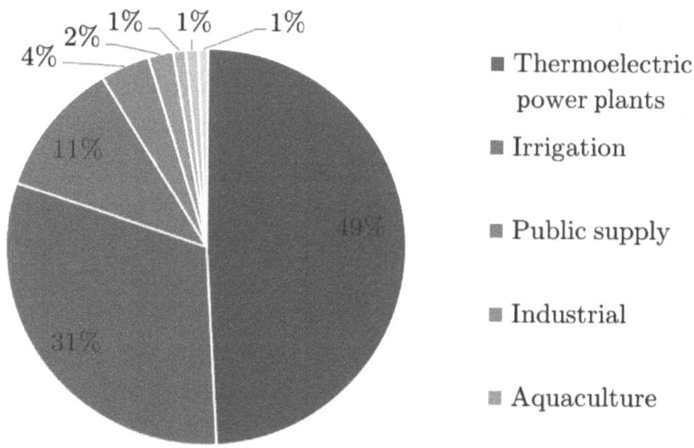

Figure 10.6 Distribution of total water use in the United States (freshwater and saline). Power plants account for nearly half of the total water use in the United States and 41% of freshwater withdrawals. *Data from Natural Resources Defense Council, Inc., Power Plant Cooling and Associated Impacts: The Need to Modernize U.S. Power Plants and Protect Our Water Resources and Aquatic Ecosystems, NRDC Issue Brief, Apr. 2014.*

decades of drought and increased demand due to rapid population growth in the area, Lake Mead's water level continues to decrease [53], heavily affecting the power generation capabilities of Hoover Powerplant, one of the country's largest hydroelectric plants. Coupled with increasing summer building loads in the area, a lack of generation capacity threatens the area's peak demand electricity supply.

Case study: Venezuela's Guri dam

The Guri dam in Venezuela, which generates hydroelectric power and provides power to 65% of the country, produces five times the power of the Hoover Powerplant. The plant struggled to provide power during a drought in 2016 and continued to struggle due to a 45% decline in rainfall since 2013. To account for the lack of generation the government implemented 4-hour daily power cuts to conserve capacity.

The relationship between energy and water also traverses the opposite direction: electricity is essential for pumping, treating, and transporting water. There exist around 200,000 drinking water treatment systems in the country and approximately 15,000 wastewater treatment plants, many of which use electricity as their main source of energy. Electric pumps can be used to draw water from a source (e.g., a lake or reservoir) and deliver water through a distribution network when a difference in gravity is not present or pressure levels are insufficient to deliver water to consumers at the specified pressure levels. In California an estimated 19% of the state's entire electricity consumption and 30% of its natural gas consumption is used for dealing with water and wastewater due to the energy requirements from pumping, treating, and more [54]. At drinking water plants, most of the energy consumed is for operating pumps; at wastewater plants, most energy is consumed by aeration, pumping, and solids processing.

At the network level, water and power networks, typically owned, operated, and maintained independently, have opportunities to coordinate for improvements in overall system efficiency [55]. Controlling electric water pumps in a grid-responsive manner has load-shifting potential which can be significant, considering operating wastewater and drinking water networks in the United States comprises roughly 4% of the national electricity consumption [56]. Variable-speed pumps, water tanks, reservoirs, and pressure valves within a water network all interact together to maintain water delivery throughout the network within pressure and flow requirements, and all have potential opportunities to shift their consumption to impact the power network, providing operational, environmental, and economic benefits [57]. Similar to power networks, water distribution networks also maintain reserves and "ancillary services" such as balancing storage, though tanks, towers, and/or reservoirs, which help balance fluctuations in water demand. During a power loss event, joint control between power and water networks can help prevent damage to the water network: power outages halt pump operation, leading to pressure changes that can damage pipes, resulting in power outages being one of the main causes of damage to water networks [58]. While these concerns are presently mitigated by backup diesel generation and pressurized surge tanks, improved joint control can improve system efficiency and prevent damage in future water networks [59,60].

Some water utilities are beginning to consider or are already participating in grid services [61], but there exist many further opportunities for these two industries to connect. For example, pumps at filtration plants at the Eastern Municipal Water District (EMWD) in California participate in DR by shutting off when requested and can stay off for multiple hours. EMWD receives $100,000 each year from SCE for providing this grid service [62]. Pennsylvania American Water also participated in a pilot program where a single 700-hp motor was used to provide frequency regulation, varying its output but never shutting off. The operators observed no notable change in service or operation, and Pennsylvania American Water received $20,000 annually for providing this service. These, along with other case studies, have proven the efficacy of coordinating these two sectors. Further potential venues for joint control within the water–energy nexus include, but are not limited to,

- performing intelligent pump scheduling to improve system efficiency, coincide with renewable generation, decrease power grid loading, and reduce electricity costs;
- installing on-site generation at water facilities, reducing demand during peak hours or potentially providing power back to the main grid; and
- adjusting the speed of variable-speed pumps to provide grid ancillary services or reduce/increase demand when needed.

While some studies state that the current lack of coordination between power and water utilities is a large barrier to their joint success [63], these two sectors have an abundance of potential for achieving joint financial and environmental benefits. Financial incentives, and increased awareness, are two of the main factors that could help facilitate future coordination within the water–energy nexus.

10.3.2 Power and gas

Natural gas is a nonrenewable fuel source that is used for heating, clothes drying, cooking, as fuel for power plants, and as transportation fuel. In 2018 natural gas production in the United States hit a record high. As seen in Fig. 10.7, natural gas used in the electric power sector comprises a larger portion of consumption compared to any other sector. This has increased over recent years; partially due to technological advancements in hydraulic fracturing (fracking), many gas-fired power plants are replacing aging and cost-ineffective coal plants—in 2015 natural gas surpassed coal as the United States' dominant fuel for power generation. Gas-fired power plants were traditionally used as "peaker plants," that is, power plants that are typically turned on for a few hours of the day to deliver power during peak grid loading, but due to decreases in fuel costs, in recent years they have also been used as base-load power plants [64]. The backbone of natural gas delivery in the commercial, residential, and industrial sectors is the natural gas piping and storage infrastructure. The natural gas pipelines in the United States, which has the largest pipeline system in the world, has its own transmission, distribution, and storage infrastructure, just like water and power networks. Also similar to the power network, but differing from water distribution networks, gas pipelines can be

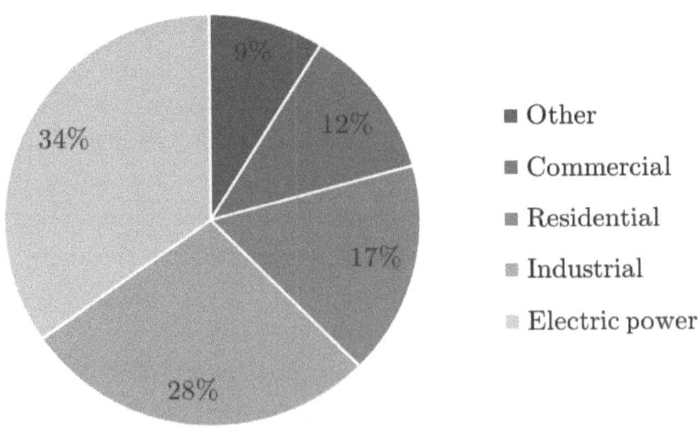

Figure 10.7 2018 US natural gas consumption. Total use in 2018 averaged over 80 billion cubic feet per day. *Data from the EIA.*

upgraded to support bidirectional flows of gas, which is useful in regions where natural gas use is seasonal, and can avoid the need of building additional pipelines [65].

Over the years, temperature fluctuations and increases in natural gas extraction have driven peak gas demand higher in the winter months, motivating a DR program for gas consumers. For example, Consolidated Edison (ConEd) in New York is piloting a 3-year, $5 million gas DR program, in attempts to manage peak demand and avoid additional pipeline installations [66] after concluding that peak gas demand in their territory will increase significantly in the next decade. Changing behavioral patterns toward electricity and gas use is a challenging problem; however, Southern California Gas Company's Smart Therm DR program saw 40% of consumers override DR requests (out of approximately 10,000 thermostats) during their 2017–18 winter program season [67]. Both engineering challenges and societal challenges continue to pose roadblocks for demand-responsiveness in power, water, and gas networks.

On the supply side, despite a tight coupling between power and gas networks, system operators for these networks rarely communicate to coordinate planning or operational decisions. Benefits from joint planning of these networks, for example, can take into account coincident loads for minimizing the costs due to expansion planning [68]. Optimizing the joint control of these networks can also help improve overall system efficiency and lower fuel costs [69,70]. Gas networks, such as water and power networks have complex, nonlinear dynamics and are not trivial to model or optimize [71]. Typical system characteristics, operational objectives, and constraint considerations for both power and gas networks can be compared in Table 10.4. Solving the joint planning and operation optimization problems is not trivial; often, integer variables are included (e.g., for on/off or include/exclude design decisions) and constraints that render the problem nonconvex (e.g., gas and power flow dynamics). Various methods of reducing problem complexity and convexifying problem constraints have been developed to help make joint gas–power problems computationally feasible [68].

10.3.3 Combined heat and power

Multigeneration systems that use gas to produce electricity, such as CHP plants (e.g., cogeneration), reduce losses due to power transmission and introduce many opportunities to coordinate electrical with thermal loads. Natural gas fueled CHP plants can have highly coupled gas–power dynamics and can offer efficiency gains over conventional gas-fired generation sources. Unlike conventional thermoelectric power plants, the waste heat from generating electricity is used to provide heating to consumers (and cooling in the case of tri-generation plants) [72]. As discussed previously in the water and energy section, heat generated by creating electricity is typically rejected into a water source or into the air. In contrast, CHP systems harness this heat, using it to deliver steam or hot water to buildings, reducing the need for on-site heat sources. This can reduce strain on the electric distribution grid due to decreased electrical loads and decrease overall electrical losses.

However, while there are many benefits from using CHP systems, they can require high capital costs due to the need for additional steam and hot water distribution infrastructure. In addition, they are most appropriate for energy districts that supply power and heat to buildings that have coincident electrical and heating loads such as hospitals and university campuses. Lastly, similar to conventional thermoelectric power plants, they require fossil fuels to operate, so while the use of CHPs can improve overall system efficiency, it is not an end-all–be-all solution for clean energy. The interconnection between gas use and power generation is evident here; however, due to this coupling, CHP plants have opportunities to participate in DR programs [73] and coordinate with energy storage [74] to provide grid services while ensuring a reliable delivery of electricity and heating to consumers. Due to their localized nature of providing electricity and heating to districts or communities of buildings, they are also often coupled with microgrids, introducing opportunities for joint microgrid and CHP coordination as well [75].

Despite many encouraging results in the literature and positive pilot studies, there is still much left to be desired for joint the coordination of water–power, gas–power, and even water–gas–power networks. Engineering roadblocks include a lack of communication infrastructure and sensor networks necessary for the coordination of these networks; roadblocks outside of engineering and technical considerations can stem from a lack of communication between sectors, risk-adverse mentalities, and a lack of consumer awareness or willingness to participate in DR programs.

10.4 Electrification of the transportation sector

Like water and gas systems, the transportation sector is also energy-intensive, although presently less linked to the electrical grid than these other systems. Transportation currently comprises about 30% of primary energy consumption in the United States, but less than 1% of electricity demand, meaning that there could be large potentials for electrification. However, as the transportation sector becomes more electrified, it means more strain on the electric grid—and a pressing need for innovation in grid planning, operations, and markets.

10.4.1 Consumer vehicles

By 2030 the IEA predicts that the number of EVs on the road will increase from roughly 3 million to 125 million [11]. Most charging of these EVs will occur at residences, as many consumers find home charging more convenient than charging at other locations, with some studies showing over 80% of charging activities occurring at home [76]. While convenient to the consumer, the associated distribution infrastructure needed to accommodate this additional increase in load is astounding, especially with the introduction of fast-charging EV stations. Level 3 charging stations (which are typically not installed at residences), for example, can charge at 50 times the rate of Level 1 chargers (which use standard 120 V residential electrical receptacles).

A comparison of Levels 1, 2, and 3 charging is seen in Fig. 10.8, with typical ranges of current draws and charging rates taken from multiple sources [78,79]. Level 1 chargers use standard 120 V AC outlets on 15–20 A branch circuits, typically pulling up to 16 A, and thus require no additional infrastructure or adapters for residential use. Level 2 chargers use standard split-phase 240 V AC outlets, similar to 240 V appliances such as electric clothes dryers, fully charging most EVs in 4–6 hours [77]. Level 3 charging is high-voltage, DC "fast-charging" that requires a 480 V DC supply and can pull up to 300 A of current, fully charging most EVs in under an hour. Considering this is a higher current than the entire rated electrical service entrance of most residences, Level 3 chargers are very expensive to install in places outside of commercial and public settings. Charging at such high currents will also create a lot of heat and may also degrade the EV battery when used frequently. Considering the slow charge rate of Level 1 chargers, most public chargers are Levels 2 and 3; the city of Boulder, CO,

Level 1	Level 2	Level 3
120 V AC	240 V AC	480 V DC
15–20 Amps	30–80 Amps	100–300 Amps
1.4 kW rate	3.6–17 kW rate	25–120 kW rate
		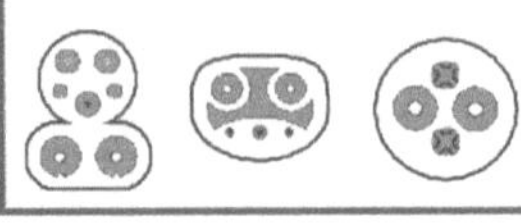

Figure 10.8 Typical ranges of voltage, current, and power ratings of Levels 1, 2, and 3 charging stations, with various plug/charger appearances for each option. Note that Level 3 chargers have vehicle-dependent charging ports (e.g., the middle EV connector under Level 3 charging is for Tesla Model S and X vehicles [77]). *EV*, Electric vehicle.

for instance, has 207 public Level 2 chargers (94%) and 13 Level 3 chargers (6%), with 31 stations being free of charge. Wireless EV charging is also an active area of research and development, whose ultimate goal is to utilize wireless, inductive charging between vehicles and the road, dynamically charging the EV as the user drives [80].

Residential distribution networks and feeders, which generally are nonmeshed, low-voltage networks that deliver power to consumers, were originally constructed for supplying power to traditional building loads such as water heating, HVAC, and lighting. In comparison, an EV with a daily commuting distance of 25 mi would require roughly 6–8 kWh to recharge, which is equivalent to the entire daily energy needs of a small household. Some consumer EVs, such as the Tesla Model S, have very large batteries—100 kWh—and have the potential of putting a very large strain on the grid. With the increasing adoption of EVs the deterioration of distribution infrastructure and overloading of distribution lines and transformers is a real and immediate problem that must be addressed. Installing higher capacity infrastructure to accommodate these loads will result in enormous costs, motivating the use of dynamic pricing schemes or other DR programs. The optimal placement of charging stations throughout a city, taking into account the distribution infrastructure and grid constraints [81], is an open and interesting research problem incorporating stochasticity and uncertainty, binary or integer variables, and traffic pattern models. Finding efficient ways of modeling and solving these problems can help cities avoid further infrastructure upgrades and costs in the near future.

10.4.2 Public transportation

The current dearth of transportation electrification is heavily impacting air quality and the environment. According to [82], the average car in the United States produces roughly 0.96 lb of CO_2 (which makes up 95% of all transportation-related emissions) per passenger mile (which is defined as one mile traveled per one passenger). Bus transit, due to the ability to transport more passengers in a smaller space, emits roughly 0.64 lb of CO_2 per passenger mile, with light rail transit at 0.36 lb/CO_2 per passenger mile. In terms of life cycle emissions the amount of CO_2 per passenger mile for any vehicle mostly comprises the operational emissions of that vehicle. Emissions from sedans, SUVs, and pickup trucks is higher than any form of public transportation but from public transportation is significant [83].

From these numbers, we can gather than electrifying the public transportation sector would make a significant impact on air pollution and emissions reduction. Some cities leverage electric trolley systems (e.g., San Francisco), and light rail systems (e.g., Chicago), but these require overhead wires to operate. Many other cities are deploying battery-electric buses, despite concerns with hill-climbing torque capabilities, range anxiety, or extreme heat and cold effects on battery efficiency. Due to various emissions-cutting goals, it is estimated that by 2045, a third of the public buses in the United States will be zero-emissions [84]. Internationally, many other countries have ambitious goals for electrification—within the Chinese city of Shenzhen, all buses are now electric, with many cities forming plans to follow suit. Switching from diesel buses to electric not only has emissions benefits but also noise pollution benefits as well.

Case study: school bus electrification

Many school buses around the United States have been converted from diesel-powered to all-electric powertrains, reducing children's exposure to air pollution. On average, school buses travel roughly 100 mi per day, making them a prime candidate for electrification. During the day, while not in use, the bus batteries can potentially be used to provide grid services such as load shifting.

A challenge for both electric public transportation infrastructure and public use is that battery-powered buses typically will not have the same range as a diesel-powered bus, meaning that buses would have to charge before fully completing their routes, or additional buses would have to be provided as alternatives. As many of these batteries will likely be charged at night, after bus service has ended, peak demand charges due to bus charging has also been identified as a major roadblock to the deployment of electric buses. Under high demand charges the cost to charge a single electric bus at night can increase by $0.24 per mile, for example [85]. In order to avoid these charges, and the associated stress and strain on the electrical distribution infrastructure, electric bus charging must be staggered, performed at different charging stations throughout the day, or, in the future, utilize wireless inductive charging.

10.4.3 Rideshare services and emerging methods of transportation

Alternative modes of electric transportation have emerged in recent years, including EV rideshare programs and applications, electric scooter-sharing systems, and electric bicycle–sharing systems. Rideshare company Lyft, for instance, offers a "Green Mode" that allows passengers to request rides from EVs, in addition to their commitment to maintaining a zero-emissions standard through the purchase of carbon credits. This further incentivizes Lyft drivers to consider purchasing or leasing EVs over gasoline-powered vehicles. An increase in EVs within rideshare programs may cause spatiotemporal changes in electrical loads throughout a city—drivers may now need to charge vehicles at a variety of hours and locations, rather than follow the traditional model of at-home EV charging [76].

Near the end of 2017, new methods of electric transportation began to arise. GPS-tracked electric scooters started to litter streets throughout the United States, seemingly popping up overnight, providing dock-less modes of transportation for a marginal fee. These electric scooter companies pay users to track and pick up the scooters with depleted batteries and charge them, which usually entails collecting multiple scooters and charging them at home with company-provided adapters. Uber's JUMP program extended this concept to dock-less electric bicycles, providing another convenient and inexpensive way to get around a city. While individual scooters and electric bikes alone may not have a significant additional stress on distribution infrastructure, en masse, their charging can be equivalent to an additional large building load. For example, if a battery on an electric bike is rated 20 Amp h at 48 V, completely charging the battery amounts to almost a kilowatt hour of energy. Simultaneously charge a few of these bikes at home, and the resulting increase in electrical load is equivalent to turning on an air conditioner or electric water heater. Further, if multiple homes on the same distribution feeder increase their load by a few kilowatt hour, the voltage in the feeder may sag further and create significant strain on the electrical infrastructure.

It is unclear what regulations will emerge in this realm, or if providing easily accessible new modes of transportation will help decrease loads in general by providing a less energy-intensive mode of transportation. Due to their nascent presence within a limited number of cities, it is difficult to predict where and how electric scooters and bicycles will impact the power and transportation sector. However, considering electric scooter companies Bird and Lime reached $1 million valuations within a year of launching and have a rapidly expanding user base, continued use of these technologies is a reasonable assumption. With that said, alternate modes of electric transportation are transpiring so quickly that by the time you read this chapter, there will likely be an entirely new technology not mentioned here.

10.4.4 Vehicle-to-grid

Typically, EV batteries, in contrast to stationary energy storage systems, draw power from the grid but do not push power back to the grid. However, because EVs have significant battery capacity, discharging these batteries through a vehicle-to-grid (V2G) system has the potential to provide grid services such as frequency regulation and peak shaving. Most car manufacturers do not allow V2G capabilities due to warranty issues and battery wear-and-tear, but in some locations such as Japan, V2G systems are being explored for resiliency benefits. Car manufacturer Nissan has developed a V2G kit for homes for this purpose: In Japan the average home uses 10–12 kWh a day. A Nissan Leaf can have a 24 kWh battery capacity, meaning that it could provide power for an entire home for 2 days during a disaster or grid failure. PJM Interconnection is one of the entities that is exploring this area in the United States; through an initiative with the University of Delaware, PJM is demonstrating how aggregate control over a fleet of EV batteries can help provide frequency regulation [86]. As EVs can be thought of as a form of distributed energy storage, the ability to use them as a grid resource can prove invaluable in times of need.

Outside of fast-timescale grid services, considering that many EVs are left unattended for multiple hours during the middle of the day at office parks or public charging stations, many opportunities arise for the batteries within these vehicles to shift daily demand patterns from commercial building loads. By studying actual arrival/departure patterns, researchers were able to determine that coordinating EV charging with distributed solar and smart building loads can provide revenue for building owners in addition to alleviating grid stressors [87]. As the power system becomes more inverted dominated, we will have fewer conventional power plants and thus large-scale inertial resources to provide automatic frequency regulation services, making EV battery storage and other energy storage systems extremely valuable for achieving fast-timescale grid services [88].

10.5 Considerations for future power systems

The previous sections in this chapter only begin to touch on the many topics that offer both challenges and opportunities for future power systems and power-dependent networks. As infrastructure ages, more renewable generation is connected to the grid and extreme weather events continue to impact the ability of generators to deliver power to loads; revolutions in traditional thinking around power systems operation must evolve to accommodate emerging challenges. Broadly speaking, there are three main considerations that the planning and operation of future power systems should take into account to address these issues and to harness the benefits from demand-side resources and interconnected system operations: physical considerations, market considerations, cyber considerations, and the overlap and intersection of the three.

10.5.1 Physical considerations

In addition to transmission-level upgrades, distribution-level, distributed upgrades such as the installation of AMI or community-scale energy storage, and distributed generation installations can prove useful to supplying power to communities when the grid has been impacted by a hazard upstream. These resources can enhance the ability to supply power locally, reducing the dependence on the main grid when it is damaged, and provide more information for grid operators through deployed AMI, which can assist operators in identifying and localizing specific outage areas and expediting repair. Region-specific grid upgrades will become more important as the threats to the grid become more geographically dependent: Burying overhead lines may be advantageous in vulnerable regions where hurricane-force winds are expected to increase the frequency of grid outages; however, in coastal regions, underground wires can be susceptible to damage from flooding, and another grid strengthening measure may be more valuable. Increases in inverter-interfaced devices such as rooftop PV and stationary energy storage will necessitate developments in power electronics and associated controls in order to maintain and preserve grid stability.

10.5.2 Market and organizational considerations

Without corresponding economic benefits and enabling markets for DR and for other entities to provide grid services, it would be nearly impossible for the coordination and collaboration between systems discussed in this chapter to occur. A two-way communication infrastructure, on relatively fast timescales (i.e., seconds or minutes), will be essential for building loads, transportation, and other networks to assist the grid and redirect load during a contingency or other grid event. Frameworks such as transactive energy markets [89], not discussed in this chapter, prove promising solutions to coordinating demand-responsive assets with grid needs through the use of pricing or carbon signals. As more building loads gain digital controls or are replaced by variable frequency drives, the fast-timescale capabilities of these assets to provide grid services is enhanced, but the markets and communication infrastructure must also be updated to facilitate these transactions. Lastly, consumer awareness and the willingness for different sectors to work together to achieve common goals is an important step that has yet to be overcome.

10.5.3 Cyber considerations

Lastly, the increasingly connected nature of modern power systems provides bountiful opportunities for increased coordination and control between multienergy networks. However, the cyber threats introduced by the Internet-of-Things and wirelessly connected grid assets is an important area of consideration when designing and operating future power systems. One weak link in the system could open further vulnerabilities in the larger power grid, causing widespread outages and damage to propagate throughout multiple networks. As systems become more interconnected, these threats become more and more pressing, and design considerations that encompass the physical and cyber systems are necessary. Thorough testing and validation through gray and white hat hacking of IoT devices, communication frameworks, transactive platforms, and connected EVs will help identify vulnerabilities that could prevent catastrophic grid failures in future power systems.

Acknowledgments

The author would like to thank Bridget Toomey, Joseph Kasprzyk, Jonathan Donadee, Kaitlyn Garifi, and the University of Colorado Boulder Spring 2019 Grid Connected Systems class for inspiring him to research this material in the first place.

References

[1] G. Bakke, The Grid: The Fraying Wires Between Americans and Our Energy Future, Bloomsbury, July 2016.
[2] Executive Office of the President, Economic Benefits of Increasing Grid Resilience to Weather Outages, August 2013.
[3] L. Clarke, J. Eom, E.H. Marten, R. Horowitz, P. Kyle, R. Link, et al., Effects of long-term climate change on global building energy expenditures, Energy Econ. 72 (2018) 667−677.
[4] E. Yamangil, R. Bent, S. Backhaus, Resilient upgrade of electrical distribution grids, in: Twenty-Ninth AAAI Conference on Artificial Intelligence, 2015.
[5] H. Nagarajan, E. Yamangil, R. Bent, P. Van Hentenryck, S. Backhaus, Optimal resilient transmission grid design, in: Power Systems Computation Conference (PSCC), 2016.
[6] State of New York Public Service Commission, Order Approving Electric, Gas and Steam Rate Plans in Accord with Joint Proposal, February, 2014.
[7] R. Campbell, Weather-Related Power Outages and Electric System Resiliency, Congressional Research Service, August 2012.
[8] American Society of Civil Engineers, Infrastructure Report Card, 2017.
[9] A. Mills, R. Wiser, J. Seel, Power Plant Retirements: Trends and Possible Drivers, Lawrence Berkeley National Laboratory, November 2017.
[10] D. Steinberg, D. Bielen, J. Eichman, K. Eurek, J. Logan, R. Mai, et al., Electrification & decarbonization: exploring U.S. energy use and greenhouse gas emissions in scenarios with widespread electrification and power sector decarbonization, in: NREL Technical Report NREL/TP-6A20-68214, July 2017.
[11] International Energy Agency, Global EV Outlook 2018, May 2018.
[12] J.D. Taft, P. De Martini, Sensing and measurement architecture for grid modernization, in: Technical Report, Pacific Northwest National Laboratory (PNNL) PNNL-25249, February 2016.
[13] A. Mulla, J. Baviskar, S. Khare, F. Kazi, The wireless technologies for smart grid communication: a review, in: 2015 Fifth International Conference on Communication Systems and Network Technologies, Gwalior, 2015.

[14] P. Yi, A. Iwayemi, C. Zhou, Developing ZigBee deployment guideline under WiFi interference for smart grid applications, IEEE Trans. Smart Grid 2 (1) (2011) 110–120.
[15] M. Mylrea, S. Nikhil, G. Gourisetti, M. Mylrea, S.N.G. Gourisetti, Cyber-Security and Optimization in Smart 'Autonomous' Buildings, 2017.
[16] L. Jia, R.J. Thomas, L. Tong, Impacts of malicious data on real-time price of electricity market operations, in: Proc. Annu. Hawaii Int. Conf. Syst. Sci., 2012,. 1907–1914.
[17] L. Xie, Y. Mo, B. Sinopoli, Integrity data attacks in power market operations, IEEE Trans. Smart Grid 2 (4) (2011) 659–666.
[18] F. Ding, B. Mather, N. Ainsworth, P. Gotseff, K. Baker, Locational sensitivity investigation on PV hosting capacity and fast track PV screening, in: IEEE PES T&D, Dallas, TX, May 2016.
[19] B. Palmintier, R. Broderick, B. Mather, M. Coddington, K. Baker, F. Ding, et al. On the path to SunShot: emerging issues and challenges in integrating solar with the distribution system, in: NREL Technical Report NREL/TP-5D00-6533, 2016.
[20] K. Horowitz, B. Palmintier, B. Mather, P. Denholm, Distribution system costs associated with the deployment of photovoltaic systems, Renew. Sustain. Energy Rev. 90 (2018) 420–433.
[21] California Public Utilities Commission (CPUC), CPUC Helps Consumers Plan for the Great Solar Eclipse, Press Release, May 2017. http://docs.cpuc.ca.gov/PublishedDocs/Published/G000/M188/K343/188343356.PDF.
[22] Nest, Solar Eclipse, Meet the Nest Thermostat. 2017. https://nest.com/blog/2017/08/10/solar-eclipse-meet-the-nest-thermostat/.
[23] A. Clark-Ginsberg, What's the Difference Between Reliability and Resilience? U.S. Department of Homeland Security, March 2016.
[24] Navigant Research, Demand response for renewables integration, in: Technical Report, 2017.
[25] P. Alstone et al., 2025 California demand response potential study, in: LBNL Report LBNL-2001113, March 2017.
[26] W. Zhang, K. Kalsi, J. Fuller, M. Elizondo, D. Chassin, Aggregate model for heterogeneous thermostatically controlled loads with demand response, in: IEEE PES General Meeting, San Diego, CA, 2012.
[27] E. Keating, Distributed Energy Resources, Presentation Slides to the California Energy Commission by Southern California Edison, August 2017.
[28] X. Jin, K. Baker, D. Christensen, S. Isley, Foresee: a user-centric home energy management system for energy efficiency and demand response, Appl. Energy 205 (2017) 1583–1595.
[29] EnerNOC, Inc., The role of demand response in integrating variable energy resources, in: Final Report, December 2013.
[30] A. Kenward, U. Raja, Blackout: extreme weather, climate change and power outages, in: Climate Central Report, 2014.
[31] J.T. Abatzoglou, A.P. Williams, Impact of anthropogenic climate change on wildfire across western US forests, Proc. Natl. Acad. Sci. U.S.A. 113 (42) (2016) 11770–11775.
[32] M. Davis, S. Clemmer, Power failure: how climate change puts our electricity at risk – and what we can do, in: Union of Concerned Scientists Report, April 2014.
[33] J.-P. Watson, R. Guttromson, C. Silva-Monroy, R. Jeffers, K. Jones, J. Ellison, et al., Conceptual Framework for Developing Resilience Metrics for the Electricity, Oil, and Gas Sectors in the United States, Sandia National Laboratories, September 2014.
[34] Y. Wang, H. Lin, W. Wang, Y. Liu, R. Wennersten, Q. Sun, Impacts of climate change on the cooling loads of residential buildings differences between occupants

with different age, in: Proceedings of the 9th International Conference on Applied Energy, 2017.
[35] S.L. Wong, K. Wan, D. Li, J. Lam, Impact of climate change on residential building envelope cooling loads in subtropical climates, Energy Build. 42 (11) (2010) 2098–2103.
[36] T. Veselka, A. Botterud, G. Conzelmann, V. Koritarov, L. Poch, J. Wang, Climate Change Impacts on the Electric Power System in the Western United States, American Geophysical Union, Fall Meeting, 2007.
[37] P. Sullivan, J. Colman, E. Kalendra, Predicting the response of electricity load to climate change, in: NREL Technical Report NREL/TP-6A20-64297, July 2015.
[38] M. Mylrea, S.N.G. Gourisetti, A. Nicholls, An introduction to buildings cybersecurity framework, in: 2017 IEEE Symp. Ser. Comput. Intell. SSCI 2017 – Proc., January 2018.
[39] J. Liu, Y. Xiao, S. Li, W. Liang, C. Chen, Cyber security and privacy issues in smart grids, IEEE Commun. Surv. Tutor. 14 (2012) 981–997.
[40] U. Greveler, P. Glosekotter, B. Justus, D. Loehr, Multimedia content identification through smart meter power usage profiles, in: Proc. Int. Conf. Inform. and Knowledge Eng., Las Vegas, NV, July 2012.
[41] U.S. Federal Bureau of Investigation (FBI), Smart Grid Electric Meters Altered to Steal Electricity, Intelligence Bulletin, Cyber Intelligence Section, May 2010.
[42] K. Baker, K. Garifi, Power signature obfuscation using flexible building loads, in: International Workshop on Non-Intrusive Load Monitoring (NILM), Austin, TX, March 2018.
[43] M. Mohammadi, Y. Noorollahi, B. Mohammadi-ivatloo, H. Yousefi, Energy hub: from a model to a concept – a review, Renew. Sustain. Energy Rev. 80 (December 2017) 1512–1527.
[44] Energy Information Administration (EIA), U.S. Electric System is Made Up of Interconnections and Balancing Authorities, Today in Energy, July 2019.
[45] University of Michigan, U.S. Water Supply and Distribution Factsheet. http://css.umich.edu/factsheets/us-water-supply-and-distribution-factsheet, August 2018.
[46] Energy Information Administration (EIA), Natural Gas Pipelines. 2019. https://www.eia.gov/energyexplained/index.php?page = natural_gas_pipelines (last accessed 6/2019).
[47] A. Hamiche, A. Stambouli, S. Flazi, A review of the water-energy nexus, Renew. Sustain. Energy Rev. 65 (2016) 319–331.
[48] J. Dai, S. Wu, G. Han, J. Weinberg, X. Xie, X. Wu, et al., Water-energy nexus: a review of methods and tools for macro-assessment, Appl. Energy 210 (2018) 393–408.
[49] E. Meyer, J. Wernau, Power Plants Releasing Hotter Water, The Chicago Tribune. https://www.chicagotribune.com/news/ct-xpm-2012-08-20-ct-met-nuclear-water-20120820-story.html, August 2012.
[50] Natural Resources Defense Council, Inc., Power Plant Cooling and Associated Impacts: The Need to Modernize U.S. Power Plants and Protect Our Water Resources and Aquatic Ecosystems, NRDC Issue Brief, April 2014.
[51] M.A. Maupin, J.F. Kenny, S.S. Hutson, J.K. Lovelace, N.L. Barber, K.S. Linsey, Estimated Use of Water in the United States in 2010, U.S. Geological Survey, Circular 1405, 2014.
[52] C. Harto, Y. Yan, Y. Demissie, D. Elcock, V. Tidwell, K. Hallettm, et al., Analysis of Drought Impacts on Electricity Production in the Western and Texas Interconnections of the United States, Argonne National Laboratory, December 2011.
[53] C. Holdren, K. Turner, Characteristics of Lake Mead, Arizona-Nevada, Lake Reserv. Manage. 26 (4) (2011) 230–239.

[54] C. Copeland, N. Carter, Energy-water nexus: the water sector's energy use, in: Congressional Research Service Report R43200, January 2017.
[55] D. Fooladivanda, J.A. Taylor, Energy-optimal pump scheduling and water flow, IEEE Trans. Control. Netw. Syst. 5 (3) (2018) 1016–1026.
[56] D. Denig-Chakroff, Reducing electricity used for water production: questions state commissions should ask regulated utilities, Water Res. Policy (2008).
[57] A. Zamzam, E. Dall'Anese, C. Zhao, J. Taylor, N. Sidiropoulos, Optimal water-power flow problem: formulation and distributed optimal solution, IEEE Trans. Control. Netw. Syst. 6 (1) (2019) 37–47.
[58] M.W. LeChevallier, R.W. Gullick, M.R. Karim, M. Friedman, J.E. Funk, The potential for health risks from intrusion of contaminants into the distribution system from pressure transients, J. Water Health 1 (1) (2003) 3–14.
[59] J.A. Taylor, R. Perryman, A. Bazylak, B. Karney, Safely landing water networks during power outages with energy storage, in: 2018 56th Annual Allerton Conference on Communication, Control, and Computing (Allerton), Monticello, IL, 2018, pp. 346–350.
[60] T. Chen, Z. Ren, C. Xu, R. Loxton, Optimal boundary control for water hammer suppression in fluid transmission pipelines, Comput. Math. Appl. 69 (4) (2015) 275–290.
[61] Water Research Foundation, Water and electric utility integrated planning, in: Report #4469, 2017.
[62] B. Sparn, R. Hunsberger, Opportunities and challenges for water and wastewater industries to provide exchangeable services, in: NREL Technical Report NREL/TP-5500-63931, November 2015.
[63] Water Research Foundation, Trends, challenges and opportunities for energy management in water and wastewater utilities, in: Report #4356, 2014.
[64] Energy Information Administration, Average Utilization for Natural Gas Combined-Cycle Plants Exceeded Coal Plants in 2015, Today in Energy. https://www.eia.gov/todayinenergy/detail.php?id = 25652, April 2016.
[65] Energy Information Administration (EIA), 32% of Natural Gas Pipeline Capacity Into the Northeast Could be Bidirectional by 2017, Today in Energy, EIA, December 2014.
[66] New York Public Service Commission, PSC Approves Con Edison's Demand Response Pilot Program, 18069/17-G-0606, August 2018.
[67] Peak Load Management Alliance (PMLA), November 2018. Available Online: https://www.peakload.org/assets/38thConf/Three-Utility-Approaches-to-Gas-Demand-Response.pdf. Three Utility Approaches to Gas Demand Response, PMLA Gas DR Panel.
[68] C. Borraz-Sanchez, R. Bent, S. Backhaus, S. Blumsack, H. Hijazi, P. van Hentenryck, Convex optimization for joint expansion planning of natural gas and power systems, in: 49th Hawaii International Conference on System Sciences (HICSS), Koloa, HI, 2016, pp. 2536–2545.
[69] A. Zlotnik, L. Roald, S. Backhaus, M. Chertkov, G. Andersson, Control Policies for Operational Coordination of Electric Power and Natural Gas Transmission Systems, American Control Conference (ACC), Boston, MA, 2016.
[70] S. An, Q. Li, T.W. Gedra, Natural gas and electricity optimal power flow, in: 2003 IEEE PES Transmission and Distribution Conference and Exposition, Dallas, TX, 2003.
[71] R.Z. Rios-Mercado, C. Borraz-Sanchez, Optimization problems in natural gas transportation systems: a state-of-the-art review, Appl. Energy 147 (2015) 536–555.
[72] Department of Energy, Combined heat and power (CHP) technical potential in the United States, in: Report DOE/EE-1328, March 2016.

[73] P. Mancarella, G. Chicco, Real-time demand response from energy shifting in distributed multi-generation, IEEE Trans. Smart Grid 4 (4) (2013) 1928–1938.
[74] J. Aghaei, M. Alizadeh, Multi-objective self-scheduling of CHP (combined heat and power)-based microgrids considering demand response programs and ESSs (energy storage systems), Energy 55 (2013) 1044–1054.
[75] J. Lu, D. Lubkeman, N. Lu, T. Sun, Z. Li, An optimal microgrid dispatch algorithm for scheduling multiple CHPs in islanding operations, in: IEEE Power & Energy Society General Meeting, Chicago, IL, July 2017.
[76] T. R. Board and N. R. Council, Overcoming Barriers to Deployment of Plug-in Electric Vehicles, The National Academies Press, Washington, DC, 2015.
[77] J. Li, Compatibility and Investment in the U.S. Electric Vehicle Market, January 2019.
[78] California Energy Commission, California plug-in electric vehicle infrastructure projections: 2017–2025, in: Staff Report CEC-600-2018-001, March 2018.
[79] K. Murthy, Electric vehicle charging stations: rate considerations for cooperatives, in: National Rural Utilities Cooperative Finance Corporation Report, August 2018.
[80] C. Qiu, K.T. Chau, C. Liu, C.C. Chan, Overview of wireless power transfer for electric vehicle charging, in: World Electric Vehicle Symposium and Exhibition (EVS27), Barcelona, 2013.
[81] Y. Xiong, J. Gan, B. An, C. Miao, A.L.C. Bazzan, Optimal electric vehicle fast charging station placement based on game theoretical framework, IEEE Trans. Intell. Transport. Syst. 19 (8) (2018) 2493–2504.
[82] U.S. Department of Transportation, Public Transportation's Role in Responding to Climate Change, January 2010.
[83] M. Chester, A. Horvath, Life-Cycle Energy and Emissions Inventories for Motorcycles, Diesel Automobiles, School Buses, Electric Buses, Chicago Rail, and New York City Rail, Center for Future Urban Transport: A Volvo Center of Excellence, UC Berkeley, 2009.
[84] D. Iaconangelo, A third of U.S. buses on path to be carbon-free – report, EnergyWire (2019).
[85] J. Gallo, T. Bloch-Rubin, J. Tomic, Peak Demand Charges and Electric Transit Buses, U.S. Department of Transportation Federal Transit Administration, October 2014.
[86] PJM, Electric Vehicles and the Grid. https://learn.pjm.com/-/media/about-pjm/newsroom/fact-sheets/electric-vehicles-and-the-grid-factsheet.ashx, May 2017.
[87] S. Yoon, K. Park, E. Hwang, Connected electric vehicles for flexible vehicle-to-grid (V2G) services, in: 2017 International Conference on Information Networking (ICOIN), Da Nang, 2017, pp. 411–413.
[88] T. Yiyun, L. Can, C. Lin, L. Lin, Research on Vehicle-to-Grid Technology, in: 2011 International Conference on Computer Distributed Control and Intelligent Environmental Monitoring, Changsha, 2011, pp. 1013–1016.
[89] C. King, Chapter 9 – Transactive energy: linking supply and demand through price signals, Distributed Generation and its Implications for the Utility Industry, Academic Press, 2014, pp. 189–204.

Index

Note: Page numbers followed by "*f*," "*t*," and "*b*" refer to figures, tables, and boxes, respectively.

A

AC induction motor, 4
AC power flow model, 24, 42–44. *See also* Decoupled linear AC power flow models, voltage magnitude
Actor-critic using Kronecker-factored trust region (ACTKR), 15–16
Adaptive boosting model (AdaBoost), 215, 245
Advanced grid operational tools, state estimation
  model validation, 165–178
    erroneous parameters from large model dataset, 165
    largest normalized Lagrange multiplier test, 166–173
    measurement errors, 165, 177–178
    model parameter errors, 165
    parameter errors, detectability and identifiability of, 177–178
  protective relaying, 192–204
    challenges of, 193
    dynamic state estimation–based protection, 193–196
    event study, 201–204, 202*f*, 203*f*
    legacy protection functions, 201
    series compensated transmission line, 197
  system monitoring, 178–192
    Bayesian dynamic state estimation, nonlinear regression, 184–192
    motivations for dynamic state estimation, 178–180
    problem formulation of dynamic state estimation, 180–183
Advanced metering infrastructure (AMI), 210, 237–238, 281–282, 293, 308–309
Aggregated microgrids, 261–262
Aging infrastructure and climate-related impacts, 277–278
  average lifespan of coal, natural gas and nuclear plants, United States, 278*t*
  power generation sources, 278
Algebraic network model, 51
AMI. *See* Advanced metering infrastructure (AMI)
AMUs. *See* Asynchronous measurement units (AMUs)
Analog ensemble model, 215–216
Ancillary services, 99–100, 298–299
  and associated timescales, 125–126, 125*t*
Artificial intelligence (AI), 12, 209–210
  CNN-LSTM deep learning architecture, 238–243
    advantage, 249–250
    convolutional neural network (CNN), 238–243
    developed CNN-LSTM model, 241–243, 242*f*
    evaluation metrics, mean of, 249*t*
    experiments, 243–245, 243*t*, 245*t*
    hyperparameters, 243–245, 244*t*
    long short-term memory network (LSTM), 239–241, 240*f*
    occupancy detection results, confusion matrices of, 248*f*
    parameters, 245*t*
    typical binary classification problem, 246–247
  deep learning gated recurrent unit (DLGRU), 231–232
  forecasting, 210
  and machine learning, 10–11
    building occupancy detection, 235–250
    forecasting, 11
    heterogeneous, 10

Artificial intelligence (AI) (*Continued*)
large volume and high velocity, 11
low value density, 11
operation and control, 11
perception, 11
structured data, 10
unstructured data, 10
smart grid, applications to, 211*f*
state estimation, 210–212
Artificial neural networks (ANNs), 213–216, 218–220, 229–231
Asynchronous advantage actor-critic (A3C), 14–16
pseudocode for, 14–15, 16*b*
Asynchronous measurement units (AMUs), 79, 81–83, 84*f*, 86–87
Asynchronous voltage phasors, 268
Augmented Lagrangian relaxation, 230–231
Automated machine learning (AutoML), 20–21
Automatic generation control (AGC), 35–36, 122, 127–129, 229–230
Automatic meter reading (AMR) systems, 281, 283*f*, 292–293
Automatic voltage regulators (AVRs), 229–230
Autoregressive integrated moving average model, 213–215
Autoregressive model, 213–214
Autoregressive moving average model, 213–214
Avista distribution system, 268–269

B

Bagging, 215
Balanced distribution systems, 41–44
Balancing Authority's area control error (ACE), 128–129
Base flexibility, 105
Bayesian dynamic state estimation, nonlinear regression, 184–192
decentralized *vs.* centralized dynamic state estimation for power system, 191–192
for nonlinear dynamic system models, 184–185
proposed unified framework, 185–188
robustifying, proposed framework for, 188–191
Bayesian inference, 137
Bayes' rule, 184–185
Behind-the-meter (BTM), 114, 119–120
Behind-the-meter (BTM) distributed energy resources (DERs), 99–100, 131–132
Bellman equations, 3–6
expectation equations, 5
optimal equations, 5–6
Bidirectional long short-term memory network (BiLSTM) layer, 240–241
Binary confusion matrix, 246–248, 246*f*
Blockchain technology in electric power systems, 138–139, 139*f*
Bluetooth, 280–281
Boosting, 215
Bootstrapping, 5
Branch power flow models, 53–54, 74–75
Bronzeville microgrid, 262–263
Brown's Ferry nuclear plant, 290
Building-to-grid, 284–285
Bulk electric power grid dispatch in inverter-based resources systems, 119–120, 120*f*
Buses in power grids, 9*f*
Bus injection power flow equations, 53, 55, 68
2-bus system, 34–35, 34*f*

C

California independent system operators (CAISO), 100–101, 105–106, 118, 126, 133–135
Energy Imbalance Market program, 130–131
typical marginal price for energy, 122–123
Cauchy distribution, 191
Centralized dynamic state estimation for power system, 191–192
Centralized *versus* distributed transactional platform, 139*f*
Chapman–Kolmogorov equation, 184

Chi-square value, 196
Cholesky decomposition, 186
Cholesky factorization, 175–176
Classification and regression trees (CARTs), 217–218
Closed-loop systems, 296–297
Close-loop cooperative control law, 59–60
Coal unit, 123
Cold-start DC power flow models, 25–26
Combined heat and power (CHP) plants, 120, 133–135, 137–138, 285, 301–302
Community choice aggregation (CCA), 130–131
Competitive ensemble methods, 215
  bagging and boosting, 215
  multi-model forecasting framework, 218–220, 219*f*
  wind speed forecasting based on, 223–227, 224*t*, 226*f*, 226*t*
Comprehensive energy system, 137–138
Consolidated Edison (ConEd) in New York, 277–278, 301
Constraint optimization method, 195
Consumer vehicles, 303–304, 304*f*
Contingency analysis (CA), 103–104
Contingency reserve, 127–129
Continuous-time models, 183
Control continuum in power industry, 129*f*
Control functions *vs.* dispatch functions for utility-scale inverter-based resources, 122*t*
Controllable load (Type II DR), 114
Conventional power flow techniques, 268–269
Convex relaxation techniques, 16–17
Convolutional neural network (CNN), 12, 237–239
Convolutional neural network-long short-term memory network deep learning architecture, 238–243
  advantage, 249–250
  convolutional neural network (CNN), 238–239
  developed CNN-LSTM model, 241–243, 242*f*
  evaluation metrics, mean of, 249*t*
  experiments, 243–245, 243*t*, 245*t*
  hyperparameters, 243–245, 244*t*
  long short-term memory network (LSTM), 239–241, 240*f*
  occupancy detection problem formulation, 238
  occupancy detection results, confusion matrices of, 248*f*
  parameters, 245*t*
  typical binary classification problem, 246–247
Cooperative control theory, 58–59
Cooperative ensemble methods, 215
  framework of, 221*f*
  probabilistic solar power forecasting, 215–216
  probabilistic wind power forecasting, 215–216, 220
  surrogate model, 220, 222
  wind power forecasting based on, 227–228, 227*t*, 228*f*, 228*t*
Cooperative real power control, 62*f*, 63
Critical parameter-measurement pair, 177
Critical parameter pair, 177
Curriculum learning, 19–20
Cybersecurity and privacy, 292–294

D

Data-acquisition system, 193
Data-driven machine learning algorithms, 237
Data-driven models, 27
Data management and analytics, 136–137
Day-ahead market clearing process, 102–103, 102*f*
Day-ahead unit commitment process, 121–122
DC power flow (DCPF) model, 9, 24–25, 28, 33–34, 40–41, 44–46
  fast computation speed, 33–34
Decentralized dynamic state estimation for power system, 191–192
Decision-making, 2
Decision tree model (DT), 237

Decoupled linear AC power flow models, voltage magnitude
balanced distribution systems, 41–44
linear power flow models for meshed transmission systems, 26–37
derivation and justification of FDLPF, 33–37
matrix formulation of DLPF model, 30–31
transformers and phase shifters, 31–33
voltage magnitude and phase angle, decoupling of, 26–29, 29*f*
linear three-phase power flow models, unbalanced distribution systems, 37–39
meshed transmission systems, 40–41
computational efficiencies, for large-scale systems, 41*t*
errors of different linear power flow models, 40*t*
IEEE 118-bus system, voltage magnitudes and branch MW flows for, 41*f*
33-node ill-conditioned system (lumped overload), 42*f*
33-node ill-conditioned system (uniform overload), 42*f*
unbalanced distribution systems, 44–46
Decoupled linearized power flow (DLPF) model, 25–26, 28, 40–41, 44–46
matrix formulation of, 30–31
voltage magnitude, 41*f*
Deep deterministic policy gradient (DDPG), 15–16
Deep distributed recurrent Q-networks-action discovery (DDRQN-AD), 230
Deep learning, 210, 216, 231, 237
Deep learning gated recurrent unit (DLGRU), 231–232, 234, 236*f*
block diagram, 234*f*
generation units' power, 236*f*
reconfiguration switching of, 236*f*
Deep Q-network (DQN), 12–14
Deep reinforcement learning (DRL), 12–22
Demand management system (DMS), 133, 137–138
Demand response (DR), 33–35, 281
CHP plants, 302
definition of, 286
emerging considerations for, 289–290
FPL's on call program, 288
incentive-based programs, 286, 287*f*
price-based programs, 286, 287*f*
Rush Hour Rewards program, 288
shape, 288–289
shed, 288–289
shift, 288–289
shimmy, 288–289
time-of-use (TOU) prices, 286, 287*t*
XCEL Energy's Windsource Program, 288
Demand response resource (DRR), 105–106, 114–117
Demand-side management (DSM), 279–280, 288
Department of Energy, 277
Dependability, 192
DERs. *See* Distributed energy resource (DER)
Deterministic forecasting methods, 213–214
DHA. *See* Dynamic hosting allowance (DHA) method
Differential and algebraic equations (DAEs), 13, 164, 180–181, 195
Direct load control (DLC), 284–285
Distance protection, 201
Distributed denial-of-service (DDoS), 292
Distributed energy resource (DER), 2, 5–8, 14–15, 21, 50, 87–88, 89*f*, 90–91, 99–102, 117–118, 131–135, 132*f*, 178–180, 182–183, 262–263, 266, 276, 283–284
abnormal frequencies, response to, 267*t*
abnormal voltages, response to
category I, 267*t*
category II, 267*t*
category III, 267*t*
grid-interactive buildings, 285–286

power losses, transmission and distribution in United States, 285*f*
reliability, 285–286
resilience, 285–286
microgrids, 5–8
photovoltaic (PV), 5–7, 6*f*
connecting with grids, 6*f*
curves, 7, 7*f*
DC output of, 6–7
Distributed generator (DGs), 138, 261, 270, 285
Distributed real power control, 63
Distributed subgradient algorithm, 58–59, 79
Distributed subgradient voltage control, 59–64
Distributed system platforms (DSPs), 101–102, 131–135
Distribution grids, 260–261
Distribution network model, 52–54, 52*f*, 54*f*
with high penetration renewables, 66*f*
with measurements, 83*f*
Distribution system operator (DSO), 87–88, 132–133, 132*f*, 135–136
functionalities in different levels, 136–138, 137*f*
advanced DMS/SCADA system, 137
comprehensive energy system, 137–138
data management and analytics, 136–137
resource dispatch system, 137
reliable and resilient operation of, 136
transmission and gas systems, connection with, 134*f*
Distribution systems, 1, 266, 270–271
errors of different linear power flow models for, 43*t*
Dragonfly 2.0 targeted power companies, 293–294
Droop control, 76, 79–80
Droop strategy, for PQ inverters, 75*f*
Dynamic distributed generation (DG) models, 51, 56–57
Dynamic hosting allowance (DHA) method, 51, 87–90, 88*f*, 89*f*
Dynamic microgrids
concept of, 263–264
definition of, 263
networked microgrids operation in, 264*f*
parallel inverters
steady-state operation, 264–266
transient-state operation, 266–273
reconfiguration operation, 272*f*
topology of, 263
Dynamic NN (DNN), 230
Dynamic security assessment (DSA), 103–104, 180
Dynamic simulations, 268–269
Dynamic state estimation (DSE), 164, 178, 184, 188
Bayesian dynamic state estimation, nonlinear regression, 184–192
decentralized *vs.* centralized dynamic state estimation for power system, 191–192
for nonlinear dynamic system models, 184–185
proposed unified framework, 185–188
robustifying, proposed framework for, 188–191
benefits of, 178–180
dependability and reliability of protection systems, improved, 179–180
enhanced reliability of system models, 180
hierarchical decentralized control, enhanced, 179
oscillations monitoring, improved, 178–179
motivations for, 178–180
problem formulation of, 180–183
Dynamic state estimation-based protection
protection zone, dynamic model of, 194
quantification of consistency, 195–196
series compensated transmission line, 197, 197*f*, 199*f*

Dynamic state estimation-based protection (*Continued*)
overall dynamic model, formulation of, 201
R, L, and C matrices, 197, 198*t*
section *k* in multisection line, 198–200, 199*f*
sequence parameters, 197, 198*t*
source impedances, 197, 197*t*
three-phase series capacitors, 200, 200*f*
trip decision, 196

E

Eastern Municipal Water District (EMWD), 299
Effective flexible capacity (EFC), 105–106
Electrical engineering, 144
Electric grid modernization, 260–261
Electricity Consumption and Occupancy (ECO) dataset, 243
Electric power system (EPS), 261
blockchain technology in, 138–139, 139*f*
Electric pumps, 298
Electric Reliability Council of Texas, 100–101
Electric storage resources, 118
Electric system topology variation, 268
Electric vehicles (EVs), 279, 303–304, 306–308
Electrification, 276
benefits, 278–279
in built environment, 278–280
of transportation sector, 279, 280*f*, 303–308
consumer vehicles, 303–304, 304*f*
public transportation, 305–306
rideshare services and emerging methods, 306–307
school bus, 305*b*
vehicle-to-grid (V2G) system, 307–308
Electromagnetic induction, 296
Element-wise rectified linear unite (ReLU), 239
Emerging customer behavior, 99–100
Emerging dynamic microgrids, 263–264
Empirical mode decomposition (EMD), 215
Energy hubs, 294
Energy Information Administration (EIA), 279
Energy storage market modeling, 110–114
capacity limits constraints with reserves, 110–111
cleared energy and reserve constraint, 111–112
downward reserve equation, 113
energy ramp-down constraints, 113
energy ramp-up constraints, 113
energy storage resources, 110–111, 114
power balance constraints, 111–112
pumped storage hydro, 110–111
ramp constraints for reserve deployment, 112
reserve activation constraints, 113–114
reserve ramp constraints, 113
resource limit constraints, 112
state-of-charge, 110–111, 113
upward reserve equation, 113
Energy storage resources (ESRs), 110–114
Energy storage systems, 76
Energy systems, reinforcement learning in, 23–40
combining prior knowledge, 38
controller training and evaluation, 38
demand response, 33–35
grid operation, 35–37
handling nonlinearity, 24
learning environment, 25–29
limitations and challenges, 39
model adaptability, 23–24
modeling uncertainty, 24
multiagent, 39
proper RL/MDP formulation, 38
real-time readiness, 23
renewable generation and battery control, 37
smart buildings, 30–33
transfer learning, 38
typical workflow, 24–25
Ensemble deterministic forecasting model, 222–223
Ensemble Kalman filter (EnKF), 184–190
Ensemble learning, 213–216

competitive ensemble learning, 215, 218–220
bagging and boosting, 215
multi-model forecasting framework, 218–220, 219*f*
wind speed forecasting based on, 223–227, 224*t*, 226*f*, 226*t*
cooperative ensemble learning, 215, 220–222
framework of, 221*f*
probabilistic solar power forecasting, 215–216
probabilistic wind power forecasting, 215–216, 220
surrogate model, 220, 222
wind power forecasting based on, 227–228, 227*t*, 228*f*, 228*t*
probabilistic solar power forecasting, 215–216
probabilistic wind power forecasting, 215–216
single machine learning algorithm models, 216–218
ERCOT, 127
EriGRID, 118
Ethernet, 148–151
Euler methods, 183
Event-data-based oscillation modal analysis, 153, 154*f*
Event replay and postevent analysis, 156, 156*f*
Evolution strategies-based reinforcement learning (ES-RL), 16–18
Exogenous variable model, 215–216
Explicit branch model of distribution network, 55–56
Extended Kalman filter (EKF), 184–186, 195

F

FACTS devices, 178–180
Fair utilization ratio method, 79–80
Fast and linear power flow model, 24
Fast decoupled linearized power flow (FDLPF) model, 26, 40–42, 44–46
computational efficiency, 40–41, 41*t*
derivation and justification of, 33–37
2-bus system, 34–35, 34*f*
numerical example, 36–37
theoretical derivation, 35–36
voltage magnitude, 41*f*
Fast-decoupled-load-flow method, 9
Fast distribution state estimation method, 136
Fast-timescale grid services, 307–308
FDR, 147–148
components, 150*t*
deployment in North America, 148, 149*f*
first generation of, 148–151, 150*f*
installation, 151, 151*f*
low installation cost and high measurement accuracy, 151
second generation of, 148–151, 150*f*
Feature learning blocks (FLB), 238–239
Federal Bureau of Investigation (FBI), 292–293
Federal Energy Regulation Commission (FERC), 118, 125–126, 278
Order 841, 110–111
Order 888, 100–101
Order 889, 100–101
Order 2000, 100–101
Feedback-based algorithms, 14–15
Feed forward neural network (FFNN), 210–212
Flexible ramp-down (FRD) capabilities, 106, 108, 110
Flexible ramp products (FRPs), 106–110
capacity limit constraints for eligible resources, 110
economic dispatch problem with, 107–108
flexible ramp-down, 106
flexible ramp-up, 106, 110
market-wide regulating down reserve requirements, 109
market-wide regulating up plus spinning plus nonspinning reserve requirement, 109

Flexible ramp products (FRPs) (*Continued*)
market-wide regulating up plus spinning reserve requirement, 109
market-wide regulating up reserve requirement, 109
ramp capacity requirement, 106–107
ramp demand curves, 107, 108*f*
resource ramping capability constraints, 110
system-wide power balance constraint, 108–109
transmission constraints, 109
Flexible ramp-up (FRU), 106, 110
Flexible resource adequacy and must offer obligation (FRAC-MOO), 105, 118
Florida Power and Light (FPL), 288
FNET/GridEye, wide-area measurement system, 147–151
architecture of, 147, 147*f*
disturbance detection and location, 152–153, 153*f*
event-data-based oscillation modal analysis, 153, 154*f*
event replay and postevent analysis, 156, 156*f*
FDR deployment in North America, 148, 149*f*
first generation of FDR, 148–151, 150*f*
hardware component, 147
interarea oscillation detection, 153, 154*f*
islanding detection, 154–155, 155*f*
machine learning–based inertia estimation, 157–159, 159*f*
model validation and parameter verification, 157, 158*f*
North American grids, 148
online ambient-data-based oscillation modal analysis, 154, 155*f*
second generation of FDR, 148–151, 150*f*
software component, 147
statistical analysis of historical data, 156–157, 157*f*
visualization, 154
worldwide power grid monitored, 148, 149*f*
Forecasting technology, 11, 210, 212–228
categorization, 212–213, 213*f*
deterministic forecasting methods, 213–214
ensemble learning, 214–216
probabilistic forecasting methods, 214
Fractional order control strategy, 230–231
Frequency control, 8, 14, 229–230
Frequency response mechanism design, 128–129, 129*f*
Front-of-the-meter (FTM), 119–120
Future power systems
cyber considerations, 309–310
market and organizational considerations, 309
physical considerations, 308–309

G

Gain matrix, 167–168, 174–176
Game theory, 22
Gas and heating/cooling (GHC) systems, 133, 135
Gas-fired power plants, 288, 300–301
Gated recurrent units (GRU), 20–21
Gaussian distribution, 169–170, 180, 183, 188–191, 243–244
Gaussian measurement errors, 169–170
Gaussian mixture model, 214
Gaussian Process Regression (GPR), 230
Gauss-Newton algorithm, 210–212
Gauss–Newton's iterative method, 186
Gauss–Seidel method, 9
General dynamic state-space model, 183
Generalized autoregressive conditional heteroskedasticity, 215
Generalized maximum (GM)-likelihood-estimator, 189–190
Generalized neurons (GN), 229–230
Generation system, 1
Generator bus, 9
Generator operators, 122
Genotypes, 17
Gird operation paradigm, 117
Global positioning system (GPS), 144–145
Global System for Mobile Communications (GSM), 280–283

Gradient boosting machine (GBM), 213–214, 217, 218*b*, 224–225
Grainger & Stevenson 4-bus system, 36, 36*f*
Green Mode, 306
Grid-connected mode, 262, 264–265
Grid-edge devices and networks, 138
Grid-edge situational inference process, 90
Grid-following inverters, 119, 119*t*
Grid-following mode, 266
Grid-forming inverters, 119, 119*t*
Grid-forming (GFM) mode, 266, 271–273
Grid-interactive buildings, 283–294
  climate and environment, 290–292
  cybersecurity and privacy, 292–294
  demand response, 281, 286–289
    definition of, 286
    emerging considerations for, 289–290
    FPL's on call program, 288
    incentive-based programs, 286, 287*f*
    price-based programs, 286, 287*f*
    Rush Hour Rewards program, 288
    shape, 288–289
    shed, 288–289
    shift, 288–289
    shimmy, 288–289
    time-of-use (TOU) prices, 286, 287*t*
    XCEL Energy's Windsource Program, 288
  distributed energy resources, 285–286
    power losses, transmission and distribution in United States, 285*f*
    reliability, 285–286
    resilience, 285–286
  2017 solar eclipse, 284*b*
Grid resiliency, 277–278

H

Hawaiian Electric Companies' Power Supply Improvement Plan, 118
HC. *See* Hosting capacity (HC)
Heating, ventilation, and airconditioning (HVAC) system, 212
Hidden Markov model (HMM), 237
High-performance computing technique, 178–179
Hoover Dam, 297–298
Hoover Powerplant, 297–298
Hosting capacity (HC), 86–88, 89*f*
Hot-start models, 25
Hydroelectricity, 297–298
Hydroelectric power plants, 294–298
Hyperparameters, 243–245, 244*t*, 245*t*
Hyper plane function, 217

I

IEEE 14-bus transmission system, 92–93, 95
IEEE 33-bus system, single line diagram of, 235*f*
IEEE 34-bus test feeder, 271–273
IEEE 57-bus systems, 210–212
IEEE 118-bus systems, 41*f*, 210–212
IEEE 123-bus system, 64, 84*f*, 86*f*, 90–91
  DHA process, 90–91
  simulation results of voltage inference on, 86*f*
  with SMUs, AMUs and PVs on bus 72, 77 and 89, 84*f*
IEEE-DC1A exciter, 180–182
IEEE 8500-node system, 64, 73*f*, 75–76, 80, 95
  frequency control of, 78*f*
  with 4 large-scale PVs, 78*f*, 80
IEEE 33-node test feeder, 44–46
  voltage magnitudes and branch MW flows for, 44*f*
IEKF, 186–187, 189–190
Importance Weighted Actor-Learner Architecture (IMPALA), 19
Incentive-based programs, 286, 287*f*
Incremental learning, 8–9
Independent power producer (IPP), 100–101
Independent system operator (ISO), 100–101, 110–112, 119–121, 126, 132, 132*f*, 135–136, 210
  security-constrained unit commitment (SCUC) formulation in, 111
Independent system operator (ISO) New England, 100–101, 127, 210
Integrated distributed electricity system (IDES), 133–136
Intelligent models, 213–214
Interarea oscillation detection, 153, 154*f*
Interconnected microgrids, 261–262

Interdependent critical networks, 294–302
  combined heat and power, 301–302
  energy hubs, 294
  power and gas, 295*t*, 300–301
  water–energy nexus, 294–300, 297*f*
    ancillary services, 298–299
    closed-loop systems, 296–297
    controlling electric water pumps, 298–299
    electricity, 298
    hydroelectricity, 297–298
    once-through cooling systems, 296
    thermoelectric power plants, 296
    Venezuela's Guri dam, 298*b*
International Energy Agency (IEA), 279
Internet, 148–151
Internet-of-Things (IoT), 32, 138, 309–310
Interruptible load (Type I DR), 114
Intrusive sensors, 236–237
Inverter-based distributed generators (DGs), 264–266, 268–273
Inverter-based resources (IBRs), 118, 120–122, 120*f*, 124, 127
  AGC design for, 130*f*
  bulk electric power grid dispatch in, 119–120, 120*f*
  plant-level control functions, 121
Islanded microgrid with high-penetration of DGs, 74–78, 74*f*
Islanded mode, 262, 264–265, 268
Islanding detection, 154–155, 155*f*

J

Jacobian matrix, 191, 195
  measurement function, 167–168, 174

K

Keras, 245
Kirchhoff law, 55–56, 68–69, 82–83
Kirchhoff's Current Laws (KCLs), 193–194
k-nearest neighbor (kNN) point forecasts, 214, 237, 245
100K-node distribution circuits, 93
Kriging-based surrogate modeling technique, 230–231
Kriging modeling, 230–231
100k synthetic circuits, 63*f*
  voltage control on, 65*f*

L

Laplace distribution, 226–227
Laplacian probability distribution, 189–191
Large-campus district cooling systems, 135
Large-scale distributed generation (DG) resources, 133–135
Large-scale distribution system, 50–51, 68
Large-scale dynamic systems, 143–144
Large-scale power systems, 165, 229–230
Largest Normalized Lagrange Multiplier (LNLM) test, 165–166
  computationally efficient implementation of, 173–177
  extraction of Lagrange multipliers, state estimation problem, 166–167
  model parameter errors, detection, identification and correction of, 171–173, 173*f*
  normalized Lagrange multipliers and hypothesis testing, 167–171
Largest normalized residual (LNR) test, 171–172
Lawrence Berkeley National Laboratory, 288–289
Learning environment
  action space, 27–28
  data-driven models, 27
  domain-specific simulators, 27
  observation/state space, 28
  reward structure, 28
  simplified physics models, 26–27
  termination condition, 29
Least squares support vector regression (LSSVR) model, 213–214
Linearized measurement model, 167–168
Linear three-phase power flow models, unbalanced distribution systems, 37–39
Line differential protection, 201–204
Load bus, 9
Load-following reserve, 127–128, 128*f*
Load forecasting, 215
Load Serving Entities (LSEs), 105
Load tap changers (LTCs), 66

Locational marginal price (LMP), 24, 103–104, 122–123
Long short-term memory (LSTM), 20–21, 231–232, 237–241, 240*f*
Low-voltage distribution networks, 276

M

Machine learning–based inertia estimation, 157–159, 159*f*
Machine learning methods, 2, 10–11
  building occupancy detection, 235–250
    CNN-LSTM deep learning architecture, 238–243
    non-intrusive occupancy detection, 236–237
    RFID-based occupancy detection system, 236–237
    WiFi-based occupancy detection system, 236–237
  forecasting, 11
  heterogeneous, 10
  large volume and high velocity, 11
  low value density, 11
  machine learning-based control, 229–230
  machine learning-based optimization, 230–231
  modern forecasting technology, 212–228
    categorization, 212–213, 213*f*
    deterministic forecasting methods, 213–214
    ensemble learning, 214–216
    probabilistic forecasting methods, 214
  network reconfiguration, 231–234
    bus voltage and angle limits, 233
    definition of, 231
    DLGRU, 231–232, 234, 234*f*
    generation units constraints, 233
    generation units features, 235*t*
    modified IEEE 33-bus test system, 234, 235*f*
    power balance constraints, 232–233
    reconfiguration constraints, 233–234
  operation and control, 11
  perception, 11
  structured data, 10
  unstructured data, 10
Machine-to-machine communication, 280–281
MA-OpenDSS, 93–95
Market clearing engine (MCE), 102–103
Markov Decision Process, 3–6, 220–222
  action set A, 3
  Bellman equations, 5
  Bellman expectation equations, 5
  Bellman optimal equations, 5–6
  definition of, 38
  discount factor γ, 4
  policy, 4
  problems, 6
  reward function R, 4
  state set S, 3
  state transition probability set P, 4
  value function, 4
Markov model, 184
MATLAB program, 42–44, 268–269
MATPOWER 5.1, 42–44
Maximum power point tracking (MPPT), 5–7, 37
Max-pooling layer, 239
Measurement errors, 165, 168–169, 172, 177–178
  detectability and identifiability of, 177–178
Merging units (MUs), 163–164
Meshed transmission systems, 26–37, 40–41
  computational efficiencies, for large-scale systems, 41*t*
  derivation and justification of FDLPF, 33–37
    2-bus system, 34–35, 34*f*
    numerical example, 36–37
    theoretical derivation, 35–36
  errors of different linear power flow models, 40*t*
  IEEE 118-bus system, voltage magnitudes and branch MW flows for, 41*f*
  matrix formulation of DLPF model, 30–31
  33-node ill-conditioned system (lumped overload), 42*f*

Meshed transmission systems (*Continued*)
33-node ill-conditioned system (uniform overload), 42*f*
transformers and phase shifters, 31–33
admittance matrix, contribution to, 31–32, 32*f*
branch MW flow, influence on, 32–33
voltage magnitude and phase angle, decoupling of, 26–29, 29*f*
Meta learning, 20–21
metric-based, 21
optimization-based, 21
RNN-based, 20–21
Microgrids, 2–3, 3*f*
clusters, 261–262
controllers, 3, 7–8
DER units, 5–8
fractional order control strategy, 230–231
frequency control reserves, 8
grid-connected mode, 3
islanded mode, 3
Midcontinent independent system operator (MISO), 106–107, 126
Mid-voltage (MV) network, 76
Midwest independent system operator, 100–101
Min-microgrid
configuration of, 271*f*
definition of, 270–271
electric boundary of, 270–271
Mixed integer linear programming (MILP), 231
M3-Laplace model, 226–227
Model-free Q-learning, 230
Model parameter errors, 165
Model predictive control (MPC), 30–31
Model time-discretization errors, 183
Monte Carlo simulations, 7–8, 7*b*, 188
Motivations for dynamic state estimation, 178–180
Multi-agent optimization method, 58–59
Multiagent system, 21–22
competitive, 22
cooperative, 21–22
mixed, 22
Multi-model probabilistic forecasting model, 222–223
Multiple microgrid system, 8
Multisector coupling, 15–17
Multi-variable regression, 214
Municipal aggregation. *See* Community choice aggregation (CCA)

N

National Energy Policy Act, 5
National Renewable Energy Laboratory, 279–280
Natural gas, 295*t*, 300–301
definition of, 300–301
2018 US natural gas consumption, 300*f*
Natural monopolies, 100–101
NERC Standard VAR-001, 122
Nested microgrids, 261–262
configuration of, 262*f*
management of, 262
Net power injection, 81
Network dynamic operation, 11–15, 12*f*
Networked microgrids with parallel inverters
advanced microgrid structures, 261–263
dynamic microgrids, concept of, 263–264
Network sensitivity, 81, 82*f*, 89*f*, 91
Neural network (NN) training, 12
computation for, 16
for RL training, 22
Neuro-fuzzy network, 213–214
Newton–Raphson solution method, 9
New York independent system operators, 100–101
33-node ill-conditioned system (lumped overload), 42*f*
33-node ill-conditioned system (uniform overload), 42*f*
Node injection power flow, 53
11,000-node system, 95
100,000-node system, 95
123-node test feeders, 44–46
Nondiscriminatory access, 5
Non-Gaussian distribution, 180
Nonintrusive load monitoring, 292–293
Non-intrusive sensors, 236–237

Nonlinear mapping-based kernel methods, 217
Nonlinear oscillator synchronization, 76
Nonlinear regression, Bayesian dynamic state estimation, 184–192
  decentralized *vs.* centralized dynamic state estimation for power system, 191–192
  for nonlinear dynamic system models, 184–185
  proposed unified framework, 185–188
  robustifying, proposed framework for, 188–191
Nonparametric approaches, 214
Nonspinning reserve, 126
Normalized Lagrange multipliers (NLMs), 170–172, 176–178
Normalized mean absolute error (nMAE), 222–224, 224*t*, 227, 227*t*
Normalized residual (NR), 171–172, 178, 204
Normalized root mean square error (nRMSE), 222–224, 225*t*, 227, 227*t*
North American grids, 148
NREL-100k system, 93

O

Occupancy-based feedback control algorithm, 212
Off-the-shelf black-box tools, 15–16
Ohm's law, 54
Olney Town Center microgrid project, 262–263
Once-through cooling systems, 296
Online ambient-data-based oscillation modal analysis, 154, 155*f*
Online learning, 8–9
Open Access Same-Time Information System, 100–101
OpenAI, 15
OpenDSS, 80
Operating reserves, 124–125
Operating thermal resource, 118–119
Optimal policy, 6
Optimal power flow (OPF), 8–10, 14–15

P

Parallel inverters
  dynamic microgrids
    steady-state operation, 264–266
    transient-state operation, 266–273
  networked microgrids with
    advanced microgrid structures, 261–263
    dynamic microgrids, concept of, 263–264
Parameter errors, 168–169
  detectability and identifiability of, 177–178
Parametric approaches, 214
Particle filter (PF), 184–186, 188–190
Peaker plants, 300–301
Peak flexibility, 105
Pearl Street DC system, 4
Pennsylvania American Water, 299
Phase locked loops (PLL), 6–7, 57–58, 76–77
Phasor, 144, 144*f*, 148
Phasor measurement unit (PMU), 144–146, 151, 163–164, 178–181, 191–192, 210, 280–281
  components of, 145*f*
Phasor measurement unit (PMU)-based wide-area measurement system (WAMS), 145–146, 146*f*
Photovoltaics (PVs), 5–7, 6*f*, 47, 265
  connecting with grids, 6*f*
  curves, 7, 7*f*
  DC output of, 6–7
Pinball loss, 222–228, 226*t*, 228*t*
PJM Interconnection, 100–101, 307
Point of common coupling (PCC), 261–262
Points of interconnection (POIs), 263–264
Points on wave measurements, 193–194
Policy, 4
  gradient theorem, 10–11
Policymakers, 131–132
Polynomial Response Surface (PRS), 230
Polynominal regression model, 215–216
Popular convex approximation, 10
Power and gas, 295*t*, 300–301
Power-dependent networks, 308

Power electronics (PEs), 2–3
Power factor control, 64–68
Power flow analysis, 24
Power flow and generator stators, 180–181
Power flow calculation, 9
Power generation, 2
Power plant cooling system, 296
Power systems, 1–3, 2*f*, 16*f*, 17
 community microgrid, 3, 4*f*
 development history in United States, 4–5
 distribution system, 1
 evolution, modernization, 260–261
 generation system, 1
 microgrids, 2–3, 3*f*
 transmission system, 1
Power system stability, 12, 13*f*
Power system stabilizer (PSS), 178–179, 229–230
Power, water and gas networks, increasing connectivity in, 280–283
PQ inverters, 76–77
 droop strategy for, 75*f*
Price-based programs, 286, 287*f*
Probabilistic forecasting methods, 214
Probabilistic solar power forecasting, 215–216
Probabilistic wind power forecasting, 215–216
Proximal policy optimization (PPO), 15–16
Public transportation, 305–306
Public Utilities Regulator Policies Act (PURPA), 5
Public Utility Holding Company Act, 5
Public Utility Regulatory Policies Act (PURPA), 100–101
Puerto Rico Energy Public Policy Act, 290–291
Pumped storage hydro (PSH), 110–111

Q

Q-learning, 8–9
Q-learning enhanced deterministic forecasting method, 220–223, 227–228
Q-learning reinforcement learning (RL) algorithm, 229–230

R

Radial base function neural network (RBFNN), 230
Radial Basis Functions (RBF), 230
Radio-frequency identification (RFID), 236–237
Ramp demand curves (RDCs), 107–108, 108*f*
Ramp-down capacity requirements, 106–107, 115–116
Ramping reserve, 126
Ramp-up capacity requirements, 106–107, 115–117
Random forest (RF), 213–214, 217–218, 224–225, 237
Reactive power control, 51, 57–58, 63–68, 67*f*, 68*f*, 75–76, 79–80
Real power control, 68, 79
Real-time closed-loop wide-area decentralized power system stabilizers (WD-PSSs), 229–230
Real-time economic dispatch processes, 121–122
Real-time market operation, 102–104, 103*f*
Real time visualization of measurement data, 152*f*
Rebound effect, 289
Receiver operating characteristic (ROC), 247–249
Recirculating cooling systems, 296–297
Recurrent neural network (RNN), 13, 210–212
Reforming the Energy Vision (REV), 131–132
Regional transmission organizations (RTOs), 100–101
Regulation down reserve, 126
Regulation mileage reserve, 125
Regulation reserves, 125–126
Reinforcement learning (RL), 210
 actor-critic, 11
 Bellman equations, 3–6
 curriculum learning, 19–20

deep reinforcement learning
A3C, 14–16
DQN, 12–14
in energy systems, 23–40
combining prior knowledge, 38
controller training and evaluation, 38
demand response, 33–35
grid operation, 35–37
limitations and challenges, 39
multiagent, 39
proper RL/MDP formulation, 38
renewable generation and battery control, 37
smart buildings, 30–33
transfer learning, 38
evolution strategies-based, 16–18, 17*b*
examples of, 3*b*
learning environment
action space, 27–28
data-driven models, 27
domain-specific simulators, 27
observation/state space, 28
reward structure, 28
simplified physics models, 26–27
termination condition, 29
Markov decision process, 3–6
action set A, 3
Bellman equations, 5
Bellman expectation equations, 5
Bellman optimal equations, 5–6
discount factor γ, 4
policy, 4
problems, 6
reward function R, 4
state set S, 3
state transition probability set P, 4
value function, 4
meta learning, 20–21
metric-based, 21
optimization-based, 21
RNN-based, 20–21
Monte Carlo, 7–8
multiagent system, 21–22
competitive, 22
cooperative, 21–22
mixed, 22
policy-based, 9–11
scalable reinforcement learning frameworks, 18–19
temporal difference, 8–9
value function approximation, 9
Reliability coordination, 122
Renewable energy credit (REC), 288
Renewable energy resources, 106
Renewable generation, 37
Residential distribution networks and feeders, 304
Resource adequacy analysis, 104–105
Resource commitment (RSC) application procedure, 102–103
Resource dispatch system, 137
Resource ramping capability constraints, 110
Retail electricity market
blockchain technology in electric power systems, 138–139, 139*f*
design of, 133–138
state-of-the-art for development, 130–132
Retail electric providers (REPs), 131–132
Retail market operators, functions, 99–100
Runge–Kutta-based methods, 183
Rural Electrification Act, 5
Rush Hour Rewards program, 288

S

Sandworm, 293–294
SARSA learning, 8–9, 8*b*
Scalable distributed feedback control algorithms, 14–15
Scalable reinforcement learning frameworks, 18–19
Scale estimator, 189–190
Scikit-learn libraries, 245
Security assessment (SA), 103–104
Security-constrained economic dispatch (SCED), 103–104, 123–124
Security-constrained optimal power flow (SCOPF), 103–104
Security-constrained unit commitment (SCUC), 24, 111–112, 115, 123–124
Semi-definite program (SDP), 10
Semidefinite relaxation, 10

Sensitivity matrix, 88
Sequential Monte Carlo (SMC) method, 136–137
Shape, demand response, 288–289
Shed, demand response, 288–289
Shift, demand response, 288–289
Shimmy, demand response, 288–289
Short-term energy, 99–100
Short-term solar forecasting method, 213–214
Short-term wind power forecasting method, 213–214
Shunt elements, 27
Siamese neural network, 21
Sigmoid and hyperbolic tangent, 216
Simulink, 268–269
Simultaneous feasibility test (SFT), 102–103
Sinusoial/cosine function, 144*f*
Slack bus, 9
Smart electricity meters, 281–282
Smart grid, 211*f*, 260, 280–281, 292–293
Smart Grids European Technology Platform, 260
Smart meters, 281
Smart switches (SSWs), 263–264, 270
Smart water meters, 281–282
SMUs. *See* Synchronous measurement and control units (SMUs)
2017 solar eclipse, 284*b*
Solar forecasting, 215
Solar photovoltaic (PV) and battery storage, 99–100
Solvable second-order cone program (SOCP), 10
Southern California Edison (SCE) territory, 289
Southern California Gas Company's Smart Therm DR program, 301
Southern Power Pool (SPP) system, 118, 126
Southwestern Power Pool, 100–101
Spatio-temporal correlation, 214
Spinning reserve, 125, 127
State estimation, 210–212
State estimation, advanced grid operational tools
  differential and algebraic equations (DAEs), 164
  dynamic state estimation (DSE) algorithms, 164
  model validation, 165–178
    erroneous parameters from large model dataset, 165
    largest normalized Lagrange multiplier test, 166–173
    measurement errors, 165, 177–178
    model parameter errors, 165
    parameter errors, detectability and identifiability of, 177–178
  protective relaying, 192–204
    challenges of, 193
    dynamic state estimation–based protection, 193–196
    event study, 201–204, 202*f*, 203*f*
    legacy protection functions, 201
    series compensated transmission line, 197
  static state estimation, 164
  system monitoring, 178–192
    Bayesian dynamic state estimation, nonlinear regression, 184–192
    motivations for dynamic state estimation, 178–180
    problem formulation of dynamic state estimation, 180–183
State evolution, probabilistic model of, 184–185
State-of-charge (SOC), 110–112
State-of-the-art artificial intelligence techniques, 231, 235–236
State-space model, 182
Static microgrids, 263–264, 264*f*
Static security assessment, 103–104
Static state estimator (SSE), 164, 178–180
Statistical methods, 213–214
Steady-state operation, 8–10, 178–180
  AC OPF, 9–10
  definition of, 8
  parallel inverter operation in microgrids
    operational constraints, 265
    physical constraints, 265
    renewable penetration, maximization of, 265

system operation reliability maximization, 265
system topology optimization problem, 266
topology constraints, 265
transmission line loss minimization, 265
unbalance constraints, 265
popular convex approximation, 10
power flow calculation, 9, 9*f*
Stiff source bus, 77*f*, 80
Stochastic gradient descent (SDG), 244–245
Stochastic sampling techniques, 137
Structured data, 10
SUMMER-GO, 210
Super peak flexibility, 105
Supervisory control and data acquisition (SCADA) system, 137–138, 143–144, 146, 146*f*, 178–180. *See also* Supervisory control and data acquisition (SCADA) system
Supplemental reserves, 126–127
Support vector machine (SVM), 210–212, 217, 230, 237, 245
Support vector regression (SVR), 213–214, 217, 224–225
Synchronized reserve. *See* Spinning reserve
Synchronized sensors, 10
Synchronous generators, 266
Synchronous measurement and control units (SMUs), 79, 81, 84*f*, 87
Synchrophasor measurement technology
advanced control, 146
phasor, 144, 144*f*
phasor measurement unit (PMU), 144–145
sinusoidal/cosine function, 144*f*
situational awareness, 146
Synthetic 100,000-node NREL System, 93
System-and cluster-level real/reactive power control, 75–76
System transient stability analysis, 14
System-wide power balance constraint, 108–109
System-wide ramp capacity requirement (RCR), 106–107

T

Temporal difference (TD) learning, 8–9
SARSA learning, 8–9, 8*b*
Tensorflow, 245
Tesla Model S, 304
Texas Public Utility Commission, 131–132
TGOV1 turbine-governor, 180–182
Thermal generators, 124, 126–127
Thermoelectric power plants, 294–296
Time-domain numerical integration, 13–14
Time-of-use (TOU) prices, 286, 287*t*, 288–289
Timescale of power system studies, 12*f*
Total installed electricity meters on residential and commercial buildings, 282*f*
Traditional hosting analysis, 86–87
Transfer learning, 33, 38
Transformers and phase shifters, meshed transmission systems, 31–33
admittance matrix, contribution to, 31–32, 32*f*
branch MW flow, influence on, 32–33
Transient-state operation, parallel inverter operation in microgrids, 266–273
advanced inverter control, 269–273, 272*f*, 273*f*
improved inverter operation performance, 268–269
inverter dynamic stability, 266–268
Transmission constraints, 109, 124
Transmission constraints management (TCM), 103–104
Transmission systems, 1, 122
errors of different linear power flow models, 40*t*
Transportation sector, electrification, 303–308
consumer vehicles, 303–304, 304*f*
public transportation, 305–306
rideshare services and emerging methods, 306–307
school bus, 305*b*
vehicle-to-grid (V2G) system, 307–308
Trapezoidal methods, 183

True negative rate, 246–247
True positive rate/recall, 246–247
Trust region policy optimization (TRPO), 15–16
Type I demand response resource, 114, 116–117
Type II demand response resource, 114–116, 115*f*

U

Uber's JUMP program, 305–306
Unbalanced distribution systems, 44–46
  linear three-phase power flow models, 37–39
Unconstraint optimization method, 195
Unscented Kalman filter (UKF), 184–187, 189–190
Unstructured data, 10
Unsynchronized sensors, 10
US Department of Energy (DOE), 260
U.S. Energy Information Administration (EIA), 30–31
Utility-scale inverter-based resources, control functions *vs.* dispatch functions for, 122*t*
Utility-scale plants, 119–120

V

Value function, 4
  approximation, 9
Vanilla policy gradient (VPG) algorithm, 30
Vehicle-to-grid (V2G) system, 307–308
Venezuela's Guri dam, 298*b*
VGGNet, 241, 243–244
Virtual leader (VL), 69–72, 70*f*
Virtual synchronous generator model, 77*f*, 80
Virtual synchronous machine (VSM), 76, 80
Voltage and frequency (VF) inverters, 76–77
Voltage inference method, 79–80, 80*f*
Voltage sensitivity, real and reactive power injections, 85*f*
Voltage source inverter (VSI), 76
Voltage/Var control, 68
Volt/VAR optimization, 138
VSM. *See* Virtual synchronous machine (VSM)

W

Water and natural gas networks, 294, 295*t*
Water–energy nexus, interdependent critical networks, 294–300, 297*f*
  ancillary services, 298–299
  closed-loop systems, 296–297
  controlling electric water pumps, 298–299
  electricity, 298
  hydroelectricity, 297–298
  once-through cooling systems, 296
  thermoelectric power plants, 296
  Venezuela's Guri dam, 298*b*
Watt-Sun, 210
Weather-related power outages, 290, 290*t*
Weighted least squares (WLS) state estimation problem, 166–168, 171–174
Wholesale electricity market
  all zero marginal cost resources, 122–123
  bulk electric power grid dispatch in inverter-based resources systems, 119–120, 120*f*
  driving forces of electricity market development, 100–102
  frequency response mechanism design, 128–129, 129*f*
  increased complexity for controls of utility-scale inverter-based resources, 121–122, 122*t*
  increased uncertainty and variability, 120–121
  motivations, 117–119, 119*f*
  new products and designs in
    considering flexibility in resource adequacy, 104–106
    demand response resources market modeling, 114–117
    energy storage market modeling, 110–114
    flexible ramping products, 106–110
  operation process of, 102–104

reserve requirements, determination of, 124–128
unit commitment and economic dispatch for up to 100% renewables, 123–124
Wide-area measurement system (WAMS), 145
FNET/GridEye, 147–151
architecture of, 147, 147*f*
disturbance detection and location, 152–153, 153*f*
event-data-based oscillation modal analysis, 153, 154*f*
event replay and postevent analysis, 156, 156*f*
FDR deployment in North America, 148, 149*f*
first generation of FDR, 148–151, 150*f*
hardware component, 147
interarea oscillation detection, 153, 154*f*
islanding detection, 154–155, 155*f*
machine learning–based inertia estimation, 157–159, 159*f*
model validation and parameter verification, 157, 158*f*
North American grids, 148
online ambient-data-based oscillation modal analysis, 154, 155*f*
second generation of FDR, 148–151, 150*f*
software component, 147
statistical analysis of historical data, 156–157, 157*f*
visualization, 154
worldwide power grid monitored, 148, 149*f*
structure, 145, 145*f*
WiFi-based occupancy detection system, 236–237
Wildfire mitigation plan, 291
Wind forecast improvement project (WFIP), 210
WindView, 210
Wireless communication technologies, 280–283

X

Xavier method, 243–244
XCEL Energy, 281–282, 286, 287*t*
Windsource Program, 288

Z

Zero marginal cost resources, 122–123
ZigBee technology, 280–283

FSC
www.fsc.org
MIX
Papier | Fördert
gute Waldnutzung
FSC® C083411

Kolibri 360 GmbH
Ferdinand-Jühlke-Straße 7
99095 Erfurt, Deutschland
produktsicherheit@kolibri360.de